Stephen I. Warshaw
U. of Calif — LLNL

Introduction to
the Quantum Theory of Scattering

PURE AND APPLIED PHYSICS
A SERIES OF MONOGRAPHS AND TEXTBOOKS

CONSULTING EDITORS

H. S. W. MASSEY
University College, London, England

KEITH A. BRUECKNER
University of California, San Diego
La Jolla, California

Volume 1. F. H. FIELD and J. L. FRANKLIN, Electron Impact Phenomena and the Properties of Gaseous Ions.

Volume 2. H. KOPFERMANN, Nuclear Moments, English Version Prepared from the Second German Edition by E. E. SCHNEIDER.

Volume 3. WALTER E. THIRRING, Principles of Quantum Electrodynamics. Translated from the German by J. BERNSTEIN. With Corrections and Additions by WALTER E. THIRRING.

Volume 4. U. FANO and G. RACAH, Irreducible Tensorial Sets.

Volume 5. E. P. WIGNER, Group Theory and Its Application to the Quantum Mechanics of Atomic Spectra. Expanded and Improved Edition. Translated from the German by J. J. GRIFFIN.

Volume 6. J. IRVING and N. MULLINEUX, Mathematics in Physics and Engineering.

Volume 7. KARL F. HERZFELD and THEODORE A. LITOVITZ, Absorption and Dispersion of Ultrasonic Waves.

Volume 8. LEON BRILLOUIN, Wave Propagation and Group Velocity.

Volume 9. FAY AJZENBERG-SELOVE (ed.), Nuclear Spectroscopy. Parts A and B.

Volume 10. D. R. BATES (ed.), Quantum Theory. In three volumes.

Volume 11. D. J. THOULESS, The Quantum Mechanics of Many-Body Systems.

Volume 12. W. S. C. WILLIAMS, An Introduction to Elementary Particles.

Volume 13. D. R. BATES (ed.), Atomic and Molecular Processes.

Volume 14. AMOS DE-SHALIT and IGAL TALMI, Nuclear Shell Theory.

Volume 15. WALTER H. BARKAS. Nuclear Research Emulsions. Part I.

Nuclear Research Emulsions. Part II. *In preparation*

Volume 16. JOSEPH CALLAWAY, Energy Band Theory.

Volume 17. JOHN M. BLATT, Theory of Superconductivity.

Volume 18. F. A. KAEMPFFER, Concepts in Quantum Mechanics.

Volume 19. R. E. BURGESS (ed.), Fluctuation Phenomena in Solids.

Volume 20. J. M. DANIELS, Oriented Nuclei: Polarized Targets and Beams.

Volume 21. R. H. HUDDLESTONE and S. L. LEONARD (eds.), Plasma Diagnostic Techniques.

Volume 22. AMNON KATZ, Classical Mechanics, Quantum Mechanics, Field Theory.

Volume 23. WARREN P. MASON, Crystal Physics of Interaction Processes.

Volume 24. F. A. BEREZIN, The Method of Second Quantization

Volume 25. E. H. S. BURHOP (ed.), High Energy Physics. In three volumes.

Volume 26. L. S. RODBERG and R. M. THALER, Introduction to the Quantum Theory of Scattering

Volume 27. R. P. SHUTT (ed.), Bubble and Spark Chambers. In two volumes.

In preparation

H. S. GREEN and R. B. LEIPNIK, The Foundations of Magnetohydrodynamics and Plasma Physics

GEOFFREY V. MARR, Photoionization Processes in Gases

Introduction to the Quantum Theory of Scattering

Leonard S. Rodberg
DEPARTMENT OF PHYSICS AND ASTRONOMY
UNIVERSITY OF MARYLAND
COLLEGE PARK, MARYLAND

R. M. Thaler
DEPARTMENT OF PHYSICS
CASE INSTITUTE OF TECHNOLOGY
CLEVELAND, OHIO

1967

ACADEMIC PRESS   *New York and London*

Copyright © 1967, by Academic Press Inc.
ALL RIGHTS RESERVED.
NO PART OF THIS BOOK MAY BE REPRODUCED IN ANY FORM,
BY PHOTOSTAT, MICROFILM, OR ANY OTHER MEANS, WITHOUT
WRITTEN PERMISSION FROM THE PUBLISHERS.

ACADEMIC PRESS INC.
111 Fifth Avenue, New York, New York 10003

United Kingdom Edition published by
ACADEMIC PRESS INC. (LONDON) LTD.
Berkeley Square House, London W.1

Library of Congress Catalog Card Number: 65-26402

PRINTED IN THE UNITED STATES OF AMERICA

To our parents

Preface

This book is intended as an introduction to scattering theory, especially as it is applied in atomic and nuclear physics and, in its more abstract and advanced forms, in high energy physics. The text is addressed principally to the graduate student in physics and to the experimental physicist, whose formal backgrounds in quantum mechanics may be limited to a first-year graduate course.

In preparing this introductory work, we have stressed the central ideas and theoretical approaches, rather than the details of specific examples and applications. To make the important points completely clear, we have often derived the same results in a number of different ways, including both physical arguments and more precise mathematical derivations. We have particularly sought to treat the underlying physical assumptions more carefully than has been usual in treatments of scattering theory. In keeping with the intended purpose of this book, we have not sought mathematical rigor when this would reduce the clarity of the physical argument. As a consequence, the presentation is highly personal. If the reader finds, after completing this book, that he is in a position to comprehend the published applications of scattering theory, this informality will have achieved its aim.

The theory discussed in this book is entirely nonrelativistic. Many of the results have been deliberately placed in a form which can be easily generalized to the relativistic domain, but we have made no attempt to do so consistently. Here, as in other situations, whenever we had to choose between clarity and simplicity on the one hand, and generality on the other, we invariably chose the former. Hence, the reader must use appropriate caution in transferring these results to relativistic problems.

In preparing this book we have attempted to provide a unified treatment of scattering theory. With this objective in mind, we have sought to use a consistent notation throughout the book, and yet to keep the notation simple and unambiguous. Since we also sought to invent as little new notation as possible, but rather to use the current conventions wherever feasible, we could not be completely successful.

The first portion of this book provides the background in conventional Schrödinger quantum mechanics for the formal theory developed in later portions. However, as much as possible, we have kept each chapter an entity to itself, with cross-references from chapter to chapter kept to a minimum. Thus, the reader should be able to use one part of the book without necessarily having to read everything which precedes it.

Because the treatment is self-contained and highly personal, we have not attempted to refer to the published origins of many of the ideas. An extensive set of references, as well as a fuller discussion of many of the topics treated herein, may be found in M. L. Goldberger and K. M. Watson, *Collision Theory* (Wiley, New York, 1964). Applications of the theory and further references are also given in T. Wu and T. Ohmura, *Quantum Theory of Scattering* (Prentice-Hall, Englewood Cliffs, New Jersey, 1962).

This book has benefited from discussions with Leslie L. Foldy. The authors are grateful also for the initiative and industry displayed by Miss Nancy Vaitkus in the preparation of the manuscript.

L. S. RODBERG
R. M. THALER

College Park, Maryland
Cleveland, Ohio
March, 1967

Contents

Preface   vii

Chapter 1. **Introduction**
1. The Schrödinger Equation 1
2. The Relation of the Boundary Condition to the Experimental Situation . 3
3. The Cross Section 6
4. The Optical Theorem 9
5. Scattering from a Collection of Scatterers 10

Chapter 2. **The Wave Packet Description of a Scattering Experiment**
1. The Free-Particle Wave Packet 14
2. Wave Packet Description of a Scattering Experiment 16

Chapter 3. **Differential Equation Methods**
1. Partial Wave Expansion 26
2. Calculation of the Phase Shift 34
3. The Low-Energy Limit 44
4. The High-Energy Limit 52
5. Elastic Scattering by a Coulomb Potential 63

Chapter 4. **Green's Functions**
1. Integral Form of the Schrödinger Equation 73
2. Properties of Green's Functions 79
3. Eigenfunction Expansions of Green's Functions 84

4. Green's Function in Three Dimensions 107
  5. Many-Particle Green's Functions 121

Chapter 5. Integral Equations of Scattering Theory and Their Solutions
  1. Integral Equations for the Wave Function 127
  2. Time-Dependent Derivation of the Integral Equation 136
  3. Methods of Solution: The Born Approximation and the Fredholm Method 141
  4. The Integral Equation for the Scattering Amplitude 157

Chapter 6. The Operator Formalism in Two-Particle Scattering Theory
  1. Operator Formalism 163
  2. Operator Form of the Scattering Equations 171
  3. The Optical Theorem 183

Chapter 7. Cross Sections for General Collision Processes
  1. General Scattering Formalism for Ordinary Scattering Processes 188
  2. Rearrangement Collisions 197
  3. Collisions Involving Identical Particles 205

Chapter 8. The Time-Dependent Approach to Scattering Theory
  1. The Schrödinger and Interaction Pictures 212
  2. Infinite Limits 217
  3. Relation to the Time-Independent Theory 222
  4. The Cross Section 225

Chapter 9. The S Matrix and the K Matrix
  1. The S Matrix 227
  2. The Optical Theorem and the K Matrix 235
  3. Diagonalization of the S Matrix 244

Chapter 10. Invariance Principles and Conservation Laws
  1. Invariance under Space Translations 249
  2. Invariance under Time Translations 255
  3. Galilean Invariance 255
  4. Rotation Invariance 261
  5. Reflection Invariance 263
  6. Time-Reversal Invariance 266

Chapter 11. **Spin and Angular Momentum**
1. System of Spin $\frac{1}{2}$ 279
2. Addition of Spin and Orbital Angular Momentum 286
3. Radial Integral Equation for Scattering by a Spin-Zero Target 289
4. Scattering Amplitudes and T Matrices 292
5. Cross Sections and Polarizations 299
6. Matrix Methods 307
7. Projection Operators 316

Chapter 12. **Applications**
1. The Two-Potential Formula 321
2. Some Examples of the Two-Potential Approach—The Distorted-Wave Approximation 327
3. The Impulse Approximation 341
4. Scattering by a Many-Body System 350
5. The Optical Potential and the Elastically Scattered Wave 363
6. Resonances 379

Index 391

Introduction to
the Quantum Theory of Scattering

Chapter

1

Introduction

In this volume we shall be concerned with the collisions of particles in accordance with the laws of quantum mechanics. Although this subject may at first appear to be a rather specialized one, the vast scope of the quantum theory of scattering is readily apparent. Scattering experiments are a principal means of studying quantum mechanical systems. If we want detailed information concerning a given particle or atomic system, for example, we have scant recourse save to scatter other particles from it. Detailed information concerning particles or atomic systems requires measurements related to their behavior at very small separations. Such measurements are in general not possible unless their kinetic energy is large and the de Broglie wavelength of the relative motion is comparable with the distances to be studied. Since nuclear dimensions are of the order of 10^{-13} cm, the importance of scattering to nuclear physics, for instance, is evident.

1. The Schrödinger Equation

We assume throughout this work that the laws of ordinary quantum mechanics hold universally. Therefore, in the nonrelativistic domain we may formulate any physical situation in terms of the solution of a Schrödinger equation using an appropriate Hamiltonian and suitable boundary conditions.

Let us consider schematically how one might describe measurements upon a system consisting of particles of type A colliding with particles of type B. The Hamiltonian for a system of two interacting particles may be written as

$$H = \frac{\mathbf{P}_A^2}{2m_A} + \frac{\mathbf{P}_B^2}{2m_B} + V_{AB}, \qquad (1.1)$$

where m_A and m_B are the masses and $\mathbf{P}_A$ and $\mathbf{P}_B$ the momenta of particles A and B, respectively, and V_{AB} is the potential energy due to their mutual interaction. We shall assume in the early chapters that the two particles interact only through a conservative two-particle potential $V_{AB}(\mathbf{r}_A - \mathbf{r}_B)$. In the steady state, where the total energy of the system is a given constant E, the Schrödinger equation is simply

$$H\,\Psi(\mathbf{r}_A, \mathbf{r}_B) = E\,\Psi(\mathbf{r}_A, \mathbf{r}_B), \tag{1.2}$$

where now the momentum $\mathbf{P}_A$ is replaced by the differential operator

$$\mathbf{P}_A = -i\hbar\nabla_A \tag{1.3}$$

and similarly for $\mathbf{P}_B$.

An important simplification results if we separate out the motion of the center of mass. We define the relative coordinate $\mathbf{r}$ by

$$\mathbf{r} \equiv \mathbf{r}_A - \mathbf{r}_B \tag{1.4}$$

and the coordinate of the center of mass by

$$\mathbf{R} \equiv \frac{m_A \mathbf{r}_A + m_B \mathbf{r}_B}{(m_A + m_B)}. \tag{1.5}$$

If the mutual interaction potential V_{AB} depends only on the relative coordinate, then Eq. (1.2) is

$$\left[-\frac{\hbar^2}{2M}\nabla_R^2 - \frac{\hbar^2}{2m}\nabla_r^2 + V(r)\right]\Psi(\mathbf{r}_A, \mathbf{r}_B) = E\,\Psi(\mathbf{r}_A, \mathbf{r}_B), \tag{1.6}$$

where the total mass M is $M = m_A + m_B$ and the reduced mass m is

$$m = \frac{m_A m_B}{m_A + m_B}. \tag{1.7}$$

The advantage of Eq. (1.6) is that it permits a separation of the variables $\mathbf{R}$ and $\mathbf{r}$. If we write the wave function $\Psi(\mathbf{r}_A, \mathbf{r}_B)$ as

$$\Psi(\mathbf{r}_A, \mathbf{r}_B) = \Phi(\mathbf{R})\,\psi(\mathbf{r}), \tag{1.8}$$

then we obtain the two separated equations

$$\frac{-\hbar^2}{2M}\nabla_R^2\,\Phi(\mathbf{R}) = E_{cm}\,\Phi(\mathbf{R}) \tag{1.9}$$

and

$$\left\{-\frac{\hbar^2}{2m}\nabla^2 + V(r)\right\}\psi(\mathbf{r}) = E_{\text{rel}}\psi(\mathbf{r}), \quad (1.10)$$

where

$$E = E_{\text{cm}} + E_{\text{rel}}. \quad (1.11)$$

If we identify particle A as a projectile initially moving in the laboratory with a velocity $\mathbf{v}$ and particle B as a target particle at rest in the laboratory prior to the collision, then the total energy is

$$E = \tfrac{1}{2}m_A v^2. \quad (1.12)$$

This relation assumes that the interaction potential $V(r)$ has a finite range beyond which it is zero, so that the total energy is the asymptotic kinetic energy. The kinetic energy associated with the motion of the center of mass of the system is given by

$$E_{\text{cm}} = \tfrac{1}{2}M\dot{R}^2 = \tfrac{1}{2}M\left(\frac{m_A \dot{\mathbf{r}}_A + m_B \dot{\mathbf{r}}_B}{M}\right)^2$$

$$= \frac{1}{2}\frac{m_A^2}{(m_A + m_B)}v^2 = \frac{m_A}{m_A + m_B}E. \quad (1.13)$$

From this it follows that

$$E_{\text{rel}} = \frac{m_B}{m_A + m_B}E = \tfrac{1}{2}mv^2. \quad (1.14)$$

2. The Relation of the Boundary Condition to the Experimental Situation

We cannot solve Eqs. (1.9) and (1.10) until we have specified the boundary conditions to be imposed on the solutions. These boundary conditions must of course reflect the physical situation in the laboratory. Before we proceed further, then, we must discuss the experimental circumstances under which the measurements of interest are performed. A schematic representation of a standard scattering experiment appears in Fig. 1.1. At the left appears the source of incident particles. This may be a modern particle accelerator or perhaps merely a sample of spontaneously decaying material. A source of particles is, of course, an indispensable element in a scattering or collision experiment, since in order for two particles to collide at least one of them must be moving. The second indispensable element of the experiment is

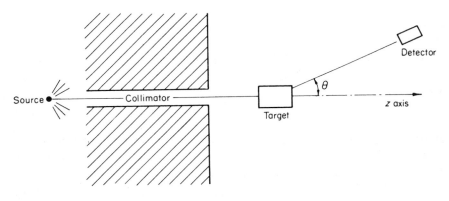

Fig. 1.1

another particle, so that there may be a collision. In Fig. 1.1 the second particle is provided by the stationary target, but it is not essential that the second particle be part of a stationary target. A perfectly feasible scattering experiment may be performed by scattering one beam of particles from another. The third indispensable element in the scattering experiment is the detector. To perform an experiment we must measure something; the detector performs this function. It simply counts the number of particles of a particular type that arrive at its position. Ideally it may be set to count only particles of a given type in a given velocity or energy.

We assume throughout that the source and detector are classical objects that have a clearly defined, precisely controllable, effect on the scattering process. The detector will be assumed to be 100% efficient and to have no effect on the scattered particle prior to the time it enters the detector.

One other element conventionally appears in all scattering experiments. This is the collimator, which shields the detector from the source. This is not essential for the scattering experiment, but, were it not for the collimator, the particle intensity striking the detector would be nearly equal to the intensity at the target. Since in general only a minute fraction of the particles arriving at the target are deflected to the detector, the presence or absence of the target, which is the essence of the experiment, would only very slightly affect the intensity received by the detector. Thus, without the collimator to shield the detector from the incident beam, the scattering experiment would be difficult to perform. With an ideal collimator no particles are received by the detector when the target is removed; the only particles received are in that portion of the incident beam that is deflected into the detector by the

2. The Boundary Condition

target. The perfect collimator thus produces a narrow beam directed at the target.

The results of the scattering experiment will vary with the energy of the incident beam. In order to simplify the analysis of the experimental data, the energy spectrum of the incident beam should be sharply peaked so that the experiment may be considered to take place at a unique energy eigenvalue. To this end, in most experiments care is taken to achieve a monochromatic incident beam. We shall assume here that the beam emerging from the collimator is both perfectly monochromatic and perfectly collimated. Of course, according to the uncertainty principle, a beam of finite cross section cannot be perfectly monochromatic and perfectly collimated as well. We may, however, assume the beam to be sufficiently well collimated that the angular divergence may be ignored in an actual experiment. In that case we must necessarily have, not a monochromatic beam represented by a plane wave, but rather a beam describable as a superposition of such waves. In the next chapter we shall discuss the scattering process using such wave packets. That discussion will justify the present idealized treatment.

Neglecting these effects for the present and assuming a monochromatic beam, we find the wave function for the motion of the center of mass to be

$$\Phi(\mathbf{R}) = e^{iKZ}, \tag{2.1}$$

where

$$K^2 = \frac{2M}{\hbar^2} E_{cm} \tag{2.2}$$

and Z is the z projection of the center-of-mass vector $\mathbf{R}$. The z axis is conventionally defined by the axis of the collimator, as shown in Fig. 1.1. Likewise, the wave function for the relative motion in the absence of the target is

$$\psi(\mathbf{r}) = e^{ikz}, \tag{2.3}$$

where

$$k^2 = \frac{2m}{\hbar^2} E_{rel}. \tag{2.4}$$

In the presence of a target, Eqs. (2.1) and (2.2) are unaltered, since the potential is assumed to depend only on the relative position of the two particles. Our interest will naturally focus, then, on the solution to the Schrödinger equation for the relative motion, Eq. (1.10). In the presence of a target that scatters the incident beam, a plane wave solution of Eq. (1.10) is no longer possible. Instead, the proper boundary condition is a condition on the wave function when the particles are far apart:

$$\psi(\mathbf{r}) \xrightarrow[r \to \infty]{} N\left[e^{ikz} + \frac{e^{ikr}}{r} f(\theta, \phi) \right]. \tag{2.5}$$

This asymptotic form is easy to understand. To simplify our thinking, it is helpful to consider the case where the target particle is infinitely heavy relative to the mass of the incident particle. In this case the center of mass remains stationary at the position of the target particle throughout the scattering process. Under these circumstances the relative coordinates represent the actual laboratory coordinates of the light particle. It is often convenient to use this limiting case, since the Schrödinger equation to be solved is then identical with Eq. (1.10), provided only that m is interpreted as the mass of the incident projectile.

In this limit we can readily visualize the physical meaning of Eq. (2.5). The origin of the coordinate system is now at the position of the target particle. We may then interpret the asymptotic form in Eq. (2.5) as the incident plane wave plus an outgoing scattered wave emanating from the target. At a very large distance from the target the only effect of the target is to remove some of the particles from the incident beam and to "reradiate" them as a spherical wave emerging from the target particle. This is, of course, a familiar property of waves, whether they be electromagnetic waves, sound waves, or "matter" waves. It is simply another form of Huygens' principle. This boundary condition will be further discussed using a wave packet treatment in the next chapter.

We must remember that Eq. (2.5) does not hold throughout all space in the experimental situation we have postulated. The incident beam is not a true plane wave, but instead is a narrow pencil. Well outside this pencil we expect to see only the spherical scattered wave, so that one may well question the correctness of Eq. (2.5) as a boundary condition. We shall show in the next chapter, using a wave packet description, that in spite of these difficulties it is in fact correct. In essence, that part of the incident beam which does not strike the target has no effect on the scattering process, and no error is caused by treating this part of the incident beam incorrectly.

3. The Cross Section

Let us now compute the currents carried by the incident and outgoing waves. Consider first the incident plane wave Ne^{ikz}. The probability current

3. The Cross Section

density is given by

$$\mathbf{j}(\mathbf{r}) = \text{Re}[\psi^\dagger (\hbar/im) \nabla \psi], \tag{3.1}$$

so that for the plane wave the current density is

$$\mathbf{j}(\mathbf{r}) = \frac{\hbar k |N|^2 \hat{\mathbf{z}}}{m} = j_{\text{inc}} \hat{\mathbf{z}}, \tag{3.2}$$

where $\hat{\mathbf{z}}$ is a unit vector in the z direction. As this indicates, the plane wave e^{ikz} represents a current flow along the z axis. In analogy with the familiar classical relation

$$\mathbf{j} = \rho \mathbf{v}, \tag{3.3}$$

we may identify the particle density as

$$\rho = |N|^2 \tag{3.4}$$

and the particle velocity as

$$\mathbf{v} = (\hbar k/m)\hat{\mathbf{z}} = (p/m)\hat{\mathbf{z}}. \tag{3.5}$$

The above relations must not be taken too literally, since neither the density nor the velocity can be uniquely defined for a plane wave function, which cannot be normalized and which, therefore, cannot be given a conventional one-particle interpretation. They are none the less suggestive.

Let us now calculate the current density for the scattered wave

$$\psi_{\text{sc}} = N(e^{ikr}/r) f(\theta, \phi). \tag{3.6}$$

From Eq. (3.1) we get

$$\mathbf{j}_{\text{sc}}(\mathbf{r}) = |N|^2 \frac{|f(\theta, \phi)|^2}{r^2} \frac{\hbar k}{m} \hat{\mathbf{r}} + \cdots = j_{\text{inc}} \frac{|f(\theta, \phi)|^2}{r^2} \hat{\mathbf{r}} + \cdots, \tag{3.7}$$

where $\hat{\mathbf{r}}$ is the radial unit vector. The higher-order terms that are neglected contain higher inverse powers of r. Since Eq. (2.5) is an asymptotic relation, valid only for large r, we may certainly ignore such terms. From Eq. (3.7) we see most clearly why the $(1/r)$ factor appears in Eq. (3.6); it is simply an expression of the usual inverse-square fall-off of the outgoing current from a source of radiation.

The function $f(\theta, \phi)$, which determines the magnitude of the current in the scattered beam, is called the scattering amplitude. If we can find the scattering amplitude, we have all the information we can possibly obtain from a scattering experiment, for then we have determined the entire wave function for large r. For the most part we are unable to make measurements except in the asymptotic region, simply because ordinary laboratory dimensions are many orders of magnitude larger than, say, nuclear or atomic dimensions.

Let us suppose that we bombard a single target particle of infinite mass with a collimated beam of particles having a current density of j_{inc} particles per unit time per unit area. We count the number of particles per unit time that are deflected into the detector. We expect the number of particles per second counted by the detector to be proportional to the incident flux and to the solid angle subtended by the detector. The detector is located along a ray specified by the angles (θ, ϕ) and subtends a solid angle $d\Omega = dA/r^2$, where dA is the projected area of the detector normal to the ray. Then the number of particles counted per unit time will be given by

$$dN_d/dt = j_{inc}\, \sigma(\theta, \phi)\, d\Omega = j_{inc}\, \sigma(\theta, \phi)\, (1/r^2)\, dA, \qquad (3.8)$$

where the proportionality factor is denoted by $\sigma(\theta, \phi)$. Clearly $\sigma(\theta, \phi)$ has the dimensions of an area; it is called the differential scattering cross section. The quantity $\sigma(\theta, \phi)\, d\Omega$ is the cross-sectional area of the incident beam that is intercepted by the target particle and scattered into the cone defined by θ, ϕ, and $d\Omega$. The total cross section,

$$\sigma_{TOT} = \int \sigma(\theta, \phi)\, d\Omega, \qquad (3.9)$$

is the total cross-sectional area of the incident beam that is intercepted and the particles therein deflected by the target particle.

From Eq. (3.8) the scattered current density is $(1/r^2) j_{inc}\, \sigma(\theta, \phi)$, whereas from Eq. (3.7) the scattered current density is $j_{inc}|f(\theta, \phi)|^2(1/r^2)$. Equating these gives the relation

$$\sigma(\theta, \phi) = |f(\theta, \phi)|^2. \qquad (3.10)$$

This relates the experimental quantity $\sigma(\theta, \phi)$, the differential cross section, to $f(\theta, \phi)$, the scattering amplitude, which characterizes the wave function at large distances from the target. It is the fundamental relation between scattering theory and scattering experiments.

From Eq. (3.8) we can see that there is an important element missing from the idealized experiment of Fig. (1.1). In order to determine the cross section, we must measure not only the scattered flux at the detector, whose position is given by the coordinates (r, θ, ϕ), but we must also know the incident flux j_{inc}. A further element of a scattering experiment, therefore, must be a device for measuring the incident current density; this is conventionally referred to as a beam monitor.

4. The Optical Theorem

We have just seen that without the target the flux along the z axis is given by Eq. (3.2), whereas with the target in place there is an additional radial flux given by Eq. (3.7). Apparently we are not conserving particles. For scattering by a conservative potential only elastic scattering is possible, that is, the incident particle can only be deflected by the target and cannot be absorbed. In a region in which there are no sources or sinks of radiation, we then expect that current is conserved:

$$\oint_{\text{surface}} \mathbf{j} \cdot \mathbf{dS} = 0. \tag{4.1}$$

It is instructive to substitute the asymptotic wave function of Eq. (2.5) into Eq. (4.1). For the current we find

$$\mathbf{j} = \frac{\hbar k}{m} \left\{ \hat{\mathbf{z}} + \frac{|f(\theta, \phi)|^2}{r^2} \hat{\mathbf{z}} + \frac{1}{r} (\hat{\mathbf{z}} + \hat{\mathbf{r}}) \operatorname{Re}[f(\theta, \phi) e^{ik(r-z)}] \right\}. \tag{4.2}$$

Choosing as the closed surface a large sphere centered on the target, we find that

$$\int r^2 \, d\Omega \, \mathbf{j} \cdot \hat{\mathbf{r}} = \left\{ \int \int |f(\theta, \phi)|^2 \, d\Omega + \operatorname{Re} \int r \, d\phi \, d\cos\theta \, f(\theta, \phi) \right.$$
$$\left. \times \exp[ikr(1 - \cos\theta)] (1 + \cos\theta) \right\} (\hbar k/m). \tag{4.3}$$

The first integral on the right-hand side is the total cross section; the second integral is the overlap of the scattered wave with the incident wave. Thus, if Eq. (4.1) is to hold, we must have

$$\sigma_{\text{TOT}} = -\operatorname{Re} \int r \, d\phi \, d\cos\theta \, f(\theta, \phi) \exp[ikr(1 - \cos\theta)] (1 + \cos\theta). \tag{4.4}$$

The integral on the right may be evaluated for large r by partial integration:

$$\int_{-1}^{1} d\cos\theta \, f(\theta, \phi) \exp[ikr(1 - \cos\theta)] (1 + \cos\theta)$$
$$= \frac{i}{kr} f(\theta, \phi) \exp[ikr(1 - \cos\theta)] (1 + \cos\theta) \Big|_{-1}^{+1}$$
$$+ \mathcal{O}\left(\frac{1}{k^2 r^2}\right) = \frac{2i}{kr} f(\theta = 0, \phi) + \mathcal{O}\left(\frac{1}{k^2 r^2}\right), \tag{4.5}$$

assuming that $f(\theta, \phi)$ can be differentiated to all orders. Since $\psi(\mathbf{r})$ is a single-valued function, $f(\theta, \phi)$ must be independent of ϕ in the forward direction. This gives the result

$$\sigma_{\text{TOT}} = \frac{4\pi}{k} \operatorname{Im} f(\theta = 0). \tag{4.6}$$

This is the optical theorem, which must be satisfied by any physically meaningful scattering amplitude. We have not derived it for the most general circumstances under which it holds. However, we do see easily the main feature of the optical theorem, namely, that it is a direct consequence of the conservation of flux.

We can see how current can be conserved when there is a scattered flux. The target casts a "shadow" in the forward direction where the intensity of the beam is reduced. The forward current is reduced by just the amount that appears in the scattered wave. This decrease in intensity, the shadow, may be viewed as resulting from the interference between the incident wave and the scattered wave in the forward direction.

Total cross sections are actually measured in just this way. The current density is measured by placing particle detectors on the z axis both before and behind the target. The difference between these two measurements yields the total flux removed from the incident plane wave by the scattering process and hence is directly related to the total cross section.

5. Scattering from a Collection of Scatterers

In the preceding we have considered the scattering from a single scattering center. This is completely general, since the scattering center may just as well be a macroscopic sample of matter as a single atom. However, if we wish to study the microscopic properties of an atomic system, the flux scattered into the detector by a macroscopic target must be related to the flux that would result from scattering by a single target atom. If we consider a sample of material containing M identical scatterers, we would expect the flux at the detector to be proportional to the number of scattering centers in the target. Since this may be of the order of Avogadro's number, the practical importance of this distinction is obvious.

Let us now treat scattering by a collection of independent equivalent scattering centers. We will neglect the "shadowing" of one target particle by another, that is, the diminution of the effective incident beam in the

5. Scattering from a Collection of Scatterers

rear of the target and all multiple scattering effects. These effects can easily be included if the target thickness is comparable with the mean free path for scattering. In order to avoid center of mass complications, we shall also assume that the target particles are massive relative to the incident particles. With these assumptions we have for the asymptotic wave function

$$\psi(r) \xrightarrow[r \to \infty]{} N\left[e^{ikz} + (\sum_{m=1}^{M} e^{i\phi_m}) \frac{e^{ikr}}{r} f(\theta, \phi)\right], \tag{5.1}$$

where the ϕ_m are the relative phases of the scattered waves from each of the target scatterers.

The phase associated with scattering by a particle at the coordinate origin is taken as zero in conformity with the boundary condition, Eq. (2.5). The coordinates of the mth scatterer are (x_m, y_m, z_m), and the relative phase of the mth scattered wave at a point with coordinates (x, y, z) is then

$$\phi_m = k\{z_m + [(x - x_m)^2 + (y - y_m)^2 + (z - z_m)^2]^{1/2} - r\}. \tag{5.2}$$

To lowest order in $1/r$ this becomes

$$\phi_m = k[-\sin\theta\,(x_m \cos\phi + y_m \sin\phi) + (1 - \cos\theta)z_m] + \mathcal{O}(1/r). \tag{5.3}$$

Notice that in the forward direction $\phi_m = 0$, and the scattering from different centers is always coherent. The sum in Eq. (5.1) is

$$S(\theta, \phi) \equiv \sum_{m=1}^{M} e^{i\phi_m}. \tag{5.4}$$

We readily see that in the forward direction

$$S(0) = M. \tag{5.5}$$

The forward scattering amplitude and hence, by the optical theorem, the total cross section are M times their single-particle values, as expected.

The differential cross section implied by Eq. (5.1) is

$$\sigma(\theta, \phi) = |f(\theta, \phi)|^2 \{M + 2 \sum_{m>n} \cos(\phi_m - \phi_n)\}. \tag{5.6}$$

The terms in the summation arise from the interference between scattered waves from different target particles. If Eq. (5.6) is integrated over all angles, this interference term will on the average contribute nothing to the total cross section; this is a necessary consequence of the optical theorem. However, the interference term does not necessarily vanish at any particular value of (θ, ϕ). The target particles might, for example, form a lattice structure. Then, if the motion of the individual particles about their mean positions could be neglected, we could easily derive the usual relations for Bragg scattering.

At the Bragg maxima the interference term is of order M^2. Nonetheless the interference term is negligible in most practical experiments. If for some reason there is an individual uncertainty in each of the phases that is $\geq 2\pi$, then the interference term will vanish. For scattering angles substantially away from the forward direction, Eq. (5.2) or Eq. (5.3) implies that the condition for the uncertainty in phase to be larger than 2π is that the uncertainty in position of the target particle be greater than the wavelength of the incident radiation. Thus, for instance, we can usually neglect the interference term in scattering from a gas. In scattering from a solid target the motion of the target particles about their equilibrium positions may be sufficient to produce a large uncertainty in the individual phases. Even at absolute zero, the zero point motion may accomplish this.

Chapter

2

The Wave Packet Description of a Scattering Experiment

In a scattering experiment a beam of particles is allowed to strike a target, and the particles that emerge from the target area are observed. From this observation one hopes to study the properties of the projectile and the target and the forces between them. We shall confine ourselves in this chapter to scattering processes in which only short-range central forces are present. In the presence of such forces the particles are not under the influence of the target when they are emitted from the source and when they enter the detector. This means that the wave function which describes the projectile when it is far from the target must be a solution of the free-particle Schrödinger equation.

If an elastic scattering experiment is described by a time-independent wave function, we asserted in the previous chapter that this function should have the asymptotic form

$$\psi(\mathbf{r}) \xrightarrow[r \to \infty]{} e^{ikz} + \frac{e^{ikr}}{r} f(\theta, \phi),$$

where the plane wave represents the beam incident upon the target and $r^{-1}e^{ikr}$ represents the outgoing wave emerging from the target. The function $f(\theta, \phi)$, the scattering amplitude, describes the detailed properties of the scattered wave. In this chapter we want to justify the use of this boundary condition and show that the time-independent theory of Chapter 1 provides a general framework on which our subsequent discussion can build.

It is not obvious that the use of a time-independent wave function is justified. A scattering event is not a stationary process but, instead, involves individual discrete particles as projectiles, and we would like to see under what conditions a time-independent description may be valid. In addition, a wave function that satisfies the above boundary condition cannot have the usual meaning of a one-particle function since it is not normalizable. We shall construct a function that *is* normalizable, and shall then show that equivalent information can in fact be obtained using the time-independent treatment.

In the following sections we shall construct a wave packet describing the motion of the projectile and shall show that the above difficulties can be overcome, while the simplicity of the time-independent treatment can nevertheless still be maintained.

1. The Free-Particle Wave Packet

We shall first discuss the free-particle wave packet. This may be constructed as a superposition of stationary-state eigenfunctions for the free particle. Such solutions have the form

$$\phi_{\mathbf{k}}(\mathbf{r}, t) = \exp(i\mathbf{k} \cdot \mathbf{r}) \exp[-i(\hbar k^2/2m)t] \tag{1.1}$$

for a particle of mass m and momentum $\hbar \mathbf{k}$. Since these eigenfunctions form a complete set, the wave packet can be written as

$$\phi(\mathbf{r}, t) = (2\pi)^{-3} \int d^3k \, A(\mathbf{k}) \exp(i\mathbf{k} \cdot \mathbf{r}) \exp[-i(\hbar k^2/2m)t]. \tag{1.2}$$

The coefficient $A(\mathbf{k})$ is the probability amplitude for finding the wave number $\mathbf{k}$ (or momentum $\hbar \mathbf{k}$) in the initial state. It may be determined by Fourier inversion if the form of the packet is known at some time, say $t = 0$, from the relation

$$A(\mathbf{k}) = \int d^3r \, \exp(-i\mathbf{k} \cdot \mathbf{r}) \, \phi(\mathbf{r}, 0). \tag{1.3}$$

The packet will describe the motion of a single free particle if it is normalized so that

$$\int d^3r \, |\phi(\mathbf{r}, t)|^2 = (2\pi)^{-3} \int d^3k \, |A(\mathbf{k})|^2 = 1. \tag{1.4}$$

The packet describing the particle as it leaves the collimator will have a width Δr of order of the size of the collimator opening. We assume that Δr is small compared to the size of the over-all region in which the experiment is performed. Thus we postulate $\Delta r \ll L$, where L is a distance that characterizes the size of the laboratory. This restriction will enable us to distinguish the scattered particles from the transmitted ones, since there will then be

1. The Free-Particle Wave Packet

sufficient room to permit placing the detector out of the path of the incident beam.

We assume that the properties of the source are such that the amplitude function $A(\mathbf{k})$ is peaked about $\mathbf{k} = \mathbf{k}_0$ with a spread in wave number Δk that is small compared to k_0, that is, $\Delta k \ll k_0$. This ensures that the kinetic energy of the particle is well defined and is an important condition in any scattering experiment where one wishes to determine the energy dependence of the scattering process. Of course, the incident beam cannot be completely monoenergetic, since we know that Δr and Δk are related by the uncertainty relation $\Delta k \, \Delta r \simeq 1$, and both cannot be arbitrarily small.

We now want to examine the motion of the wave packet. In doing this it is convenient to expand the energy-dependent phase factor about the median energy. This is easily done using the relation

$$k^2 = (\mathbf{k}_0 + \mathbf{k} - \mathbf{k}_0)^2 = k_0^2 + 2\mathbf{k}_0 \cdot (\mathbf{k} - \mathbf{k}_0) + (\mathbf{k} - \mathbf{k}_0)^2$$
$$= -k_0^2 + 2\mathbf{k} \cdot \mathbf{k}_0 + (\mathbf{k} - \mathbf{k}_0)^2. \quad (1.5)$$

On the assumption that the spread in k is small, the last term may be neglected and Eq. (1.2) will become

$$\phi(\mathbf{r}, t) = (2\pi)^{-3} \exp[i(\hbar k_0^2/2m)t] \int d^3k \, A(\mathbf{k}) \exp[i\mathbf{k} \cdot (\mathbf{r} - \mathbf{v}_0 t)], \quad (1.6)$$

where we have used the relation $(\hbar k_0/m) = (p_0/m) = v_0$. We can readily see that Eq. (1.6) may be written as

$$\phi(\mathbf{r}, t) = \exp[i(\hbar k_0^2/2m)t] \, \phi(\mathbf{r} - \mathbf{v}_0 t, 0). \quad (1.7)$$

If the packet $\phi(\mathbf{r}, t)$ is centered about the origin $r = 0$ at $t = 0$, the packet at time t will have exactly the same shape, but is centered about $\mathbf{r} = \mathbf{v}_0 t$. Thus the wave packet moves with the classical velocity v_0, and there is no change in its shape.

This result will be valid provided $\{\hbar(\mathbf{k} - \mathbf{k}_0)^2/2m\}t \ll 1$. Since $(\mathbf{k} - \mathbf{k}_0)^2 \lesssim (\Delta k)^2$ and $t \lesssim (2L/v_0)$, this is equivalent to the condition $\{(\Delta k)^2/k_0\}L \ll 1$. This requirement is easily fulfilled since, according to the uncertainty relation, it is equivalent to the condition $(\Delta r/\lambda_0) \ll L/\Delta r$. This implies that there must be many wavelengths contained within the packet, a condition that is easily satisfied. For instance, thermal electrons, with energies of the order of $1/40$ eV, have $\lambda_0 \sim 10^{-7}$ cm, while $\Delta r \sim 1$ cm and $L \sim 100$ cm; for heavier or faster particles, the condition is satisfied still better.

To see how the shape changes if this condition is not fulfilled, it is instructive to examine the Gaussian wave packet for which the integral in Eq. (1.2) may be performed analytically. In this case

$$A(\mathbf{k}) = (4\pi/\sigma)^{3/4} \exp[-(\mathbf{k} - \mathbf{k}_0)^2/2\sigma] \tag{1.8}$$

and

$$\phi(\mathbf{r}, t) = (\sigma/\pi)^{3/4} (1 + i\hbar\sigma t/m)^{-3/2} \exp[i(\hbar k_0^2 t/2m)]$$
$$\times \exp[-(\sigma/2)(\mathbf{r} - \mathbf{v}_0 t)^2/(1 + i\hbar\sigma t/m)] \exp[i\mathbf{k}_0 \cdot (\mathbf{r} - \mathbf{v}_0 t)]. \tag{1.9}$$

If the quantity $(1 + i\hbar\sigma t/m)$ were set equal to unity wherever it appears in Eq. (1.9), this equation would have the form of Eq. (1.7); that is, the wave packet would be represented by a function centered about $\mathbf{r} = \mathbf{v}_0 t$, whose shape is independent of time.

The probability density corresponding to Eq. (1.9) is

$$|\phi(\mathbf{r}, t)|^2 = \left[\frac{1}{\pi} \frac{\sigma}{\{1 + (\hbar\sigma t/m)^2\}}\right]^{3/2} \exp\left[\frac{-\sigma(\mathbf{r} - \mathbf{v}_0 t)^2}{\{1 + (\hbar\sigma t/m)^2\}}\right] \tag{1.10}$$

This is a Gaussian function centered about $\mathbf{r} = \mathbf{v}_0 t$ and having a size $\Delta r \simeq \sigma^{-1/2}$ at $t = 0$, increasing with time as $\Delta r \simeq \sigma^{-1/2}[1 + (\hbar\sigma t/m)^2]^{1/2}$. This result has a simple interpretation. Because there is a range of velocities within the packet, the wave packet will gradually expand. The change in size of the packet $\delta(\Delta r)$ is related to the momentum spread by $\delta(\Delta r) \sim \Delta v t \sim (\hbar \Delta k/m)t$. If the fractional change in the size of the packet is to be small, we must have

$$\{\delta(\Delta r)/\Delta r\} \sim \{\hbar(\Delta k)^2 t/m\} \sim \{(\Delta k)^2 L/k_0\} \ll 1. \tag{1.11}$$

This is the condition introduced earlier.

2. Wave Packet Description of a Scattering Experiment

In a typical scattering experiment a beam of particles is emitted from a collimated source. Thereafter, the experimenter has no control over the particles until they reach his detector. Thus, once the beam has left the source, its propagation is controlled solely by the laws of quantum mechanics and the Hamiltonian of the projectile-target system. In later chapters we shall study in some detail the effect of the short-range interaction between projectile and target on the scattering measurements. In this section we shall describe the beam as it leaves the source and then obtain a description of the scattered particles as they arrive in the region of the detector. The discussion that follows will concentrate on those general aspects of the scattering ex-

2. Wave Packet Description of a Scattering Experiment

periment for which a knowledge of the detailed interaction in the region between the source and the detector is not necessary.

The individual particles in the beam will each be described by a wave packet. Since we shall find that the detailed structure of this packet does not affect the results of the experiment, we shall for convenience assume that each particle is described by the same packet. We further assume that the individual packets that compose the beam do not overlap.

We shall in addition impose the conditions discussed in Section 1. First, in order that the scattered particles may be distinguished from the unscattered ones, we require that the packets be small compared to typical dimensions of the laboratory; i.e.,

$$\Delta r \ll L. \quad \text{(collimation)} \tag{2.1}$$

For convenience, the packet should be large compared to the size of the target. This is easily achieved since typical atomic scatterers have dimensions of the order of 10^{-8} cm.

Second, we require that the energy be well defined, i.e.,

$$\Delta k \ll k_0, \tag{2.2}$$

since one usually wishes to investigate the scattering as a function of energy. Because of the uncertainty relation, this is equivalent to the condition $\lambda_0 \ll \Delta r$. Third, we impose the further condition that the packet not spread appreciably during the course of the experiment, or

$$(\Delta k)^2 L/k_0 \ll 1. \tag{2.3}$$

This will ensure that the change in the packet's shape is not confused with scattering by the target.

We now construct a wave packet that is prepared at the source so as to be compatible with these conditions, Eqs. (2.1)–(2.3), and which then propagates under the influence of a total Hamiltonian that includes the projectile-target interaction. It must therefore satisfy the time-dependent Schrödinger equation

$$i\hbar \frac{\partial \psi(\mathbf{r}, t)}{\partial t} = \left\{ -\frac{\hbar^2}{2m} \nabla^2 + V(r) \right\} \psi(\mathbf{r}, t). \tag{2.4}$$

As in the free-particle case, we write the packet as a superposition of the complete set of stationary-state solutions of this Schrödinger equation. We shall place the target at the origin of the coordinate system ($r = 0$) and

18   2. THE WAVE PACKET DESCRIPTION OF A SCATTERING EXPERIMENT

choose these solutions $\psi_\mathbf{k}(\mathbf{r})$ to satisfy the boundary condition of Chapter 1,

$$\psi_\mathbf{k}(\mathbf{r}) \xrightarrow[r \to \infty]{} \exp(i\mathbf{k}\cdot\mathbf{r}) + f_\mathbf{k}(\theta)\exp(ikr)/r. \qquad (2.5)$$

Later we will show under what conditions the scattering amplitude $f_\mathbf{k}(\theta)$ is independent of the azimuthal angle; in this section we assume cylindrical symmetry as a matter of convenience, since none of our arguments depend in any essential way on this assumption. Only the boundary condition Eq. (2.5) is needed in our present discussion; the form of this solution in the neighborhood of the target will not be needed.

We write the packet as

$$\psi(\mathbf{r}, t) = (2\pi)^{-3} \int d^3k\, A(\mathbf{k})\, \psi_\mathbf{k}(\mathbf{r}) \exp[-i(\hbar k^2/2m)t]. \qquad (2.6)$$

The energy eigenvalue corresponding to the eigenfunction $\psi_\mathbf{k}(\mathbf{r})$ is simply $E_\mathbf{k} = (\hbar^2 k^2/2m)$, since the forces are assumed to have a short range and can have no effect on the energy when the target and projectile are well separated. The forces are also assumed to be conservative, so that a potential can be defined and the total energy is conserved.

The condition, Eq. (2.1), and the assumption of short-range forces guarantee that the packet will not overlap the target when their centers are widely separated. Therefore, we may replace $\psi_\mathbf{k}(\mathbf{r})$ by its asymptotic form, Eq. (2.5), when the packet is far from the target. If this is done, $\psi(\mathbf{r}, t)$ breaks up into two terms:

$$\psi(\mathbf{r}, t) = \phi(\mathbf{r}, t) + \psi_{sc}(\mathbf{r}, t), \qquad (2.7)$$

with $\phi(\mathbf{r}, t)$ identical with Eq. (1.2) and

$$\psi_{sc}(\mathbf{r}, t) = (2\pi)^{-3} \int d^3k\, A(\mathbf{k}) f_\mathbf{k}(\theta)\, r^{-1} \exp(ikr) \exp[-i(\hbar k^2/2m)t]. \qquad (2.8)$$

We will choose $A(\mathbf{k})$ so that $\phi(\mathbf{r}, t)$ satisfies conditions (2.1)–(2.3) and is centered about $\mathbf{r} = \mathbf{v}_0 t$ at time t. An important consequence of this form for $\psi(\mathbf{r}, t)$ is then that the scattered wave $\psi_{sc}(\mathbf{r}, t)$ vanishes for very early times, that is, until the incident wave packet $\phi(\mathbf{r}, t)$ reaches the target.

To show this, we will assume that the scattering amplitude is slowly varying over the spread of wave numbers Δk, so that $f_\mathbf{k}(\theta) \simeq f_{\mathbf{k}_0}(\theta)$. This assumption makes the calculation simpler and implies no significant restriction on the results, since it is very easy to satisfy in realistic cases. We then find

$$\psi_{sc}(\mathbf{r}, t) = (2\pi)^{-3} r^{-1} f_{\mathbf{k}_0}(\theta) \int d^3k\, A(\mathbf{k}) \exp(ikr) \exp[-i(\hbar k^2/2m)t]. \qquad (2.9)$$

The integral is just $\phi(r\hat{z}, t)$, where $\hat{z}$ is a unit vector in the direction of $\mathbf{k}_0$. To see this we observe that

$$k = |\mathbf{k}| = |\mathbf{k}_0 + \mathbf{k} - \mathbf{k}_0| = \{k_0^2 + 2\mathbf{k}_0 \cdot (\mathbf{k} - \mathbf{k}_0) + (\mathbf{k} - \mathbf{k}_0)^2\}^{1/2}$$
$$= k_0 + \mathbf{k}_0 \cdot (\mathbf{k} - \mathbf{k}_0)/k_0 + \mathcal{O}\{(\mathbf{k} - \mathbf{k}_0)^2/k_0\} \quad (2.10)$$

Thus, to order $(\Delta k)^2/k_0^2$, we have

$$k = (\mathbf{k}_0 \cdot \mathbf{k})/k_0; \quad (2.11)$$

to this order, k is equal to its projection along $\mathbf{k}_0$. By condition (2.3) we can neglect terms of order $(\Delta k)^2 r/k_0$, so that Eq. (2.9) becomes, upon comparison with Eq. (1.2),

$$\psi_{sc}(\mathbf{r}, t) = r^{-1} f_{\mathbf{k}_0}(\theta) \phi(r\hat{z}, t). \quad (2.12)$$

The amplitude of the free-particle packet at the position $r\hat{z}$ at time t is $\phi(r\hat{z}, t)$. The point $r\hat{z}$ lies a distance r from the target along the *positive* z axis, while at time t the center of the wave packet $\phi(\mathbf{r}, t)$ is at $\mathbf{v}_0 t$, which is on the *negative* z axis for $t < 0$. We see that, as long as the free packet $\phi(\mathbf{r}, t)$ has not advanced far enough to overlap the target, $\psi_{sc}(\mathbf{r}, t)$ will be essentially zero. When $|v_0 t|$ becomes of the order of the packet size Δr, this overlap will occur. Our choice for $\psi(\mathbf{r}, t)$, as a superposition of time-independent eigenfunctions $\psi_\mathbf{k}(\mathbf{r})$ satisfying *outgoing-wave* boundary conditions, therefore satisfies the condition

$$\psi(\mathbf{r}, t) \xrightarrow[t \to -\infty]{} \phi(\mathbf{r}, t), \quad (2.13)$$

which only says that for very early times the wave packet reduces to the incident free-particle wave packet.

We now allow the wave packet to proceed through the region of interaction and to reach the force-free region where the asymptotic form is again valid. At this time $\psi_{sc}(\mathbf{r}, t)$ is no longer zero, since the center of the free-particle packet has moved onto the positive z axis. Equation (2.12) is still valid, and it informs us that the scattered wave lies on a sphere of radius $r = v_0 t$ for $t \gg 0$.

After the incident packet has passed the target a spherical scattered wave "shell" centered on the origin and having a radius $r = v_0 t$ emerges from the target. From Eq. (2.12) this shell has a thickness Δr equal to the size of the packet. The wave amplitude decreases as r^{-1}, since the total probability in the scattered packet must remain constant as the packet moves outward from the target. The scattered wave amplitude is distributed over the sphere according to the function $f_{\mathbf{k}_0}(\theta)$.

Let us now suppose that the detector is located at a distance r from the target, and at an angle θ to the incident beam, as in Fig. 1.1. The detector counts all particles that pass through its face, with a projected area dA, which we write as $dA = r^2 \, d\Omega$. The differential is used to indicate that the detector is small enough that $f_{k_0}(\theta)$ may be considered to be constant over its face. Because of condition (2.1), we can place the detector outside the incident beam, that is to say, we have chosen the angular spread of the incident beam to be small enough that, for all but a very small range of angles near $\theta = 0$, the detector will only register scattered particles.

For simplicity we shall first consider the case of a "beam" consisting of one particle and ask for the probability that the particle passes through the detector. Since the incident packet is normalized to unity, this can be obtained by integrating $|\psi_{sc}(\mathbf{r}, t)|^2$ over the volume swept out by the detector as this wave moves through it.

The packet moves with a speed v_0 with no change in shape, so that in a time dt the volume swept out is $v_0 \, dt \, dA$. Thus the probability of detection is simply

$$dP_d(\theta, \phi) = v_0 \, dA \int_{-\infty}^{\infty} dt \, |\psi_{sc}(\mathbf{r}, t)|^2$$
$$= v_0 \, d\Omega \, |f_{k_0}(\theta)|^2 \int_{-\infty}^{\infty} dt \, |\phi(r\hat{\mathbf{z}}, t)|^2. \quad (2.14)$$

This is, however, not a useful form, since it depends both on the details of the scattering process, through $|f_{k_0}(\theta)|^2$, and on the shape of the wave packet. The amplitude of the scattered wave is proportional to $\phi(r\hat{\mathbf{z}}, t)$, which in turn is the amplitude of the incident packet at the position $r\hat{\mathbf{z}}$. This is a function of the shape of the packet because of the normalization condition, and hence is dependent on the detailed properties of the source.

In order to obtain a characterization of the scattering process which is independent of the details of the source, we must be able to eliminate the offending integral over the incident packet. This can easily be done if we recognize that

$$\mathbf{j}_{inc} = \hat{\mathbf{z}} v_0 \int_{-\infty}^{\infty} dt \, |\phi(r\hat{\mathbf{z}}, t)|^2 \quad (2.15)$$

is the probability that the incident particle crosses a unit area located on the z axis. The dependence on the aperture of the detector can also be eliminated. We ask then for the ratio of the probability of detection per unit solid angle at $\mathbf{r}$ to the probability that the incident projectile crosses a unit area located

2. Wave Packet Description of a Scattering Experiment

on the z axis. This ratio is independent of details concerning the source. The ratio so defined has the dimensions of area; it is the differential cross section $\sigma_{\mathbf{k}_0}(\theta)$.

We thus have the fundamental result

$$\sigma_{\mathbf{k}_0}(\theta) \equiv \frac{1}{|\mathbf{j}_{inc}|} \frac{d\, P_d(\theta, \phi)}{d\Omega} = |f_{\mathbf{k}_0}(\theta)|^2. \qquad (2.16)$$

Although the expression "cross section" is used to describe the result of this experimental measurement, in the definition above we have established no relation between the cross section and the classical size, or mass distribution, in the target. We shall see later under what circumstances the cross section so defined is roughly equal to the classical cross-sectional area of the target. In very many situations the physical size of the target particle bears very little relation to the cross section as defined above.

The arguments leading to Eq. (2.16) have two very important implications. First, the differential cross section is a measurable quantity which is independent of the source-detector characteristics. This independence came about because through the use of Eq. (2.12) we were able to relate the shape of the scattered wave packet to the shape of the incident wave packet in such a way that the ratio of measurable quantities defined as $\sigma_{\mathbf{k}_0}(\theta)$ did not depend in any way on the shape of the incident wave packet. We can therefore see the great importance of condition (2.3). If the packet were to change its shape over the period of the experiment, this cancellation would not occur and the observation of the scattering would of necessity involve source characteristics which would be very difficult to eliminate. The second conclusion we may draw from this discussion is that the differential cross section gives the experimental result in terms of the scattering amplitude $f_{\mathbf{k}_0}(\theta)$, which may be determined by solving the *time-independent* Schrödinger equation. Because of this we may carry out our further investigation of the scattering process completely in the time-independent framework.

Before we leave this subject, it is worthwhile to restate the definition of $\sigma_{\mathbf{k}_0}(\theta)$ in terms of a more realistic incident beam. Suppose the source emits a total of N particles. The differential cross section is then the ratio of the number of particles detected at $\mathbf{r}$, per unit solid angle, to the number that cross a unit area located along the incident beam. If the particles are emitted at a constant rate, $\sigma_{\mathbf{k}_0}(\theta)$ is the ratio of the number detected per unit time per unit solid angle to the *incident flux*, that is, to the number per unit time per unit area in the center of the incident beam. This may be taken as the definition of the experimental differential cross section.

3. Conservation of Probability and the Optical Theorem

Although we might realistically discuss all scattering by means of wave packets, we shall use the wave packet description to explore just one more problem. We shall investigate the conditions imposed on the scattering amplitude by the condition that probability must be conserved.

The wave function $\psi(\mathbf{r}, t)$ is a solution of the time-dependent Schrödinger equation

$$i\hbar \frac{\partial \psi}{\partial t} = H\psi. \tag{3.1}$$

In the standard way it follows from this and the Hermitian nature of H that the total probability represented by $\int d^3r \, |\psi(\mathbf{r}, t)|^2$ is conserved. Thus

$$\frac{\partial}{\partial t} \int d^3r \, |\psi(\mathbf{r}, t)|^2 = \int d^3r \, \{\psi^*(\mathbf{r}, t) \frac{\partial \psi}{\partial t}(\mathbf{r}, t) + \frac{\partial \psi^*(\mathbf{r}, t)}{\partial t} \psi(\mathbf{r}, t)\}$$

$$= (1/i\hbar) \int d^3r \, \{\psi^*(\mathbf{r}, t) H \psi(\mathbf{r}, t)$$
$$- (H \psi^*(\mathbf{r}, t)) \psi(\mathbf{r}, t)\} = 0. \tag{3.2}$$

For very early times we have seen that

$$\psi(\mathbf{r}, t) \xrightarrow[t \to -\infty]{} \phi(\mathbf{r}, t).$$

For very late times we found that

$$\psi(\mathbf{r}, t) \xrightarrow[t \to +\infty]{} \phi(\mathbf{r}, t) + \psi_{\text{sc}}(\mathbf{r}, t).$$

If probability is to be conserved, it is necessary that these be equal:

$$\int d^3r \, |\phi(\mathbf{r}, t) + \psi_{\text{sc}}(\mathbf{r}, t)|^2 = \int d^3r \, |\phi(\mathbf{r}, t)|^2, \tag{3.3}$$

or

$$\int d^3r \, |\psi_{\text{sc}}(\mathbf{r}, t)|^2 = -\int d^3r \, \{\phi^*(\mathbf{r}, t) \psi_{\text{sc}}(\mathbf{r}, t)$$
$$+ \phi(\mathbf{r}, t) \psi_{\text{sc}}^*(\mathbf{r}, t)\}. \tag{3.4}$$

If probability is to be conserved while a scattered wave is produced, there must be a decrease in the intensity of the part of the beam that continues along the z axis. This "shadow" is produced by a destructive interference between the incident wave and the scattered wave. The interpretation of

3. Conservation of Probability and the Optical Theorem

Eq. (3.4) is that ϕ and ψ_{sc} interfere in such a way that the loss of probability in the transmitted beam is just equal to the total probability of finding the particle in the scattered beam.

We now evaluate the integrals in Eq. (3.4). We recall that $\psi_{sc}(\mathbf{r}, t)$ is a packet that spreads out radially from the target located at the coordinate origin, while $\phi(\mathbf{r}, t)$ is a packet of size Δr traveling along the z axis with its center at $\mathbf{r} = \mathbf{v}_0 t$. Thus $\psi_{sc}(\mathbf{r}, t)$ and $\phi(\mathbf{r}, t)$ overlap only where $\mathbf{r}$ lies near the z axis, and there is negligible contribution to the right-hand side of Eq. (3.4) except from the neighborhood of $z = v_0 t$ (see Fig. 2.1). As the packets move

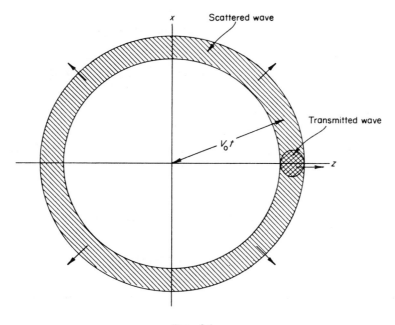

FIG. 2.1

away from the origin, i.e., as $t \to \infty$, the solid angle subtended by $\phi(\mathbf{r}, t)$ goes to zero. Therefore, for large times we may write the right-hand side of Eq. (3.4) as

$$-2 \operatorname{Re}\left\{\int d^3 r\, \phi^*(\mathbf{r}, t) \frac{f(\theta)}{r} \phi(r\hat{\mathbf{z}}, t)\right\}$$

$$\xrightarrow[t \to \infty]{} -2 \operatorname{Re}\left\{f(0) \int d^3 r\, \phi^*(\mathbf{r}, t) \frac{1}{r} \phi(r\hat{\mathbf{z}}, t)\right\}. \quad (3.5)$$

For very late times only the forward-scattering wave can interfere with the incident pulse, so that only the forward-scattering amplitude enters in the interference term.

In the resulting integral only $\phi^*(\mathbf{r}, t)$ depends on the angle of $\mathbf{r}$. If we use Eq. (1.2) for the initial wave packet and the familiar integral

$$\int d\Omega \exp(-i\mathbf{k}\cdot\mathbf{r}) = 4\pi \frac{\sin kr}{kr}, \tag{3.6}$$

we find

$$\int d\Omega \, \phi^*(\mathbf{r}, t) = (2\pi)^{-3} \int d\Omega \, d^3k \, A^*(\mathbf{k}) \exp(-i\mathbf{k}\cdot\mathbf{r}) \exp[i(\hbar k^2/2m)t]$$

$$= (4\pi)(2\pi)^{-3} \int d^3k \, A^*(\mathbf{k}) \frac{\sin kr}{kr} \exp[i(\hbar k^2/2m)t]. \tag{3.7}$$

Through use of condition (2.2), which implies $k^{-1} \simeq k_0^{-1}$, Eq. (3.7) becomes

$$\int d\Omega \, \phi^*(\mathbf{r}, t) = \frac{4\pi}{2ik_0 r} (2\pi)^{-3} \int d^3k \, A^*(\mathbf{k}) \{e^{ikr} - e^{-ikr}\} \exp[i(\hbar k^2/2m)t]. \tag{3.8}$$

If this is compared with Eq. (1.2), we see that this is just

$$\int d\Omega \, \phi^*(\mathbf{r}, t) = \frac{4\pi}{2ik_0 r} \{\phi^*(-r\hat{\mathbf{z}}, t) - \phi^*(r\hat{\mathbf{z}}, t)\}. \tag{3.9}$$

The first term is the amplitude of the incident packet at a point on the negative z axis; this is zero for $t \to +\infty$. Thus, for large t Eq. (3.7) reduces to

$$\int d\Omega \, \phi^*(\mathbf{r}, t) \xrightarrow[t \to +\infty]{} -\frac{4\pi}{2ik_0 r} \phi^*(r\hat{\mathbf{z}}, t), \tag{3.10}$$

and Eq. (3.5) becomes

$$-2 \, \mathrm{Re}\left\{\int d^3r \, \phi^*(\mathbf{r}, t) \, r^{-1} f(\theta) \, \phi(r\hat{\mathbf{z}}, t)\right\}$$

$$\xrightarrow[t \to +\infty]{} -2 \, \mathrm{Re}\{(2\pi i/k_0) f(0)\} \int_0^\infty dr \, |\phi(r\hat{\mathbf{z}}, t)|^2. \tag{3.11}$$

3. Conservation of Probability and the Optical Theorem

On the other hand, for large t the left-hand side of Eq. (3.4) becomes

$$\int d^3r\, |\psi_{\text{sc}}(\mathbf{r}, t)|^2 \xrightarrow[t \to +\infty]{} \int_0^\infty dr \int d\Omega\, |f(\theta)|^2\, |\phi(r\hat{\mathbf{z}}, t)|^2$$

$$= \sigma_{\text{TOT}} \int_0^\infty dr\, |\phi(r\hat{\mathbf{z}}, t)|^2, \tag{3.12}$$

where $\sigma_{\text{TOT}} \equiv \int d\Omega\, |f(\theta)|^2$ is the total scattering cross section. Combining Eqs. (3.4), (3.11), and (3.12), we obtain

$$\frac{4\pi}{k_0} \operatorname{Im} f(0) = \sigma_{\text{TOT}}. \tag{3.13}$$

This important consequence of probability conservation is the *optical theorem*. It not only shows that the scattering amplitude is always complex—since σ_{TOT} is always positive unless $f(\theta)$ is identically zero—but that the real and imaginary parts of the scattering amplitude are not completely independent of each other. It will play a fundamental role in much of the succeeding discussion.

Chapter

3

Differential Equation Methods

In this chapter we shall describe those aspects of scattering theory that are manifested in elastic scattering by a central, spherically symmetric potential. Many of these same results will be derived more generally and more easily in later chapters. However, we will use this special but important case to illustrate the physical ideas that underlie the scattering analysis. The time-independent Schrödinger equation will be used as the basis for the development. In the previous chapters we saw that the time-independent approach is quite general and completely correct.

1. Partial Wave Expansion

The Schrödinger equation describing the relative motion of two particles is

$$\left[\frac{-\hbar^2}{2m}\nabla^2 + V(r)\right]\psi(\mathbf{r}) = E\,\psi(\mathbf{r}). \tag{1.1}$$

In spherical coordinates the Laplacian operator ∇^2 may be written as

$$\nabla^2 = \frac{1}{r^2}\frac{\partial}{\partial r}\left(r^2\frac{\partial}{\partial r}\right) - \frac{1}{\hbar^2 r^2}L^2(\theta, \varphi), \tag{1.2}$$

where

$$L^2(\theta, \varphi) = -\hbar^2\left[\frac{1}{\sin\theta}\frac{\partial}{\partial\theta}\left(\sin\theta\frac{\partial}{\partial\theta}\right) + \frac{1}{\sin^2\theta}\frac{\partial^2}{\partial\varphi^2}\right]. \tag{1.3}$$

The operator L^2 given by Eq. (1.3) is just

$$\mathbf{L}^2 = (\mathbf{r} \times \mathbf{p}) \cdot (\mathbf{r} \times \mathbf{p}) = -\hbar^2(\mathbf{r} \times \nabla) \cdot (\mathbf{r} \times \nabla) \tag{1.4}$$

and represents the square of the orbital angular momentum. The spherical harmonics $Y_{lm}(\theta, \phi)$ are defined to be eigenfunctions of L^2 and the z component of **L**,

$$L_z = (\mathbf{r} \times \mathbf{p}) \cdot \hat{z} = -i\hbar \frac{\partial}{\partial \varphi}, \tag{1.5}$$

such that

$$L^2 Y_{lm}(\theta, \varphi) = l(l+1)\hbar^2 Y_{lm}(\theta, \varphi) \tag{1.6}$$

and

$$L_z Y_{lm}(\theta, \varphi) = m\hbar Y_{lm}(\theta, \varphi). \tag{1.7}$$

The function

$$\psi_{lm}(\mathbf{r}) \equiv r^{-1} u_l(r) Y_{lm}(\theta, \varphi) \tag{1.8}$$

is then a solution of Eq. (1.1), provided $u_l(r)$ is a solution of the radial equation

$$\frac{d^2 u_l(r)}{dr^2} + \left[k^2 - U(r) - \frac{l(l+1)}{r^2} \right] u_l(r) = 0, \tag{1.9}$$

where $k^2 = (2m/\hbar^2)E$ and $U(r) = (2m/\hbar^2) V(r)$.

The energy E can be chosen to be positive and equal to the kinetic energy of the projectile when it is far from the scattering center. The potential $V(r)$ will be assumed to vanish sufficiently rapidly with increasing r that it may be neglected beyond some radius $r = R$. This assumption does not hold for a potential that vanishes at infinity as r^{-n}. The most important example of such potential, the Coulomb potential, will be discussed separately.

We shall also require that $u_l(r)$ vanish at the origin. This boundary condition at the origin follows from the requirement that the wave function $\psi(\mathbf{r})$ and its gradient be finite everywhere, in particular at $r = 0$. This condition is sufficient to determine the radial wave function, to within an over-all normalization constant that will be fixed later.

It is evident that any linear combination of the functions $\psi_{lm}(\mathbf{r})$, for different l and m, is also a solution of Eq. (1.1). The physical situation is symmetric about the z axis, so we shall seek a solution of Eq. (1.1) which has this symmetry and is therefore independent of ϕ. We may write such a solution as

$$\psi(\mathbf{r}) = (kr)^{-1} \sum_{l=0}^{\infty} u_l(r) Y_{l,0}(\theta), \tag{1.10}$$

where the radial function $u_l(r)$ satisfies the radial equation, Eq. (1.9).

Much of what follows is, of necessity, an exercise in the solution of second-order differential equations. However, we should not lose sight of the physical significance of the results we obtain, since we are interested primarily in the physical scattering problem.

Asymptotic Behavior of the Radial Wave Function

Let us first examine the form of $u_l(r)$ beyond the range of the potential. For $r > R$, $u_l(r)$ may be expressed in terms of the solutions of the differential equation

$$\left[\frac{d^2}{dr^2} + k^2 - \frac{l(l+1)}{r^2}\right] w_l(kr) = 0. \tag{1.11}$$

The functions $w_l(kr)$ are related to the well-known spherical Bessel functions, that is, $w_l(kr)$ may be any linear combination of the regular and irregular solutions

$$F_l(kr) = kr\, j_l(kr) = (\tfrac{1}{2}\pi kr)^{1/2} J_{l+\frac{1}{2}}(kr) \tag{1.12}$$

and

$$G_l(kr) = -kr\, n_l(kr) = (\tfrac{1}{2}\pi kr)^{1/2}(-1)^l J_{-l-\frac{1}{2}}(kr). \tag{1.13}$$

As examples, $F_0(kr) = \sin kr$ and $G_0(kr) = \cos kr$. The function $G_l(kr)$ defined in Eq. (1.13) is conventionally referred to as the irregular free solution, but of course any solution of Eq. (1.11) that is independent of $F_l(kr)$ is irregular. Thus any linear combination of $F_l(kr)$ and $G_l(kr)$ can be taken as the irregular solution to Eq. (1.11).

For our purposes the most important properties of these functions are their behavior for small and large values of r. Near the origin

$$F_l(kr) \xrightarrow[kr \ll l]{} \frac{(kr)^{l+1}}{1\cdot 3\cdot 5 \cdots (2l+1)} \tag{1.14}$$

and

$$G_l(kr) \xrightarrow[kr \ll l]{} \frac{1\cdot 3\cdot 5 \cdots (2l-1)}{(kr)^l}. \tag{1.15}$$

This behavior determines the number of partial waves that will make a significant contribution to the scattering process. In particular, the behavior of $F_l(kr)$ reflects the familiar angular momentum "barrier" which causes the wave function to decrease rapidly as r decreases for $kr \ll l$. Thus, for values

1. Partial Wave Expansion

of $l \gg kR$, we would not expect the incident free-particle wave to penetrate the potential and would, correspondingly, expect little scattering.

At large distances these functions behave as

$$F_l(kr) \xrightarrow[kr \gg l]{} \sin(kr - \tfrac{1}{2}l\pi) \tag{1.16}$$

and

$$G_l(kr) \xrightarrow[kr \gg l]{} \cos(kr - \tfrac{1}{2}\pi l). \tag{1.17}$$

For $r > R$, $u_l(r)$ is a linear combination of the free functions $F_l(kr)$ and $G_l(kr)$:

$$u_l(r) = A\, F_l(kr) + B\, G_l(kr). \tag{1.18}$$

Although both A and B may be complex, the ratio (B/A) must be real if $V(r)$ is real. We can see this most easily by noting that, since the operator in Eq. (1.9) is real, a regular solution that is real can always be found, and any other regular solution of Eq. (1.9) can differ from this solution only by a complex multiplicative constant. At large distances the radial wave function given by Eq. (1.18) then has the form

$$\begin{aligned} u_l(r) \xrightarrow[r \to \infty]{} & A \sin(kr - \tfrac{1}{2}\pi l) + B \cos(kr - \tfrac{1}{2}\pi l) \\ = & (A^2 + B^2)^{1/2} \sin(kr - \tfrac{1}{2}\pi l + \delta_l), \end{aligned} \tag{1.19}$$

where we have introduced the real phase shift

$$\delta_l = \tan^{-1}(B/A). \tag{1.20}$$

We have noted that the boundary condition of regularity at the origin determines the solution of the radial equation [Eq. (1.9)] to within a multiplicative constant. Beyond the range of the force, the radial function is given by Eq. (1.18), and it is clear that the proportion of regular to irregular solutions is independent of r for all values of $r > R$. In the asymptotic region, $r \to \infty$, where the regular and irregular functions take on their asymptotic values as given by Eqs. (1.16) and (1.17), the radial function is given by Eqs. (1.19) and (1.20).

Comparing this result with Eq. (1.16), the effect of the potential on the lth partial wave in the asymptotic region can be seen to be a shift in the phase of the radial function relative to the phase of $F_l(kr)$, the regular solution in the absence of the potential. This is illustrated in Fig. 3.1, in which the effects of a weak attractive and a weak repulsive potential are compared.

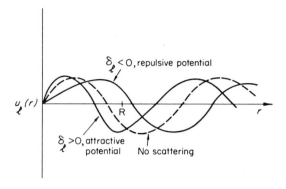

Fig. 3.1

Normalization of the Radial Wave Function

The normalization of $u_l(r)$ is determined by the requirement that $\psi(\mathbf{r}) = \psi(r, \theta)$, given by Eq. (1.10), have the asymptotic behavior

$$\psi(\mathbf{r}) \xrightarrow[r \to \infty]{} e^{ikz} + \frac{e^{ikr}}{r} f(\theta) = \phi(\mathbf{r}) + \psi_{sc}(\mathbf{r}). \quad (1.21)$$

The work of the previous chapters was primarily concerned with establishing Eq. (1.21) as the desired boundary condition.

The plane wave $e^{i\mathbf{k}\cdot\mathbf{r}}$ is a solution of the free Schrödinger equation

$$-(\hbar^2/2m)\nabla^2 \phi(\mathbf{r}) = E\,\phi(\mathbf{r}). \quad (1.22)$$

If we take **k** to be along the z axis, the plane wave is independent of azimuth and hence can be expressed as a sum of products of spherical harmonics $Y_{l0}(\theta)$ and the regular solutions $F_l(kr)$ of Eq. (1.11). The resulting expansion is

$$e^{ikz} = (kr)^{-1} \sum_{l=0}^{\infty} i^l [4\pi(2l+1)]^{1/2}\, F_l(kr)\, Y_{l0}(\theta). \quad (1.23)$$

The general plane wave describing motion in an arbitrary direction $(\theta_\mathbf{k}, \varphi_\mathbf{k})$ may be expanded as

$$e^{i\mathbf{k}\cdot\mathbf{r}} = 4\pi(kr)^{-1} \sum_{l=0}^{\infty} \sum_{m=-l}^{l} i^l\, F_l(kr)\, Y_{lm}^*(\theta_\mathbf{k}, \varphi_\mathbf{k})\, Y_{lm}(\theta, \varphi). \quad (1.24)$$

It is of interest to note that the asymptotic wave function [Eq. (1.21)] is not a solution to the force-free equation. The plane wave part of Eq. (1.21), $\phi(\mathbf{r})$, is such a solution, but the outgoing spherical wave part $\psi_{sc}(\mathbf{r})$ is not.

1. Partial Wave Expansion

By applying the kinetic energy operator to $\psi_{sc}(\mathbf{r}) = r^{-1} e^{ikr} f(\theta)$, we find

$$-\frac{\hbar^2}{2m} \nabla^2 \psi_{sc}(\mathbf{r}) = \frac{e^{ikr}}{r} \left[E - \frac{1}{2mr^2} L^2(\theta, \varphi) \right] f(\theta). \tag{1.25}$$

If $\psi_{sc}(\mathbf{r})$ were to satisfy the force-free equation, then $f(\theta)$ would satisfy the equation

$$L^2(\theta, \varphi) f(\theta) = 0, \tag{1.26}$$

which is Legendre's equation of order zero [cf. Eq. (1.6)]. This would imply that $f(\theta)$ is a linear combination of $P_0(\theta)$ and $Q_0(\theta)$, but since $P_0(\theta) = 1$ and $Q_0(\theta) = \frac{1}{2} \log[(\cos\theta + 1)/(\cos\theta - 1)]$, a scattering amplitude that satisfied Eq. (1.26) could not satisfy the optical theorem, to single out but one of the many fatal defects of such a scattering amplitude. Thus, we see that $\psi_{sc}(\mathbf{r})$ is indeed only an *asymptotic* solution of the force-free equation, and beyond the range of the force we have

$$\psi(\mathbf{r}) \underset{r > R}{=} \phi(\mathbf{r}) + \frac{e^{ikr}}{r} f(\theta) + \mathcal{O}\left(\frac{1}{r^3}\right). \tag{1.27}$$

We have now seen that a regular solution of the radial Schrödinger equation, Eq. (1.9), has the asymptotic form

$$u_l(r) \underset{r \to \infty}{\longrightarrow} N_l \sin(kr - \tfrac{1}{2}l\pi + \delta_l), \tag{1.28}$$

and we wish to determine the normalization constant N_l from the asymptotic boundary condition, Eq. (1.21). To do this we use Eq. (1.10) to obtain

$$\begin{aligned}
\psi(\mathbf{r}) \underset{r \to \infty}{\longrightarrow} &\ (kr)^{-1} \sum_{l=0}^{\infty} N_l \sin(kr - \tfrac{1}{2}l\pi + \delta_l) Y_{l0}(\theta) \\
= &\ (kr)^{-1} \sum_{l=0}^{\infty} N_l \{\cos\delta_l \sin(kr - \tfrac{1}{2}\pi l) \\
&\ + \sin\delta_l \cos(kr - \tfrac{1}{2}\pi l)\} Y_{l0}(\theta) \\
= &\ (kr)^{-1} \sum_{l=0}^{\infty} N_l (\cos\delta_l - i\sin\delta_l) \sin(kr - \tfrac{1}{2}l\pi) Y_{l0}(\theta) \\
&\ + (kr)^{-1} e^{ikr} \sum_{l=0}^{\infty} N_l \sin\delta_l\, i^{-l} Y_{l0}(\theta) \\
\equiv &\ e^{ikz} + r^{-1} e^{ikr} f(\theta) \\
= &\ (kr)^{-1} \sum_{l=0}^{\infty} i^l [4\pi(2l+1)]^{1/2} \sin(kr - \tfrac{1}{2}l\pi) Y_{l0}(\theta) \\
&\ + r^{-1} e^{ikr} f(\theta).
\end{aligned} \tag{1.29}$$

From this one obtains the normalization constant

$$N_l = i^l e^{i\delta_l} [4\pi(2l + 1)]^{1/2}. \tag{1.30}$$

We recall that

$$Y_{l0}(\theta) = [(2l + 1)/4\pi]^{1/2} P_l(\cos \theta), \tag{1.31}$$

so that the scattering amplitude can then be identified as

$$f(\theta) = k^{-1} \sum_{l=0}^{\infty} (2l + 1) e^{i\delta_l} \sin \delta_l P_l(\cos \theta). \tag{1.32}$$

It is convenient now to write the wave function as

$$\psi(\mathbf{r}) = (kr)^{-1} \sum_{l=0}^{\infty} (2l + 1) i^l u_l(r) P_l(\cos \theta) \tag{1.33}$$

and to normalize $u_l(r)$ such that

$$u_l(r) \xrightarrow[r \to \infty]{} e^{i\delta_l} \sin(kr - \tfrac{1}{2}l\pi + \delta_l) \tag{1.34}$$

or

$$u_l(r) \underset{r > R}{=} e^{i\delta_l} \{\cos \delta_l F_l(kr) + \sin \delta_l G_l(kr)\}. \tag{1.35}$$

If $\delta_l = 0$, $u_l(r)$ reduces to $F_l(kr)$ and Eq. (1.33) reduces to the plane wave expansion.

Equation (1.35) can be written in several alternative ways which show the relation of the phase shift to the scattering process. In this original form the asymptotic behavior of the wave function is

$$\begin{aligned}
u_l(r) \underset{r > R}{=}\ & e^{i\delta_l} [\cos \delta_l F_l(kr) + \sin \delta_l G_l(kr)] \\
\xrightarrow[r \to \infty]{}\ & e^{i\delta_l} [\cos \delta_l \sin(kr - \tfrac{1}{2}\pi l) + \sin \delta_l \cos(kr - \tfrac{1}{2}\pi l)] \\
=\ & e^{i\delta_l} \sin(kr - \tfrac{1}{2}\pi l + \delta_l).
\end{aligned} \tag{1.36}$$

The presence of the potential introduces into the wave function beyond the range of the force a contribution from the irregular free-particle solution $G_l(kr)$. The relative amplitude of this contribution is conventionally denoted by

$$K_l = \tan \delta_l. \tag{1.37}$$

Alternatively, the wave function can be expressed as

$$u_l(r) \underset{r > R}{=} F_l(kr) + e^{i\delta_l} \sin \delta_l \{G_l(kr) + i F_l(kr)\}$$
$$\underset{r \to \infty}{\longrightarrow} \sin(kr - \tfrac{1}{2}\pi l) + \exp[i(kr - \tfrac{1}{2}\pi l)] e^{i\delta_l} \sin \delta_l. \quad (1.38)$$

The scattering process adds to the free-particle regular wave function an outgoing wave whose amplitude is conventionally denoted by

$$T_l = e^{i\delta_l} \sin \delta_l. \quad (1.39)$$

Even though this outgoing wave is added to the incident plane wave function, the total number of particles must be conserved. We have already investigated in previous chapters whether the additional outgoing spherical wave can be compatible with current conservation, and we found that the scattering amplitude had to satisfy the optical theorem. We now wish to see whether the wave function we have just obtained is compatible with the conservation of current.

Instead of simply reproducing the earlier demonstrations, let us this time calculate the asymptotic radial current from a wave function of the form

$$\psi_l(\mathbf{r}) \underset{r \to \infty}{\longrightarrow} r^{-1} [A_l \sin(kr - \tfrac{1}{2}\pi l) + B_l \cos(kr - \tfrac{1}{2}\pi l)] Y_{l0}(\theta). \quad (1.40)$$

The result is

$$\mathbf{j}(\mathbf{r}) \underset{r \to \infty}{\longrightarrow} (\hbar k/m) r^{-1} \operatorname{Im}(A_l B_l^*) Y_{l0}(\theta) \hat{\mathbf{r}}, \quad (1.41)$$

so that there is no net current if the requirement $\operatorname{Im} A_l B_l^* = 0$ is met. This condition is equivalent to requiring that the ratio $B/A = \tan \delta_l$ be real. For scattering by a central potential, we have already seen that the phase shift is real if the potential is real. If the phase shifts are real, the optical theorem is satisfied, for the total cross section is then

$$\sigma_{\text{TOT}} = \int d\Omega |f(\theta)|^2 = 4\pi k^{-2} \sum_l (2l+1) \sin^2 \delta_l = 4\pi k^{-1} \operatorname{Im} f(\theta = 0), (1.42)$$

where we have used Eq. (1.32) and the fact that $P_l(\theta = 0) = 1$. The reader may notice also that, through any surface completely enclosing the origin, there is no net current arising from the interference of different partial waves, even if the phase shifts are not real. The vanishing of this interference requires only that the potential be spherically symmetric so that the expansion in spherical harmonics, Eq. (1.10), is valid.

Current conservation can be seen explicitly in a third form of $u_l(r)$ using ingoing and outgoing waves. This can be obtained by expressing the radial wave function beyond the range of the force as

$$u_l(r) \underset{r>R}{=} \tfrac{1}{2}i[\{G_l(kr) - i F_l(kr)\} - e^{2i\delta_l}\{G_l(kr) + i F_l(kr)\}]$$
$$\underset{r\to\infty}{\longrightarrow} \tfrac{1}{2}i[\exp(-i(kr - \tfrac{1}{2}\pi l)) - e^{2i\delta_l}\exp(i(kr - \tfrac{1}{2}\pi l))]. \quad (1.43)$$

Here we can see that the incoming spherical wave is unaffected by the scattering process, while the outgoing wave is multiplied by the quantity

$$S_l = e^{2i\delta_l}. \quad (1.44)$$

Since $|S_l| = 1$ for a real phase shift, it is apparent, as noted above, that the net flux of particles is zero. Only the phase, and not the amplitude, of the outgoing spherical wave is affected by the presence of the potential.

We may summarize this discussion in a slightly different way. Instead of choosing $F_l(kr)$ and $G_l(kr)$ as the linearly independent solutions of the force-free equation, in terms of which to express the radial wave function beyond the range of force, we might use two other linearly independent solutions of Eq. (1.11). These new functions are necessarily linear combinations of $F_l(kr)$ and $G_l(kr)$. If we choose them to be $F_l(kr)$ and $H_l^{(+)}(kr) \equiv G_l(kr) + i F_l(kr)$, then the radial function is

$$u_l(r) \underset{r>R}{=} A_l' F_l(kr) + B_l' H_l^{(+)}(kr), \quad (1.45)$$

and $B_l'/A_l' = e^{i\delta_l}\sin\delta_l = T_l$. If the two functions are taken to be $H_l^{(+)}(kr)$ and $H_l^{(-)}(kr) \equiv G_l(kr) - i F_l(kr)$, then the radial wave function is

$$u_l(r) \underset{r>R}{=} A_l'' H_l^{(-)}(kr) + B_l'' H_l^{(+)}(kr) \quad (1.46)$$

and $B_l''/A_l'' = e^{2i\delta_l} \equiv S_l$. It will be remembered that our original choice,

$$u_l(r) \underset{r>R}{=} A_l F_l(kr) + B_l G_l(kr), \quad (1.47)$$

gave $B_l/A_l = \tan\delta_l = K_l$.

We shall see later that the three modes of expression we have used, involving the quantities K_l, T_l, and S_l, are the one-particle analogs of the K or reaction matrix, the T or transition matrix, and the S or scattering matrix, respectively, in the general scattering formalism.

2. Calculation of the Phase Shift

The computation of the phase shift can be illustrated by means of the

2. Calculation of the Phase Shift

square well potential, the so-called "potential hole." Let $V(r)$ be given by

$$V(r) = \begin{cases} V_0, & r < R, \\ 0, & r > R. \end{cases} \qquad (2.1)$$

If U_0 is defined to be $(2m/\hbar^2)V_0$, the radial Schrödinger equation for the square well becomes

$$\frac{d^2 u_l(r)}{dr^2} + \left[k^2 - U_0(r) - \frac{l(l+1)}{r^2} \right] u_l(r) = 0 \quad \text{for} \quad r < R \qquad (2.2)$$

and

$$\frac{d^2 u_l(r)}{dr^2} + \left[k^2 - \frac{l(l+1)}{r^2} \right] u_l(r) = 0 \quad \text{for} \quad r > R. \qquad (2.3)$$

The regular solution of Eq. (2.2) which vanishes at the origin is

$$u_l(r) = C\, F_l(Kr), \qquad (2.4)$$

where $K^2 = k^2 - U_0$. The solution to Eq. (2.3) is written, as in Section 1, as

$$u_l(r) = A\, F_l(kr) + B\, G_l(kr). \qquad (2.5)$$

The condition that the wave function and its derivative be continuous at $r = R$, the point at which the potential is discontinuous, is

$$\lim_{\varepsilon \to 0} u_l(r = R - \varepsilon) = \lim_{\varepsilon \to 0} u(r = R + \varepsilon) \qquad (2.6)$$

and

$$\lim_{\varepsilon \to 0} L(r = R - \varepsilon) = \lim_{\varepsilon \to 0} L(r = R + \varepsilon), \qquad (2.7)$$

where the logarithmic derivative L is defined by

$$L(r) = \frac{\frac{d}{dr}\{u_l(r)\}}{u_l(r)} = \frac{u_l'(r)}{u_l(r)}. \qquad (2.8)$$

The continuity of the logarithmic derivative at $r = R$ gives

$$L(R) = \frac{F_l'(kR) + (B/A)\, G_l'(kR)}{F_l(kR) + (B/A)\, G_l(kR)} = \frac{F_l'(KR)}{F_l(KR)}, \qquad (2.9)$$

and the phase shift is obtained by solving this:

$$\tan \delta_l = \frac{B}{A} = -\frac{F_l'(kR) - L(R)\, F_l(kR)}{G_l'(kR) - L(R)\, G_l(kR)}. \qquad (2.10)$$

where $L(R)$ is taken to be

$$L(R) = \frac{F_l'(KR)}{F_l(KR)}. \qquad (2.11)$$

[Note that $F_l'(KR) \equiv (d/dr) F_l(KR)$.]

This result was derived for a square well potential. However, Eq. (2.10) is far more general than this particular application might suggest. If $V(r)$ is any potential that can be assumed to vanish for r larger than R, $u_l(r)$ will have the free-particle form, Eq. (2.5), for $r > R$. Assuming that we have obtained a regular solution of the radial Schrödinger equation, Eq. (1.9), Eq. (2.10) can be used to obtain the phase shift with $L(R)$ determined from Eq. (2.8).

We shall here use Eq. (2.10) to calculate just one example. Let us take the case of scattering by a hard sphere, where the potential is repulsive and infinitely strong, and calculate the S wave or zero angular momentum phase shift. In the limit as $K^2 \to -\infty$, $L(R)$ goes to infinity and Eq. (2.10) becomes

$$\tan \delta_0 = -F_0(kR)/G_0(kR) = -\tan(kR), \qquad (2.12)$$

and hence

$$\delta_0 = -kR. \qquad (2.13)$$

Since the hard sphere is impenetrable, the wave function must be zero for $r < R$. Outside the sphere it must be $\sin(kr + \delta_0)$. By continuity, then, we find Eq. (2.13).

At low energies only S-wave, $l = 0$, scattering occurs, so that if $kR \ll 1$,

$$f(\theta) = -k^{-1} e^{-ikR} \sin kR, \qquad (2.14)$$

and the total cross section for hard-sphere scattering is

$$\sigma_{\text{TOT}} = 4\pi k^{-2} \sin^2 kR \sim 4\pi R^2 \left[1 + \mathcal{O}(k^2 R^2)\right]. \qquad (2.15)$$

The low-energy cross section for hard-sphere scattering is thus four times the classical cross section.

INTEGRAL EXPRESSIONS FOR THE PHASE SHIFT

We can obtain from the Schrödinger equation expressions for the phase shift in terms of integrals over the wave function $u_l(r)$, as well as differential relations such as Eq. (2.10). These integral expressions are useful in making approximate calculations and are important in the formal theory discussed in later chapters.

2. Calculation of the Phase Shift

We begin with the Schrödinger equation describing the scattering,

$$u_l''(r) + \left[k^2 - \frac{l(l+1)}{r^2} - U(r)\right] u_l(r) = 0, \tag{2.16}$$

and the differential equation for a free particle,

$$w_l''(r) + \left[k^2 - \frac{l(l+1)}{r^2}\right] w_l(r) = 0. \tag{2.17}$$

Multiplication of Eq. (2.16) by $w_l(r)$ and of Eq. (2.7) by $u_l(r)$, followed by subtraction of the resulting equations, yields

$$w_l(r) u_l''(r) - u_l(r) w_l''(r) = w_l(r) U(r) u_l(r). \tag{2.18}$$

We now integrate this from the origin to infinity and obtain

$$\{w_l(r) u_l'(r) - u_l(r) w_l'(r)\}\Big|_0^\infty = \int_0^\infty dr\, w_l(r) U(r) u_l(r). \tag{2.19}$$

By choosing $w_l(r)$ to be the regular solution of Eq. (2.17),

$$w_l(r) = F_l(kr), \tag{2.20}$$

and further recognizing that, for $r > R$,

$$u_l(r) = A\, F_l(kr) + B\, G_l(kr), \tag{2.21}$$

we obtain, from Eq. (2.19),

$$B = -k^{-1} \int_0^\infty dr\, F_l(kr) U(r) u_l(r). \tag{2.22}$$

The reader will observe that Eq. (2.22) depends on the normalization of $u_l(r)$. We have insisted that $u_l(r)$ be the regular solution of the radial Schrödinger equation but, if its normalization remains unspecified, then B as given by Eq. (2.22) is also unspecified. If the radial wave function is normalized such that

$$u_l(r) \underset{r>R}{=} e^{i\delta_l} \{\cos \delta_l\, F_l(kr) + \sin \delta_l\, G_l(kr)\}, \tag{2.23}$$

then Eq. (2.22) becomes

$$e^{i\delta_l} \sin \delta_l = T_l = -k^{-1} \int_0^\infty dr\, F_l(kr) U(r) u_l(r). \tag{2.24}$$

On the other hand, if the radial wave function is normalized such that

$$u_l(r) \underset{r>R}{=} F_l(kr) + \tan \delta_l \, G_l(kr), \tag{2.25}$$

then Eq. (2.22) becomes

$$\tan \delta_l = K_l = -k^{-1} \int_0^\infty dr \, F_l(kr) \, U(r) \, u_l(r). \tag{2.26}$$

Numerically, it is usually much easier to work with the normalization of Eq. (2.25), and hence Eq. (2.26), since for a real potential we may then work in a representation in which the wave function $u_l(r)$ is entirely real.

In actual numerical calculations the radial differential equation is integrated from the origin. In that case the normalization of the numerical solution is fixed by the derivative of the wave function at the origin and thus could be quite arbitrary. For the regular solution of Eq. (2.16) one might choose $u_l(r)$ such that, for small r,

$$u_l(r) \underset{r \to 0}{\longrightarrow} \Lambda \, F_l(kr) \underset{r \to 0}{\longrightarrow} \Lambda \, \frac{k^{l+1} r^{l+1}}{1 \cdot 3 \cdot 5 \cdots (2l+1)}. \tag{2.27}$$

If we assume that $V(r)$ is less singular at the origin than $r^{-1-|\varepsilon|}$, then Eq. (2.27) follows immediately from Eq. (2.16). But since the Schrödinger equation is not soluble at all if $V(r)$ is more singular than r^{-1}, this is completely general.

With this choice the differential equation can be integrated numerically, if necessary. For large r the solution will have the form of Eq. (2.21). If in Eq. (2.19) the solution of the free radial equation $w_l(r)$ is chosen to be

$$w_l(r) = G_l(kr), \tag{2.28}$$

then evaluation of the left-hand side of Eq. (2.19) gives

$$\{w_l(r) \, u_l'(r) - u_l(r) \, w_l'(r)\} \Big|_0^\infty = (A - \Lambda) k$$

$$= \int_0^\infty dr \, G_l(kr) \, U(r) \, u_l(r), \tag{2.29}$$

where

$$\Lambda \equiv [u_l(r)/F_l(kr)]_{r=0} = \frac{1 \cdot 3 \cdot 5 \cdots (2l+1)}{k^{l+1}} \left[\frac{u_l(r)}{r^{l+1}}\right]_{r=0}$$

$$= \frac{1 \cdot 3 \cdot 5 \cdots (2l+1)}{(l+1)! k^{l+1}} \left[\frac{\partial^{l+1} u_l(r)}{\partial r^{l+1}}\right]_{r=0}. \tag{2.30}$$

2. Calculation of the Phase Shift

Thus we may write an integral expression for the amplitude of the wave function

$$A = \Lambda + k^{-1} \int_0^\infty dr\, G_l(kr)\, U(r)\, u_l(r)$$

$$= \lim_{r \to 0} \left\{ \frac{1 \cdot 3 \cdot 5 \cdots (2l+1)\, u_l(r)}{(kr)^{l+1}} \right\} + k^{-1} \int_0^\infty dr\, G_l(kr)\, U(r)\, u_l(r). \quad (2.31)$$

Now we are in a position to achieve an equation for K_l which is independent of the normalization of $u_l(r)$, viz.,

$$B/A = K_l = \tan \delta_l$$

$$= \frac{-k^{-1} \int_0^\infty dr\, F_l(kr)\, U(r)\, u_l(r)}{\lim_{r \to 0}\{1 \cdot 3 \cdot 5 \cdots (2l+1)(kr)^{-l-1} u_l(r)\} + k^{-1} \int_0^\infty dr\, G_l(kr)\, U(r)\, u_l(r)}. \quad (2.32)$$

The expression for T_l,

$$T_l = e^{i\delta_l} \sin \delta_l$$

$$= \frac{-k^{-1} \int_0^\infty dr\, F_l(kr)\, U(r)\, u_l(r)}{\lim_{r \to 0}\{1 \cdot 3 \cdot 5 \cdots (2l+1)(kr)^{-l-1} u_l(r)\} + k^{-1} \int_0^\infty dr\, H_l^{(+)}(kr)\, U(r)\, u_l(r)}, \quad (2.33)$$

may be obtained in exactly the same way, with the substitution of $H_l^{(+)}(kr) \equiv G_l(kr) + i F_l(kr)$ for $G_l(kr)$ in the above derivation. Similarly, the expression for S_l,

$$S_l = e^{2i\delta_l}$$

$$= \frac{\lim_{r \to 0}\{1 \cdot 3 \cdot 5 \cdots (2l+1)(kr)^{-l-1} u_l(r)\} + k^{-1} \int_0^\infty dr\, H_l^{(-)}(kr)\, U(r)\, u_l(r)}{\lim_{r \to 0}\{1 \cdot 3 \cdot 5 \cdots (2l+1)(kr)^{-l-1} u_l(r)\} + k^{-1} \int_0^\infty dr\, H_l^{(+)}(kr)\, U(r)\, u_l(r)}, \quad (2.34)$$

is readily obtained. Since these expressions are independent of normalization, we must be able to relate one to another. Thus, for example, the use of Eq. (2.34) and the identity $T_l = (2i)^{-1}(S_l - 1)$ immediately gives Eq. (2.33).

We may observe that the expressions for T_l and K_l given earlier by Eqs. (2.24) and (2.26) are precisely Eqs. (2.32) and (2.33), with the normalization so chosen in each case as to make the denominator unity. While it is entirely possible to choose a normalization for $u_l(r)$ such that the denominator of Eq. (2.34) is unity also, we shall see later that such a choice of normalization is not very useful.

At very low energies, where $kR << l$, we see from either Eq. (2.24) or Eq. (2.26) that the phase shift goes as

$$T_l \sim K_l \sim \delta_l \sim k^{2l+1}, \qquad kR << l. \tag{2.35}$$

This follows from the low-energy properties of the spherical Bessel functions and the fact that, in order for the denominator to be unity in Eq. (2.32) or in Eq. (2.33) as $k \to 0$, $u_l(r)$ must be normalized such that

$$u_l(k, r) \xrightarrow[k \to 0]{} k^{l+1} \eta_l(r), \tag{2.36}$$

where $\eta_l(r)$ is independent of k. At the origin, of course, $\eta_l(r) \sim r^{l+1}$. It is *not* correct, however, to assume that we may replace $u_l(r)$ by $F_l(kr)$ either in Eqs. (2.24) and (2.26) or in Eqs. (2.32) and (2.33). We shall see later, in the discussion of the perturbation expansion, under what circumstances so doing may prove to be a good approximation.

From Eq. (2.35) we see that at low energies only S-wave scattering will contribute to the scattering amplitude. Moreover, for any finite energy, provided only that the range of the force is finite, the angular momentum expansion for the scattering amplitude will effectively terminate, since for $l >> kR$ the phase shift vanishes.

At very high energies, where $kR >> l$, we see from either Eq. (2.24) or Eq. (2.26) that the phase shift behaves as

$$\delta_l \sim k^{-1}, \qquad kR >> l. \tag{2.37}$$

At high energies it does become a good approximation to replace $u(r)$ by $F(kr)$ in Eqs. (2.24) and (2.26) or in Eqs. (2.32) and (2.33). This point will be discussed at greater length in Chapter 5.

The Two-Potential Formula

Our treatment of scattering has concerned itself primarily with determining how the solutions of the free Schrödinger equation are affected by the presence of the interaction. Thus, if H_0 is the kinetic energy operator, our investigations have concerned themselves with the comparison of the solutions for the Hamiltonian $H = H_0 + V$ with the solutions for H_0. It is not necessary, however, that H_0 be the free-particle Hamiltonian, and the methods we have used are readily adapted to other choices for H_0. On occasion physically meaningful questions can thus be framed more appropriately. For example, in the scattering of charged nuclear particles, we are usually interested in the deviations from the purely electromagnetic effects caused by the specifically nuclear forces. Sometimes, however, the decomposition of the Hamiltonian is dictated by purely practical calculational considerations. Thus, if the Hamiltonian H differs only slightly from another Hamiltonian H_0 whose solutions are known, or are easily found, it may prove very convenient to express the Hamiltonian H as $H = H_0 + H'$.

Since we are here treating scattering by a spherically symmetric potential, we decompose $V(r)$ as

$$V(r) = V_0(r) + V_1(r), \tag{2.38}$$

and

$$H = H_0 + V_1(r) \tag{2.39}$$

with

$$H_0 = -(\hbar^2/2m)\nabla^2 + V_0(r). \tag{2.40}$$

The Schrödinger equation corresponding to H_0 may be separated by means of the angular momentum expansion. The radial wave function, denoted in this case by $v_l(r)$, obeys the radial wave equation

$$v_l''(r) + \left[k^2 - \frac{l(l+1)}{r^2} - U_0(r)\right]v_l(r) = 0, \tag{2.41}$$

and the normalization of $v_l(r)$ is chosen such that

$$v_l(r) \xrightarrow[r\to\infty]{} \exp(i\delta_l^{(0)})\sin(kr - \tfrac{1}{2}\pi l + \delta_l^{(0)}), \tag{2.42}$$

where $\delta_l^{(0)}$ is the phase shift due to $V_0(r)$.

The Schrödinger equation corresponding to H may be similarly decomposed. The radial wave function, denoted by $u_l(r)$, obeys the radial equation

$$u_l''(r) + \left[k^2 - \frac{l(l+1)}{r^2} - U_0(r) - U_1(r)\right]u_l(r) = 0. \tag{2.43}$$

From Eqs. (2.41) and (2.43), we get

$$v_l(r)u_l''(r) - u_l(r)v_l''(r) = \frac{d}{dr}\left[v_l(r)u_l'(r) - u_l(r)v_l'(r)\right]$$
$$= -v_l(r)U_1(r)u_l(r). \tag{2.44}$$

Integration of Eq. (2.44), with $u_l(r)$ normalized so that

$$u_l(r) \xrightarrow[r \to \infty]{} e^{i\delta_l} \sin(kr - \tfrac{1}{2}\pi l + \delta_l), \tag{2.45}$$

gives the result

$$T_l^{(1)} = \exp(2i\delta_l^{(0)}) \exp(i\delta_l^{(1)}) \sin \delta_l^{(1)} = -k^{-1} \int_0^\infty dr\, v_l(r)\, U_1(r)\, u_l(r), \tag{2.46}$$

where the relative phase shift $\delta_l^{(1)}$ is defined as

$$\delta_l^{(1)} = \delta_l - \delta_l^{(0)}. \tag{2.47}$$

The quantity $T_l^{(1)}$ given by Eq. (2.46) is the additional contribution to the partial wave scattering amplitude T_l from the potential $U_1(r)$. The partial wave amplitude T_l may be rewritten as

$$T_l = e^{i\delta_l} \sin \delta_l = \frac{e^{2i\delta_l} - 1}{2i} = \frac{\exp(2i\delta_l^{(0)}) - 1}{2i} + \frac{e^{2i\delta_l} - \exp(2i\delta_l^{(0)})}{2i}$$

$$= \exp(i\delta_l^{(0)}) \sin \delta_l^{(0)} + \exp(2i\delta_l^{(0)}) \exp(i\delta_l^{(1)}) \sin \delta_l^{(1)} = T_l^{(0)} + T_l^{(1)}. \tag{2.48}$$

In the light of Eq. (2.48), the result given by Eq. (2.46) appears very natural. Just as Eq. (2.24) for the partial wave amplitude T_l is an integral over the perturbing potential, the unperturbed-state function and the final-state function, so is Eq. (2.46) also just such an integral. However, it is most important to recognize that from Eq. (2.24) we have

$$T_l = -k^{-1} \int_0^\infty dr\, F_l(kr)\, U_0(r)\, u_l(r) - k^{-1} \int_0^\infty dr\, F_l(kr)\, U_1(r)\, u_l(r), \tag{2.49}$$

whereas from Eqs. (2.46) and (2.48) we have

$$T_l = -k^{-1} \int_0^\infty dr\, F_l(kr)\, U_0(r)\, v_l(r) - k^{-1} \int_0^\infty dr\, v_l(r)\, U_1(r)\, u_l(r), \tag{2.50}$$

where in Eqs. (2.49) and (2.50) $u_l(r)$ and $v_l(r)$ are normalized as in Eqs. (2.45) and (2.42), respectively. The equality of the right-hand sides of Eqs. (2.49) and (2.50) is by no means obvious. However, they are, as we have now seen, two different descriptions of the same scattering process. The two terms on the right of Eq. (2.49) together represent the partial wave amplitude in the entire scattering process due to the interactions $V_0(r)$ and $V_1(r)$, but they have no significance separately. In Eq. (2.50) the first term is the scattering amplitude in another scattering process, namely, one in which $V_1(r)$ is switched off;

the second term then represents the additional scattering generated by switching on $V_1(r)$.

The full scattering amplitude $f(\theta)$ is given by the formula

$$f(\theta) = k^{-1} \sum_{l=0}^{\infty} (2l+1) e^{i\delta_l} \sin \delta_l \, P_l(\cos \theta). \tag{2.51}$$

The scattering amplitude may also be expressed in terms of $\delta_l^{(0)}$ and $\delta_l^{(1)}$ via

$$f(\theta) = f_0(\theta) + k^{-1} \sum_{l=0}^{\infty} (2l+1) \exp(2i\delta_l^{(0)}) \exp(i\delta_l^{(1)}) \sin \delta_l^{(1)} P_l(\cos\theta), \tag{2.52}$$

where $f_0(\theta)$ is the scattering amplitude for $V_1(r) = 0$, and only the second term, in which $T_l^{(1)}$ appears, depends upon the additional potential $V_1(r)$. It is clear from this that $f(\theta)$ is not simply the sum of the scattering amplitude due to $V_0(r)$ in the absence of $V_1(r)$ and the scattering amplitude due to $V_1(r)$ in the absence of $V_0(r)$ but, instead, involves the scattering amplitude due to $V_1(r)$ in the presence of $V_0(r)$.

One may also derive three-dimensional formulas for the scattering amplitude $f(\theta)$ analogous to these integral expressions for one dimension. The result for a single potential is

$$f(\theta) = -(4\pi)^{-1} \int d^3r \, e^{-i\mathbf{k}\cdot\mathbf{r}} U(r) \psi(\mathbf{r}). \tag{2.53}$$

This relation will be obtained directly from the integral equation for $\psi(\mathbf{r})$ in Chapter 5. Here we shall just show that it is equivalent to the partial wave expansion of $f(\theta)$, Eq. (2.51). Introducing the partial wave expansion of $\psi(\mathbf{r})$, Eq. (1.33), and the expansion

$$e^{-i\mathbf{k}\cdot\mathbf{r}} = (4\pi/kr) \sum_{l=0}^{\infty} \sum_{m=-l}^{l} i^l F_l(kr) Y_{lm}(\theta_k, \phi_k) Y_{lm}^*(\theta, \phi), \tag{2.54}$$

we obtain

$$f(\theta) = -k^{-2} \sum_{l=0}^{\infty} (2l+1) \int_0^{\infty} dr \, F_l(kr) U(r) u_l(r) P_l(\cos \theta)$$

$$= k^{-1} \sum_{l=0}^{\infty} (2l+1) e^{i\delta_l} \sin \delta_l \, P_l(\cos \theta). \tag{2.55}$$

In this result we have used the integral expression of Eq. (2.24) for T_l in which $u_l(r)$ has the proper normalization consistent with its presence in the partial wave expansion of $\psi(\mathbf{r})$. Equation (2.55) is the familiar Legendre expansion for the scattering amplitude.

Thus both the partial wave expressions and the full scattering amplitude can be expressed as integrals of a free-particle wave function and the full scattering function over the region of the potential well. As we shall see, this result is of fundamental importance in scattering theory.

A result in three dimensions analagous to the two-potential equation, Eq. (2.50), is easy to develop. However, because this result is easy to misinterpret, we postpone discussion of this matter until Chapters 4 and 5, where we will study this question in detail.

3. The Low-Energy Limit

In this section the behavior of the phase shift at low energies will be considered. For potentials that can be assumed to vanish beyond $r = R$, the earlier expressions for $\tan \delta_l$, Eqs. (2.10) and (2.26), can be used to explore the low-energy behavior of the phase shift. When the external kinetic energy E is small compared to the depth of the potential, the wave function inside the potential will not depend sensitively on E. The total kinetic energy at any radius is $E + |V(r)|$, which is very nearly equal to $|V(r)|$ at low energies, so we can to a first approximation consider the logarithmic derivative $L(R)$ to be independent of energy.

If we introduce the low-energy behavior of $F_l(kr)$ and $G_l(kr)$ from Eqs. (1.14) and (1.15), we find from Eq. (2.10)

$$\tan \delta_l \xrightarrow[k \to 0]{} \frac{(l+1) - R L_0(R)}{l + R L_0(R)} \frac{(kR)^{2l+1}}{[1 \cdot 3 \cdot 5 \cdots (2l-1)]^2 (2l+1)}, \quad (3.1)$$

where $L_0(R)$ is the zero-energy logarithmic derivative. Thus, as the energy approaches zero, the tangent of the phase shift also approaches zero as

$$\tan \delta_l \propto k^{2l+1}. \quad (3.2)$$

For $l = 0$ (S waves), Eq. (3.1) yields

$$\tan \delta_0 \xrightarrow[k \to 0]{} \frac{1 - R L_0(R)}{L_0(R)} k = -ka. \quad (3.3)$$

The quantity a is usually called the zero-energy scattering length, or simply the scattering length. The scattering length is defined by

$$a = \frac{R L_0(R) - 1}{L_0(R)}. \quad (3.4)$$

If the wave function is small at $r = R$, $RL_0(R)$ will be large and $(a/R) \simeq 1$; if, instead, $u_l'(R)$ is nearly zero, $RL_0(R)$ will be small and $a \simeq -L_0(R)^{-1}$.

3. Low-Energy Limit

The scattering length a has a simple geometric interpretation. In the low-energy limit the external wave function is

$$u_0(r) \underset{r>R}{=} e^{i\delta_l} \sin(kr + \delta_0) \underset{k \to 0}{\longrightarrow} k(r - a). \tag{3.5}$$

Thus a is the point nearest the origin at which the external wave function, or its extrapolation toward the origin, vanishes. The scattering length also has an important physical significance. In the low-energy limit only the S wave makes a nonzero contribution to the cross section, so that the angular distribution of the scattering is spherically symmetric and the total cross section is

$$\sigma_{\text{TOT}} = (4\pi/k^2) \sin^2 \delta_0 \underset{k \to 0}{\longrightarrow} 4\pi a^2. \tag{3.6}$$

This is exactly the result we obtained in Eq. (2.15) for the low-energy scattering of a hard sphere of radius a. Thus the scattering length is the "effective radius" of the target at zero energy.

The Effective Range Expansion

To go to higher energies one may expand the interior wave function in powers of the energy. This is most easily done by comparing the Schrödinger equations for two different energies. To make explicit the energy dependence of the wave function, we write the radial wave function as $u_l(k, r)$. The radial equation for $E = (\hbar^2 k^2/2m)$ is

$$u_l''(k, r) + \left[k^2 - \frac{l(l+1)}{r^2} - U(r) \right] u_l(k, r) = 0, \tag{3.7}$$

while for $\bar{E} = (\hbar^2 \bar{k}^2/2m)$ the radial equation is

$$u_l''(\bar{k}, r) + \left[\bar{k}^2 - \frac{l(l+1)}{r^2} - U(r) \right] u_l(\bar{k}, r) = 0. \tag{3.8}$$

If we multiply the first equation by $u_l(\bar{k}, r)$ and the second by $u_l(k, r)$ and subtract, we obtain in the usual way

$$\frac{d}{dr} \left[u_l(\bar{k}, r) u_l'(k, r) - u_l'(\bar{k}, r) u_l(k, r) \right] = (\bar{k}^2 - k^2) u_l(\bar{k}, r) u_l(k, r). \tag{3.9}$$

At this point we might expect to integrate Eq. (3.9) as we did in Section 2. However, if we integrated from $r = 0$ to $r \to \infty$, the integral on the right-hand side of Eq. (3.9) would not converge. We can avoid this difficulty by defining a comparison function $w_l(k, r)$ such that, for $r > R$, it is identical with the radial wave function:

$$u_l(k, r) = w_l(k, r), \qquad r > R. \tag{3.10}$$

For $r > R$ the function $u_l(k, r)$ obeys the free-particle radial Schrödinger equation, and we can define $w_l(k, r)$ as that solution of the free equation which satisfies the boundary condition, Eq. (3.10). The functions $w_l(k, r)$ and $w_l(\bar{k}, r)$ also obey an equation exactly like Eq. (3.9). Combining the two equations for u_l and w_l and integrating, we obtain

$$[w_l(\bar{k}, r) w_l'(k, r) - w_l'(\bar{k}, r) w_l(k, r)]$$
$$- [u_l(\bar{k}, r) u_l'(k, r) - u_l'(\bar{k}, r) u_l(k, r)]$$
$$= (k^2 - \bar{k}^2) \int_r^\infty dr' \, [w_l(\bar{k}, r') w_l(k, r') - u_l(\bar{k}, r') u_l(k, r')] \tag{3.11}$$

We have not taken the lower limit in Eq. (3.11) to be $r = 0$ since the integral would diverge in that limit for $l > 0$. For $l = 0$ we can take the limit $r \to 0$ without further ado. In that case the second term on the left-hand side of Eq. (3.11) vanishes because of the regularity of the u's, and we are left with

$$[w_0(\bar{k}, 0) w_0'(k, 0) - w_0'(\bar{k}, 0) w_0(k, 0)]$$
$$= (k^2 - \bar{k}^2) \int_0^\infty dr \, [w_0(\bar{k}, r) w_0(k, r) - u_0(\bar{k}, r) u_0(k, r)]. \tag{3.12}$$

The expression on the left-hand side is, of course, dependent on the normalization of the wave function. We shall here normalize $u_l(k, r)$ such that

$$u_l(k, r) \underset{r > R}{=} w_l(k, r) = \cot \delta_l \, F_l(kr) + G_l(kr). \tag{3.13}$$

For $k = 0$ and $l = 0$ the external wave function so normalized has the form

$$u_0(0, r) = w_0(0, r) = (1 - r/a), \qquad r > R. \tag{3.14}$$

For $u_0(k, r)$ normalized in this way we find that Eq. (3.12) becomes

$$k \cot \delta_0 - \bar{k} \cot \bar{\delta}_0$$
$$= (k^2 - \bar{k}^2) \int_0^\infty dr \, [w_0(\bar{k}, r) w_0(k, r) - u_0(\bar{k}, r) u_0(k, r)]. \tag{3.15}$$

We may now, if we wish, take the limit as $\bar{k} \to 0$. From Eq. (3.3) we have

$$\lim_{\bar{k} \to 0} \bar{k} \cot \bar{\delta}_0 = -(1/a), \tag{3.16}$$

where a is the scattering length for zero orbital angular momentum. Thus Eq. (3.15) becomes

$$k \cot \delta_0 = -(1/a) + k^2 \int_0^\infty dr \, [w_0(0, r) \, w_0(k, r) - u_0(0, r) \, u_0(k, r)]. \tag{3.17}$$

We shall show below that both $w_0(k, r)$ and $u_0(k, r)$ can be expanded in powers of k^2, so that we can obtain from this the effective range expansion in its most usual form,

$$k \cot \delta_0 = -(1/a) + \tfrac{1}{2} \rho k^2 - P\rho^3 k^4 + \mathcal{O}(k^6), \tag{3.18}$$

where

$$\rho = 2 \int_0^\infty dr \, [w_0^2(0, r) - u_0^2(0, r)] \tag{3.19}$$

is the so-called "effective range" and P is the "shape-dependent parameter." This result provides an expansion for the $l = 0$ phase shift in which the leading terms are directly expressed in terms of the zero-energy wave function. In general, Eq. (3.15) implies that, for any two energies E and $\bar{E}$,

$$k \cot \delta_0 = \bar{k} \cot \bar{\delta}_0 + \tfrac{1}{2}(k^2 - \bar{k}^2) \rho(E, \bar{E}), \tag{3.20}$$

where

$$\rho(E, \bar{E}) = 2 \int_0^\infty dr \, [w_0(\bar{k}, r) \, w_0(k, r) - u_0(\bar{k}, r) \, u_0(k, r)] \tag{3.21}$$

is the generalized effective range.

For $l > 0$ a complication arises in the derivation of the effective range expansion because of the existence of singularities in Eq. (3.11) at the origin. Nonetheless, an expansion in powers of k^2 can be obtained by subtracting an expression similar to Eq. (3.9) in which $G_l(kr)$ and $G_l(\bar{k}r)$ replace $u_l(k, r)$ and $u_l(\bar{k}, r)$. This eliminates the singularity at the origin and allows one to derive the effective range expansion

$$k^{2l+1} \cot \delta_l = -(a_l)^{-2l-1} + k^2 \Big\{ \int_0^R dr \, [\hat{w}_l^2(0, r) - \hat{u}_l^2(0, r) - \hat{G}_l^2(0, r)]$$

$$- R(2l - 1)^{-1} \, \hat{G}_l^2(0, R) \Big\} + \mathcal{O}(k^4), \tag{3.22}$$

where the functions with roofs are defined by $\hat{f}_l(0, r) \equiv \lim_{k \to 0} k^l f_l(k, r)$, and R, may take on any value beyond the range of the force. Other expressions for $k^{2l+1} \cot \delta_l$ may also be found. Since the practical utility of the effective range expansion for $l > 0$ is not great, we shall devote no further attention to this topic.

To complete this development of the effective range expansion, we must investigate the energy dependence of the wave function $u_l(k, r)$, and especially its behavior for small k. We might be tempted to think that, because k^2 rather than k enters the Schrödinger equation, $u_l(k, r)$ can only depend on k^2, but this is not so. The function $g(k)\,e^{ikr}$, which is a solution of the free radial equation for $l = 0$, provides a striking counterexample.

To investigate the energy dependence of $u_l(k, r)$, we attempt an expansion of the radial wave function in powers of k by writing

$$u_l(k, r) = N(k) \sum_{n=0}^{\infty} y_l^{(n)}(r)\, k^n. \tag{3.23}$$

Inserting this into the radial Schrödinger equation, Eq. (3.7), and examining of each power of k we find that $y_l^{(0)}(r)$ and $y_l^{(1)}(r)$ both obey the $k = 0$ radial equation and that the functions $y_l^{(n)}(r)$ obey a recurrence equation

$$\frac{d^2\, y_l^{(n)}(r)}{dr^2} - \left[\frac{l(l+1)}{r^2} + V(r)\right] y_l^{(n)}(r) = -k^2\, y_l^{(n-2)}(r). \tag{3.24}$$

Because of the requirement of regularity at the origin, $y_l^{(0)}(r)$ and $y_l^{(1)}(r)$ must be the regular solution and can differ at most by a multiplicative factor, which may be a function of k. Since $y_l^{(0)}(r)$ and $y_l^{(1)}(r)$ are the same functions of r, we now see from Eq. (3.24) that we need only the even, or the odd powers, in Eq. (3.23). Thus we write, generally,

$$u_l(k, r) = N(k) \sum_{n=0}^{\infty} y_l^{(2n)}(r)\, k^{2n}. \tag{3.25}$$

Now the normalization of $u_l(k, r)$ given by Eq. (3.13) implies that $B = 1$ in Eq. (2.21), so that by Eq. (2.22)

$$1 = -k^{-1} \int_0^{\infty} dr\, F_l(kr)\, U(r)\, u_l(k, r). \tag{3.26}$$

Together with the known low-energy behavior of $F_l(kr)$, Eq. (1.14), this implies $N(k) \sim k^{-l}$ as $k \to 0$. Now since $F_l(kr)$ can also be represented by a series such as Eq. (3.23) with $N(k)$ replaced by k^{l+1}, we immediately conclude that, in order to satisfy Eq. (3.26), $N(k)$ must be of the form

$$N(k) = k^{-l} \sum_{n=0}^{\infty} \gamma_{2n} k^{2n}. \tag{3.27}$$

Putting all these results together, we find that we are able to write $u_l(k, r)$, when normalized according to Eq. (3.13), as

$$u(k, r) = k^{-l} \sum_{n=0}^{\infty} z_l^{(2n)}(r) k^{2n}, \qquad (3.28)$$

which is the desired result.

VALIDITY OF THE EFFECTIVE RANGE EXPANSION

The effective range expansion is of value only when the first few terms of the series are sufficient to give a good approximation. We shall ask under what circumstances this is the case. Let us first examine the integral for the effective range, Eq. (3.19). Since $w_0(0, r)$ has the form given by Eq. (3.14), we have

$$\begin{aligned}\rho &= 2 \int_0^R dr \, [w_0^2(0, r) - u_0^2(0, r)] \\ &= 2R[1 - (R/a) + \tfrac{1}{3}(R/a)^2] - 2 \int_0^R dr \, u_0^2(0, r), \end{aligned} \qquad (3.29)$$

where R may have any value beyond the range of the force. The function $u_0(0, r)$ vanishes at the origin and is equal to $w_0(0, r)$ beyond the range of the force. Although it is possible in certain extraordinary conditions for the effective range integral to be negative, we shall not consider these situations. For positive ρ, Eq. (3.29) may be written as

$$\begin{aligned}\rho &= 2 \int_0^\infty dr \, [w_0^2(0, r) - u_0^2(0, r)] = 2 \int_0^R dr \, w_0^2(0, r) \\ &= 2\bar{R}\,[1 - (\bar{R}/a) + \tfrac{1}{3}(\bar{R}/a)^2], \end{aligned} \qquad (3.30)$$

where $\bar{R}$ is some radius less than R. If in addition $|R/a| \ll 1$, then the effective range $\rho \simeq 2\bar{R}$, and ρ is of the order of the range of the force under these circumstances. If, on the other hand, a is very small relative to the range of the force, then ρ can be very much larger than the range of the force.

Let us now examine the next term in the expansion in the same way. The shape-dependent parameter defined in Eq. (3.18) is easily seen to be

$$\begin{aligned}P &= -\frac{1}{2\rho^3(0, 0)} \left[\frac{\partial \rho(k^2, 0)}{\partial (k^2)} \right]_{k^2=0} \\ &= -\frac{1}{\rho^3(0, 0)} \int_0^\infty \left\{ w(0, r) \left[\frac{\partial w(k, r)}{\partial (k^2)} \right]_{k=0} - u(0, r) \left[\frac{\partial u(\,, r)}{\partial (k^2)} \right] \right\} dr \\ &= -\frac{1}{2\rho^3} \left[\frac{\partial}{\partial k^2} \int_0^\infty dr \, \{w_0^2(k, r) - u_0^2(k, r)\} \right]_{k^2=0}. \end{aligned} \qquad (3.31)$$

If we use the same approach here as in Eq. (3.30), we find

$$P = -\left[\frac{1}{4}\left(\frac{\bar{R}}{\rho}\right)^2 - \frac{1}{6}\left(1 + \frac{\rho}{a}\right)\left(\frac{\bar{R}}{\rho}\right)^3 + \frac{1}{6}\left(\frac{\bar{R}}{a}\right)\left(\frac{\bar{R}}{\rho}\right)^3 - \frac{1}{30}\left(\frac{\bar{R}}{a}\right)^2\left(\frac{\bar{R}}{\rho}\right)^3\right]. \quad (3.32)$$

Although $\bar{R}$ is not independent of k^2, we have ignored any contribution from the energy dependence of $\bar{R}$ in obtaining Eq. (3.32). In the approximation $|\rho/a| \ll 1$, we see from Eq. (3.30) that $\bar{R} \simeq \frac{1}{2}\rho$, so that the shape-dependent parameter P in this approximation is

$$P \approx -\frac{1}{24}\left(1 - \tfrac{1}{4}\frac{\rho}{a}\right). \quad (3.33)$$

Thus we see that, for $|\rho/a| \ll 1$, the effective range is of the order of the range of the force, and the expansion is good so long as $k\rho < 1$. In the case of scattering by a hard sphere, $\delta_0 = -kR$, where R is the radius of the sphere, giving $a = R$, $\rho = 2/3a$, and $P = -(3/40)$.

We have seen that the effective range is generally of the order of the range of the force. On the other hand, the scattering length may assume any value, and the effective range expansion for $k \cot \delta_0$ will be useful, as we have noted, when $|a|$ is large. This condition in turn implies that the potential is strong, since it implies a large phase shift at low energies. In fact, the scattering length is infinite when there is a bound state at zero energy, so the condition $|a| > |\rho|$ will hold only when there is a bound state slightly below zero energy or a scattering resonance slightly above zero energy. A resonant energy is here defined as an energy at which the phase shift passes through $\frac{1}{2}\pi$ plus some multiple of π, so that the tangent of the phase shift is infinite and the scattering is as large as possible. Such a resonance is sometimes also referred to as a virtual level.

It is easy to see why a large scattering length implies a level near zero energy. A large scattering length implies a large admixture of the irregular function in the external wave function. At positive energies an infinite admixture of the irregular function gives a resonance. At negative energies the external wave function for a bound state is a decreasing exponential falling off as the inverse square root of the binding energy. In the limit as the binding energy goes to zero, the external wave function tends to a constant, implying an infinite scattering length (since the zero-energy wave function never vanishes). A large negative scattering length implies a resonance near zero energy; a large positive scattering length implies a bound state near zero energy.

3. Low-Energy Limit

When the potential does not produce a level, real or virtual, near zero energy, the effective range expansion of $k^{2l+1} \cot \delta_l$ will break down. That is to say, it will fail when $|\rho/a| \gtrsim 1$, or when there is little scattering at low energy. When this condition holds, we may instead perform an expansion of $(1/k)^{2l+1} \tan \delta_l$.

An expansion in powers of k^2 for $(1/k)^{2l+1} \tan \delta_l$ can be obtained if the wave function is normalized so that

$$u_l(k, r) \underset{r>R}{=} w_l(kr) = F_l(kr) + \tan \delta_l\, G_l(kr) \tag{3.34}$$

instead of as in Eq. (3.13). With this normalization, manipulations exactly as before lead to the result, for $l = 0$,

$$k^{-1} \tan \delta_0 = -a - \tfrac{1}{2}\rho a^2 k^2 + \rho' \rho^3 a^2 k^4 + \cdots, \tag{3.35}$$

with a and ρ defined as before. This result is entirely as expected, since the power series expansion for the inverse of Eq. (3.18) is

$$k^{-1} \tan \delta_0 = a^2[-1/a - \tfrac{1}{2}\rho k^2 + (P - \tfrac{1}{4}a/\rho)\rho^3 k^4 + \cdots], \tag{3.36}$$

which is just Eq. (3.35).

Study of Eq. (3.35) quickly shows that this expansion is useful for $k\rho \lesssim 1$ and $|a/\rho| \ll 1$. The two expressions Eq. (3.18) and Eq. (3.35) are thus complementary and should be employed as $|\rho/a|$ is smaller or larger than unity.

All of these considerations apply specifically to scattering by a potential having a finite range. For an infinite-range potential, such as the Coulomb force or the r^{-4} potential between an electron and a neutral atom, these expansions are no longer valid. The existence of the effective range ρ depends on the assumption that $u_l(k, r)$ behaves as a free-particle function at a sufficiently large distance from the scattering center. This condition is not satisfied for such long-range potentials and, while expansions analogous to the effective range expansion can be derived, they will not be simple power series in k^2.

The expansion for a potential that behaves for large r as (e^2/r) has the form

$$2k\eta K = -(1/a) + \tfrac{1}{2}\rho k^2 + \cdots, \tag{3.37}$$

with K defined to be

$$K = \frac{\pi \cot \delta_0}{e^{2\pi\eta} - 1} - \log \eta - 0.5772 + \eta^2 \sum_{n=1}^{\infty} \frac{1}{n(n^2 + \eta^2)}, \tag{3.38}$$

where $\eta = e^2/\hbar v$ and v is the classical velocity. The phase shift that appears in Eq. (3.37) is not the total phase shift but is just that part of the phase shift which arises from the part of the potential which differs from the Coulomb potential. This point will be discussed in Section 5.

For a potential that behaves asymptotically as (β^2/r^4), the appropriate expansion is

$$k \cot \delta_0 = -(1/a) + (\pi/\beta^2 3a^2)k + ck^2 \log(\tfrac{1}{4}\beta k) + \tfrac{1}{2}\rho' k^2 + \cdots. \quad (3.39)$$

As these results show, the infinite range of the force leads to an energy dependence that does not permit an expansion in powers of k^2 in the neighborhood of zero energy.

4. The High-Energy Limit

Let us now examine the solutions of the radial Schrödinger equation when the incident energy is large compared to the potential energy. In this situation the wave function will not differ very greatly from the wave function for a free particle. In what follows we will exploit this similarity.

THE PHASE-AMPLITUDE METHOD

The radial Schrödinger equation may be written as

$$u_l'' + \left[k^2 - \frac{l(l+1)}{r^2} - U(r) \right] u_l = u_l'' + k_l^2(r) u_l = 0, \quad (4.1)$$

where $k_l(r)$ is the "local wave number" defined to be

$$k_l(r) = \left[k^2 - \frac{l(l+1)}{r^2} - U(r) \right]^{1/2}. \quad (4.2)$$

The corresponding "local wavelength" is $2\pi/k_l(r)$. At high kinetic energies the wave number will change little over distances of the order of this wavelength, except near points where it is very long, that is, where $k_l(r) = 0$. These are the classical turning points, where the radial momentum vanishes. The singularity of the angular momentum barrier at the origin ensures that there will be at least one such turning point for $l > 0$, no matter how high the energy. Away from these points we have $k^2 << \{l(l+1)/r^2\} + U(r)$, and the wave number will be slowly varying.

If $k_l(r)$ could be considered to be a constant in the neighborhood of r, the

4. The High-Energy Limit

wave function $u_l(r)$ would be $\exp[\pm i r k_l(r)]$. The total phase in this region would depend upon the variation of $k_l(r)$ as $r \to \infty$, since the form of $u_l(r)$ is fixed by the boundary condition at infinity. The change of the phase of this function over a distance Δr in this neighborhood would be $\Delta \phi_l = k_l(r) \Delta r$, so that we expect that at high energies the derivative of the phase will be just $k_l(r)$.

Since the phase should have this simple behavior, it is convenient to write the two linearly independent solutions of the Schrödinger equation as

$$u_l^{(\pm)}(r) = A_l(r) \exp[\pm i \phi_l(r)], \tag{4.3}$$

where the amplitude $A_l(r)$ and the phase $\phi_l(r)$ are real functions. Substituting this into the Schrödinger equation gives

$$A_l'' - A_l(\phi_l')^2 \pm i(2A_l'\phi_l' + A_l\phi_l'') + k_l^2(r) A_l = 0. \tag{4.4}$$

The real and the imaginary parts of this equation must each be zero, giving the two equations

$$A_l'' - A_l(\phi_l')^2 + k_l^2(r) A_l = 0 \tag{4.5}$$

and

$$2A_l'\phi_l' + A_l\phi_l'' = 0. \tag{4.6}$$

So long as the amplitude $A_l(r)$ is not zero, Eq. (4.6) implies that

$$A_l^2 \phi_l' = \text{constant} = W, \tag{4.7}$$

which is just the condition that the Wronskian of the two independent solutions of Eq. (4.1) be a constant.

Either A_l or ϕ_l' can now be eliminated from Eq. (4.5). If we choose to eliminate the phase, we obtain a nonlinear differential equation for the amplitude:

$$A_l'' - W^2 A_l^{-3} + k_l^2(r) A_l = 0. \tag{4.8}$$

Alternatively, the differential equation satisfied by the phase is

$$\phi_l''' - \frac{3}{2} \frac{(\phi_l'')^2}{\phi_l'} + 2(\phi_l')^3 - 2k_l^2(r) \phi_l' = 0. \tag{4.9}$$

Either of these equations, together with Eq. (4.7), will give both the phase and the amplitude, when the appropriate boundary conditions are applied.

Let us choose the regular and irregular solutions as

$$u_l^{(1)}(r) = A_l(r) \sin \phi_l(r) \quad \text{and} \quad u_l^{(2)}(r) = A_l(r) \cos \phi_l(r). \quad (4.10)$$

According to Section 1 we may normalize the regular solution so that it satisfies the boundary condition

$$A_l(r) \sin \phi_l(r) \xrightarrow[r \to \infty]{} \sin(kr - \tfrac{1}{2}\pi l + \delta_l), \quad (4.11)$$

from which we infer that the phase behaves asymptotically as

$$\phi_l(r) \xrightarrow[r \to \infty]{} kr - \tfrac{1}{2}\pi l + \delta_l \quad (4.12)$$

or

$$\phi_l'(r) \xrightarrow[r \to \infty]{} k, \quad (4.13)$$

and the amplitude behaves as

$$A_l(r) \xrightarrow[r \to \infty]{} 1. \quad (4.14)$$

This choice of normalization clearly gives $W = k$ for the Wronskian appearing in Eq. (4.7). Likewise, the form of Eq. (4.10) implies that

$$\phi_l(r) \xrightarrow[r \to 0]{} 0, \quad (4.15)$$

while the amplitude $A(r)$ becomes infinite at the origin for angular momenta greater than zero.

The phase-amplitude method seems precisely suited to scattering problems, in which we are primarily interested in the phase at large distances. However, the difficulties associated with nonlinear equations are so great that only in special circumstances is this method useful. The high energy limit is one of these.

Since we are primarily interested in the phase shift, we will work with Eq. (4.9). We have seen that, for large values of r, $\phi_l'(r)$ approaches the constant value k. Thus at large distances the higher derivatives ϕ_l'' and ϕ_l''' become negligible. Let us then try to solve Eq. (4.9) by an iterative process, in which we obtain a first approximation by neglecting these derivatives. Neglecting ϕ_l'' and ϕ_l''', we find from Eq. (4.9)

$$\phi_l'(r) \simeq k_l(r). \quad (4.16)$$

In this approximation, therefore, the rate of change of the phase at the radius

r is given by the local momentum or wave number, as we conjectured at the beginning of this section.

Inserting this result into the higher-derivative terms of Eq. (4.9), we obtain for the next approximation

$$\phi_l'(r) \simeq \left[k_l^2(r) - \frac{1}{2} \frac{k_l''(r)}{k_l(r)} + \frac{3}{4} \frac{(k_l'(r))^2}{(k_l(r))^2} \right]^{1/2}. \qquad (4.17)$$

This process may be continued to higher orders. It is clear that this procedure cannot converge at a classical turning point where $k_l(r) = 0$; it is equally clear that the method is quite satisfactory when the change in the local wave number over a wavelength is small, that is, when $k_l'(r) \ll k_l^2(r)$. We will see shortly that in practice this condition need not be satisfied, but that a weaker condition will suffice. These conditions are most easily satisfied at high energies where $k_l(r)$ becomes sufficiently large; they will also be satisfied at moderate energies if the potential is slowly varying.

THE JWKB APPROXIMATION

We shall henceforth assume that the first approximation, Eq. (4.16), is valid. This is the so-called JWKB approximation. We will specialize further to the case where $k_l(r)$ has only one zero, that is, where there is only one classical turning point. For our scattering problem this becomes the distance of closest approach for a particle of angular momentum l. Outside this radius the derivatives of $k_l(r)$ will become small if the potential is smoothly varying, even though the potential itself may still be appreciable.

In the JWKB approximation, then, the phase is

$$\phi_l(r) \simeq \phi_l(r_1) + \int_{r_1}^{r} dr \left[k^2 - \frac{l(l+1)}{r^2} - U(r) \right]^{1/2}, \qquad (4.18)$$

where r_1 is the classical turning point defined by

$$k_l^2(r_1) = k^2 - \frac{l(l+1)}{r_1^2} - U(r_1) = 0. \qquad (4.19)$$

The scattering phase shift δ_l is defined relative to the phase of the free-particle wave function, the spherical Bessel function. By analogy with Eq. (4.18), the free-particle phase is

$$\hat{\phi}_l(r) = \hat{\phi}_l(r_0) + \int_{r_0}^{r} dr \left[k^2 - l(l+1)/r^2 \right]^{1/2}, \qquad (4.20)$$

where the classical turning point is

$$r_0 = [l(l+1)/k^2]^{1/2}. \tag{4.21}$$

As a result, the scattering phase shift is

$$\delta_l = \lim_{r \to \infty} (\phi_l(r) - \hat{\phi}(r)) = \hat{\phi}_l(r_1) - \hat{\phi}_l(r_0)$$

$$+ \lim_{r \to \infty} \left[\int_{r_1}^{r} dr \left(k^2 - \frac{l(l+1)}{r^2} - U(r) \right)^{1/2} \right.$$

$$\left. - \int_{r_0}^{r} dr \left(k^2 - \frac{l(l+1)}{r^2} \right)^{1/2} \right]. \tag{4.22}$$

Although the approximation, Eq. (4.16), is not very good in the neighborhood of the classical turning point, the integration in Eqs. (4.18) and (4.20) can nevertheless be performed and will yield finite results. For Eq. (4.20) to be a good approximation to the phase shift, we must merely require that the largest contribution must come from the region well beyond the turning point. Thus the potential must "stick out" well beyond this point.

The phase $\phi_l(r_1) - \hat{\phi}_l(r_0)$ is the contribution to the phase shift from the region inside the classical turning points. Classically, the particles are excluded from this region, so that this contribution will be small. In the JWKB approximation it is set equal to zero, giving

$$\delta_l \cong \lim_{r \to \infty} \left[\int_{r_1}^{r} dr \left(k^2 - \frac{l(l+1)}{r^2} - U(r) \right)^{1/2} \right.$$

$$\left. - \int_{r_0}^{r} dr \left(k^2 - \frac{l(l+1)}{r^2} \right)^{1/2} \right]. \tag{4.23}$$

A more careful treatment shows that this result is somewhat improved if $l(l+1)$ is replaced by $(l+\tfrac{1}{2})^2$; this is usually only significant for very small l.

The phase $\phi_l(r_1) - \hat{\phi}_l(r_0)$ is small, and its neglect is a less drastic approximation than was made in arriving originally at Eq. (4.18). This phase can be estimated if we know the form of $U(r)$. For instance, suppose $U(r)$ were constant within the classical turning point:

$$U(r) = U(r_1), \quad r \le r_1. \tag{4.24}$$

In this region the regular and irregular solutions are known to be

$$u_l^{(1)}(r) = F_l(Kr) \quad \text{and} \quad u_l^{(2)}(r) = G_l(Kr), \tag{4.25}$$

where

$$K = [k^2 - U(r_1)]^{1/2} = \left[\frac{l(l+1)}{r_1^2}\right]^{1/2}. \tag{4.26}$$

According to Eq. (4.10) the phase angle is

$$\phi_l(r) = \tan^{-1}\left(\frac{u_l^{(1)}(r)}{u_l^{(2)}(r)}\right) = \tan^{-1}\left(\frac{F_l(Kr)}{G_l(Kr)}\right), \quad r \le r_1. \tag{4.27}$$

In an analogous way the phase angle for the free particle is

$$\hat{\phi}_l(r) = \tan^{-1}\left(\frac{F_l(kr)}{G_l(kr)}\right). \tag{4.28}$$

Since $Kr_1 = (l(l+1))^{1/2} = kr_0$, we therefore have the result

$$\phi_l(r_1) = \hat{\phi}_l(r_0) = \tan^{-1}\left(\frac{F_l(kr_0)}{G_l(kr_0)}\right). \tag{4.29}$$

For this case, then, the phase difference is identically zero.

We can see the effect of a variation in the potential, and at the same time determine the error due to the breakdown in Eq. (4.16) near the turning point, by using the conventional connection formulas of the JWKB approximation. One performs a Taylor expansion of $k_l^2(r)$ about the turning point and keeps the first term:

$$k_l^2(r) = \lambda^2(r - r_1). \tag{4.30}$$

Inserting this into the Schrödinger equation, Eq. (4.1), gives an equation that can be reduced to Bessel's equation for functions of order $\frac{1}{3}$. Applying the boundary condition at infinity, the regular solution is

$$u_l(r) \sim \left(\frac{\chi(r)}{k_l(r)}\right)^{1/2}[J_{1/3}(\chi(r)) + J_{-1/3}(\chi(r))]$$

$$\xrightarrow[\text{large } r]{} \left(\frac{2}{\pi k_l(r)}\right)^{1/2} 2\cos(\pi/6)\sin(\chi(r) + \pi/4), \tag{4.31}$$

where the argument is

$$\chi(r) = (2/3)\lambda(r - r_1)^{3/2} = \int_{r_1}^{r} dr \, k_l(r). \tag{4.32}$$

This same form is obtained for both the scattered wave and the free-particle wave, with r_1 replaced by r_0 in the latter case. Thus one finds by using a correct treatment in the neighborhood of the turning point that even a linear variation of the potential about that point still yields Eq. (4.23) for the scattering phase shift, implying that

$$\phi_l(r_1) - \hat{\phi}_l(r_0) = 0. \tag{4.33}$$

This also shows that our earlier incorrect treatment of $\phi_l'(r)$ in the neighborhood of the turning point has no effect on the phase shift, provided $U(r)$ is slowly varying in that neighborhood.

While it has appeared in the preceding that the JWKB approximation is valid only if the potential is everywhere slowly varying, in fact even this condition is not necessary. Let us consider a square well that has a discontinuity at $r = R$,

$$U(r) = \begin{cases} -U_0, & r < R, \\ 0, & r > R. \end{cases} \tag{4.34}$$

The JWKB approximation gives for the $l = 0$ phase shift

$$\delta_0 \cong \int_0^R dr \, [(k^2 - U_0)^{1/2} - k] = (K - k)R. \tag{4.35}$$

From Eqs. (2.10) and (2.11) we see that the exact result is

$$\tan \delta_0 = \frac{\cos kR \sin KR - (K/k) \sin kR \cos KR}{\sin kR \sin KR + (K/k) \cos kR \cos KR}. \tag{4.36}$$

On the other hand, Eq. (4.35) gives

$$\tan \delta_0 \cong \frac{\sin(K-k)R}{\cos(K-k)R} = \frac{\cos kR \sin KR - \sin kR \cos KR}{\sin kR \sin KR + \cos kR \cos KR}. \tag{4.37}$$

The difference between these is a measure of the contribution of the discontinuity. At high energies, when $k^2 \gg U_0$, these become identical. Thus the change in the potential over a distance of the order of a wavelength must be small compared to the total energy, but it is not necessary that the derivatives of $U(r)$ be small.

We now can see that the JWKB approximation neglects the reflection and diffraction due to the discontinuity in the potential. These effects are just the wave properties of quantum mechanics making themselves evident. The JWKB approximation, in neglecting such effects, should be a classical approximation. It is indeed, and is often discussed as a classical limit. In fact, the JWKB method is frequently presented as the zeroth-order term in the expansion of the wave function in powers of $\hbar$. We have chosen not to emphasize this aspect, as we are interested primarily in it as an approximation method in quantum mechanics.

THE CLASSICAL LIMIT

At high energies the potential energy will be small compared to the radial part of the kinetic energy, except near the classical turning point. Since we have argued that the region near the turning point does not contribute significantly to the phase shift, we may therefore assume that $|U(r)| << k^2 - l(l+1)/r^2$ in the high-energy limit. Then Eq. (4.23) becomes, in that limit,

$$\delta_l \xrightarrow[k\to\infty]{} -\tfrac{1}{2}\int_{r_0}^{\infty} dr\, U(r)[k^2 - l(l+1)/r^2]^{-1/2}. \quad (4.38)$$

If we introduce the impact parameter

$$b = k^{-1}[l(l+1)]^{1/2} = k^{-1}[(l+\tfrac{1}{2})^2 - \tfrac{1}{4}]^{1/2} \simeq (l+\tfrac{1}{2})/k \quad (4.39)$$

and the coordinate $z = (r^2 - b^2)^{1/2}$, Eq. (4.38) assumes the suggestive form

$$\delta_l \equiv \delta(b) \xrightarrow[k\to\infty]{} -(1/4k)\int_{-\infty}^{\infty} dz\, U((b^2+z^2)^{1/2}). \quad (4.40)$$

The coordinate z may be interpreted as the coordinate of the incident particle in the direction of its motion, as measured from the center of the target (see Fig. 3.2). This expression clearly represents the phase shift as the particle

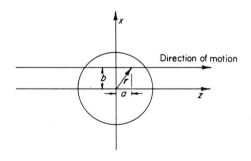

FIG. 3.2

moves along a line parallel to the z axis. It neglects any deflection of the particle from its incident direction; this is consistent with our expectation that a high-energy particle will suffer little deflection as a result of the scattering process.

Let us now examine the scattering amplitude in the classical limit, that is, at a sufficiently high energy that the particle may be localized. At these energies many angular momentum waves will contribute, so that the partial wave expansion for $f(\theta)$ may be approximated by an integral. In addition, most of the scattering will result from high partial waves, so that we can assume that $l << 1$. This follows from the fact that the phase shift will, for "normal" potentials, behave as shown in Fig. 3.3, as a function of l. The region where

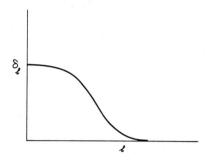

FIG. 3.3

δ_l is relatively constant will make little contribution, as we will see below. As a result, we have

$$f(\theta) = \sum_{l=0}^{\infty} (2l+1) \frac{e^{2i\delta_l} - 1}{2ik} P_l(\cos\theta)$$

$$\simeq \int_0^\infty l\, dl\, \frac{e^{2i\delta_l} - 1}{ik} P_l(\cos\theta). \qquad (4.41)$$

For large l and small angles, the Legendre polynomial may be replaced by a Bessel function using the expansion

$$P_l(\cos\theta) \simeq J_0(2(l+\tfrac{1}{2})\sin\tfrac{1}{2}\theta) + \tfrac{1}{4}\sin^2\tfrac{1}{2}\theta + \cdots. \qquad (4.42)$$

This greatly simplifies the integration, since it avoids integrating over the order of the Legendre polynomial. It is also convenient to use $b = l/k$ and to recognize that the momentum transferred to the scattered particle, which we shall denote by q, is

$$q = |\mathbf{k} - \mathbf{k}'| = (k^2 + k'^2 - 2kk' \cos \theta)^{1/2}$$
$$= k(2(1 - \cos \theta))^{1/2} = 2k \sin \tfrac{1}{2}\theta. \quad (4.43)$$

Then Eq. (4.41) may be rewritten to obtain the semiclassical approximation

$$f(\theta) = (k/i) \int_0^\infty b \, db \, (e^{2i\delta(b)} - 1) J_0(qb). \quad (4.44)$$

This formula is reminiscent of very similar formulas in the classical theory of diffraction. Its physical implications can be seen most strikingly by assuming that $\delta(b)$ has a limiting form suggested by Fig. 3.3, namely, that it is constant for $b < R$ and zero for $b > R$. In this simple case the integral in Eq. (4.44) can be performed analytically giving

$$f(\theta) \propto \frac{J_1(qR)}{qR}, \quad (4.45)$$

where $J_1(qR)$ is the first-order Bessel function. This result, familiar from the theory of Fraunhofer diffraction, gives the cross section shown in Fig. 3.4. As expected, the cross section is sharply peaked in the forward direction and is concentrated within the region having $\theta \lesssim (1/kR)$.

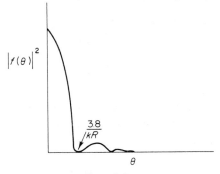

FIG. 3.4

At high energies we can extract the physical content of Eq. (4.44) by using the method of "steepest descent." For large q the integrand will oscillate rapidly as b varies. Asymptotically the Bessel function has the form

$$J_0(qb) \xrightarrow[qb \to \infty]{} (2\pi qb)^{-1/2} [\exp(i(qb - \tfrac{1}{4}\pi)) + \exp(-i(qb - \tfrac{1}{4}\pi))] \quad (4.46)$$

so that the dominant contribution to the integral will come from those values of b for which $2\delta(b) \pm qb$ is nearly constant. The term in Eq. (4.44) that is independent of the phase shift does not contribute to the scattering away from the forward direction, as we may see by returning to Eq. (4.41) and noting that $\sum_{l=0}^{\infty} (l + \tfrac{1}{2}) P_l(\cos \theta) = \delta(1 - \cos \theta)$. Thus the important region of b is determined, for fixed θ, by the relation

$$\frac{d\delta(b)}{db} = \pm \tfrac{1}{2} q = \pm k \sin \tfrac{1}{2}\theta. \tag{4.47}$$

It may be seen from this, as we pointed out earlier, that values of b for which $\delta(b)$ is a constant will not contribute to scattering out of the forward direction $(q > 0)$.

This equation relates the scattering angle θ to the angular momentum $l = kb$, which classically can be determined by localizing the incident particle. It can be evaluated if we have specific forms for the phase shift. If we introduce the high-energy approximation for $\delta(b)$, Eq. (4.40), we find

$$\frac{d\delta(b)}{db} = -(1/4k) \int_{-\infty}^{\infty} dz \, b(b^2 + z^2)^{-1/2} \left[\frac{\partial U(r)}{\partial r} \right]_{r=(b^2+z^2)^{1/2}}$$

$$= -(1/2\hbar v) \int_{-\infty}^{\infty} dz \, F_x(z), \tag{4.48}$$

where we have introduced $F_x(z)$, the force in the x direction (see Fig. 3.2). Using $dz = v \, dt$, Eqs. (4.47) and (4.48) imply that

$$\int_{-\infty}^{\infty} dt \, F_x(t) = \pm \hbar q \tag{4.49}$$

or, in words, the impulse in the x direction is equal to the momentum transfer. This is just the high-energy approximation for the scattering of a classical particle by a fixed target.

A more exact answer can be obtained by using the original JWKB formula for the phase shift, Eq. (4.23). When l is large this may be written as

$$\delta_l = \delta(b) \cong \lim_{r \to \infty} \int_{r_1}^{r} dr \, [(k^2 - (kb/r)^2 - U(r))^{1/2} - kr + \tfrac{1}{2}\pi kb] \tag{4.50}$$

which, inserted into Eq. (4.47), yields

$$-2bk^2 \int_{r_1}^{\infty} r^{-2} \, dr \, [k^2 - (kb/r)^2 - U(r)]^{-1/2} + \pi k = \pm q. \tag{4.51}$$

Expressing this in terms of the energy $E = \hbar^2 k^2 / 2m$ and the angular momentum $L = \hbar k b$, and using the fact that for small angles $q = k\theta$, this result is

$$-2L \int_{r_1}^{\infty} r^{-2} \, dr \, [2m(E - V(r)) - L^2/r^2]^{-1/2} + \pi = \pm \theta. \quad (4.52)$$

This is just the familiar classical scattering formula, which is readily derived directly from Newton's second law.

5. Elastic Scattering by a Coulomb Potential

The oldest and most familiar problem in the theory of scattering, that of the scattering of a charged particle by a Coulomb potential, is, paradoxically, outside the bounds of our previous discussion. A completely satisfactory mathematical treatment consistent with the required physical boundary conditions has yet to be found. The difficulty arises from the infinitely long range of the Coulomb interaction. In our previous discussion we limited ourselves to interactions that decreased for increasing r more rapidly than $1/r$, in order that we might be able to write the asymptotic solution of the Schrödinger equation as a solution of the force-free equation. In the case of the Coulomb interaction this cannot be done, and an entirely different treatment is required.

The fact that a problem exists can be seen immediately in the JWKB approximation. If we write $U(r) = (e^2/r)$, this implies

$$\begin{aligned} \delta_l &\simeq -\tfrac{1}{2} \lim_{r \to \infty} \int_{r_0}^{r} dr \, \frac{U(r)}{[k^2 - l(l+1)/r^2]^{1/2}} \\ &= \frac{-e^2}{2k} \lim_{r \to \infty} \{\log 2kr - \log[l(l+1)]^{1/2}\}. \end{aligned} \quad (5.1)$$

This limit does not exist, and the usual definition of the phase shift fails. This failure is immediately traceable to the fact that the Coulomb potential does not decrease sufficiently fast for increasing r, so that the phase of $u_l(r)$ relative to that of the free-particle wave function does not approach a constant at infinity.

SOLUTION OF THE SCHRÖDINGER EQUATION

Let us consider the scattering of a particle of charge $Z_1 e$ by a particle of charge $Z_2 e$. For the moment we shall assume that the only force between the particles is the electrostatic force, with potential energy given by

$$V(r) = Z_1 Z_2 e^2 / r. \quad (5.2)$$

The Schrödinger equation is then

$$-\frac{\hbar^2}{2m} \nabla^2 \psi(\mathbf{r}) + \frac{Z_1 Z_2 e^2}{r} \psi(\mathbf{r}) = E \psi(\mathbf{r}). \tag{5.3}$$

Proceeding as before, we make the usual angular momentum expansion

$$\psi(\mathbf{r}) = (kr)^{-1} \sum_{l=0}^{\infty} (2l+1) i^l u_l(r) P_l(\cos\theta). \tag{5.4}$$

The radial wave function $u_l(r)$ satisfies the radial equation

$$u_l''(r) + \left[k^2 - \frac{2m}{\hbar^2} \frac{Z_1 Z_2 e^2}{r} - \frac{l(l+1)}{r^2} \right] u_l(r) = 0, \tag{5.5}$$

or

$$u_l''(\rho) + \left[1 - \frac{2\eta}{\rho} - \frac{l(l+1)}{\rho^2} \right] u_l(\rho) = 0. \tag{5.6}$$

The prime denotes differentiation with respect to the dimensionless variable

$$\rho \equiv kr \tag{5.7}$$

and

$$\eta \equiv \frac{m Z_1 Z_2 e^2}{\hbar^2 k} = \frac{Z_1 Z_2 e^2}{\hbar v}, \tag{5.8}$$

where $v \equiv (\hbar k/m) = (p/m)$ is the classical velocity of a particle with momentum $\hbar k$.

Equation (5.6) may be transformed into a standard mathematical form by the substitution

$$u_l(\rho) = \rho^{l+1} e^{i\rho} \Phi_l(\rho), \tag{5.9}$$

which yields as the differential equation for $\Phi_l(\rho)$

$$\rho \Phi_l''(\rho) + (2l + 2 + 2i\rho) \Phi_l'(\rho) + 2i(l + 1 + i\eta) \Phi_l(\rho) = 0. \tag{5.10}$$

This is readily identified as a hypergeometric equation whose regular solution is

$$\Phi_l(\rho) = C_l \, _1F_1(l + 1 + i\eta, 2l + 2; -2i\rho), \tag{5.11}$$

where the confluent hypergeometric function $_1F_1(a, b; z)$ is defined by

$$_1F_1(a, b; z) \equiv \sum_{n=1}^{\infty} \frac{\{\Gamma(a+n)/\Gamma(a)\} z^n}{\{\Gamma(b+n)/\Gamma(b)\} n!}. \tag{5.12}$$

5. Elastic Scattering by a Coulomb Potential

The asymptotic expression for the regular hypergeometric function $_1F_1(a, b; z)$, as $|z|$ becomes infinite, is

$$_1F_1(a, b; z) \xrightarrow[|z|\to\infty]{} \frac{\Gamma(b)\, e^{z/2}}{\Gamma(a)\,\Gamma(b-a)} \{\Gamma(a)\, e^{-z/2}\, (-z)^{-a} + \Gamma(b-a)\, e^{z/2}\, z^{a-b}\}. \quad (5.13)$$

Substitution of the values $a = l + 1 + i\eta$, $b = 2l + 2$, $z = -2i\rho$ into Eq. (5.13) gives after some manipulation

$$_1F_1(l + 1 + i\eta, 2l + 2; -2i\rho)$$
$$\xrightarrow[\rho\to\infty]{} 2\frac{(2l+1)!\, e^{-i\rho}\, (2\rho)^{-(l+1)}\, e^{\frac{1}{2}\pi\eta}}{|\Gamma(l+1+i\eta)|} \sin(\rho - \tfrac{1}{2}\pi l + \sigma_l - \eta \log 2\rho), \quad (5.14)$$

or

$$u_l(r) \xrightarrow[r\to\infty]{} \frac{C_l(2l+1)!\, e^{\frac{1}{2}\pi\eta}}{|\Gamma(l+1+i\eta)|\, 2^l} \sin(kr - \tfrac{1}{2}\pi l + \sigma_l - \eta \log 2\, kr), \quad (5.15)$$

where

$$\sigma_l \equiv \arg \Gamma(l + 1 + i\eta) = \mathrm{Im}\, \log \Gamma(l + 1 + i\eta) \quad (5.16)$$

is called the Coulomb phase shift. The troublesome logarithmic phase factor shows itself here in the radial wave function, but let us proceed as if our earlier development were still applicable.

We choose the normalization constant C_l so that the wave function behaves asymptotically as

$$u_l(r) \xrightarrow[r\to\infty]{} e^{i\sigma_l} \sin(kr - \tfrac{1}{2}\pi l + \sigma_l - \eta \log 2kr). \quad (5.17)$$

If we define the regular radial Coulomb function $F_l(\eta; kr)$ by

$$F_l(\eta; kr) = e^{-i\sigma_l}\, u_l(r), \quad (5.18)$$

our solution may then be expressed as

$$F_l(\eta; kr) = \frac{e^{-\frac{1}{2}\pi\eta}\, |\Gamma(l+1+i\eta)|\, (2kr)^{l+1}\, e^{ikr}}{2(2l+1)!}$$
$$\times\, _1F_1(l + 1 + i\eta, 2l + 2; -2ikr). \quad (5.19)$$

Were it not for the additional phase factor $-\eta \log 2kr$ in Eq. (5.17), the theory would proceed exactly as previously, and we would have the partial wave expansion

$$\psi(\mathbf{r}) = (kr)^{-1} \sum_{l=0}^{\infty} (2l+1) i^l \, e^{i\sigma_l} \, F_l(\eta; kr) \, P_l(\cos\theta). \tag{5.20}$$

However, under these circumstances, we have not shown that this is a physically meaningful solution. Ordinarily we would check this by showing that the resulting three-dimensional wave function $\psi(\mathbf{r})$ satisfies the appropriate boundary conditions at large distances, but in this case no choice for the normalization of $u_l(\mathbf{r})$ can provide a solution that has the form

$$\psi(\mathbf{r}) \xrightarrow[r\to\infty]{} e^{ikz} + \frac{e^{ikr}}{r} f(\theta). \tag{5.21}$$

We can see this by examining the exact solution for $\psi(\mathbf{r})$ in three dimensions. The Coulomb potential is one of the few potentials for which such a solution can be obtained in terms of known functions. We introduce the parabolic coordinates

$$\begin{aligned} \zeta &= r - z = r(1-\cos\theta) \\ \lambda &= r + z = r(1-\cos\theta) \\ \phi &= \phi \end{aligned} \tag{5.22}$$

and seek a solution of the form

$$\psi(\mathbf{r}) = e^{ikz} \chi(\zeta). \tag{5.23}$$

We then find that the Schrödinger equation, Eq. (5.3), becomes

$$\zeta \chi'' + (1-ik\zeta)\chi' - \eta k \chi = 0. \tag{5.24}$$

This is again the hypergeometric equation, of the same form as Eq. (5.10), with the regular solution

$$\chi(\zeta) = C_1 F_1(-i\eta, 1; ik\zeta). \tag{5.25}$$

If we choose $C = e^{\frac{1}{2}\pi\eta} \Gamma(1 + i\eta)$, we have the solution

$$\psi(\mathbf{r}) = e^{-\frac{1}{2}\pi\eta} \Gamma(1 + i\eta) \, e^{ikz} \, {}_1F_1(-i\eta; 1; ikr(1-\cos\theta)). \tag{5.26}$$

The sum in Eq. (5.20) can be shown to be equal to this solution of the Schrödinger equation.

Through the use of Eq. (5.13) we see that this has the asymptotic form

$$\psi(\mathbf{r}) \xrightarrow[r\to\infty]{} \exp\{i[kz + \eta \log k(r - z)]\}$$

$$+ r^{-1} \exp[i(kr - \eta \log kr)] f_{\text{Coul}}(\theta), \tag{5.27}$$

5. Elastic Scattering by a Coulomb Potential

where the "Coulomb scattering amplitude" is

$$f_{\text{Coul}}(\theta) = \frac{-\eta \exp[2i\sigma_0 - i\eta \log(1 - \cos\theta)]}{k(1 - \cos\theta)}. \tag{5.28}$$

Again, we find logarithmic phase factors which prevent $\psi(\mathbf{r})$ from assuming the desired asymptotic form, Eq. (5.21). Thus we have no assurance that this wave function provides a correct description of the physical situation, although, since the logarithmic terms produce no net current at infinity, we do know that $f_{\text{Coul}}(\theta)$ can be interpreted as a scattering amplitude with $\sigma_{\text{Coul}}(\theta) = |f_{\text{Coul}}(\theta)|^2$.

The logarithmic terms modify both the "plane" wave and the "scattered" wave. Their presence is a direct consequence of the infinite range of the Coulomb force. No matter how far away from the scattering center we may be, we can never consider the Coulomb force to be negligible, and can never write the asymptotic solution as a linear combination of solutions of the force-free equation. Hence we can never get a true plane wave or a true spherical outgoing wave. This is true in the classical treatment as well, where the trajectory cannot be considered to have a straight-line portion even at infinity.

A second feature that attracts our immediate attention is the infinite cross section in the forward direction. That is, if we write $\sigma_{\text{Coul}}(\theta) = |f_{\text{Coul}}(\theta)|^2$, then $\sigma_{\text{Coul}}(0)$ is infinite. If we consider the total cross section to be $\sigma_{\text{TOT}} = \int \sigma_{\text{Coul}}(\theta)\,d\Omega$, the total cross section is also infinite. This is also to be expected from the infinite range of the Coulomb force and appears in the classical treatment too. No particle can pass so far from the scattering center that it escapes some deflection. The infinities arise from the scattering of the very distant particles in the forward direction.

As we have noted, we have no real assurance that Eq. (5.26) represents the solution of the three-dimensional Schrödinger equation we seek. In our previous work we were guided to a unique solution of the three-dimensional Schrödinger equation by seeking a solution that was nowhere infinite and that asymptotically took the form of a plane wave plus a scattered wave. In the present case we can only satisfy the first of these conditions. Thus we are assured that Eq. (5.4), with $u_l(r)$ given by Eqs. (5.18) and (5.19), is a solution but the normalization of $u_l(r)$ remains indoubt.

There can be no doubt that Eq. (5.4), with the normalization of $u_l(r)$ undefined, gives the entire class of regular solutions to Eq. (5.3). Further, there can be no doubt that Eq. (5.26) is one such solution; in fact, it is the only regular solution of the form of Eq. (5.23). However, we may question that this solution is the physically necessary solution.

THE SHIELDED COULOMB POTENTIAL

This problem can be approached by recognizing that in real experiments the Coulomb potential does not extend to infinity, but is shielded or neutralized by neighboring charge centers. Thus, nuclei are shielded by the surrounding electron cloud, and ions are shielded by attracting other ions or electrons around them. The shielding radius will generally be of atomic dimensions, whereas the field point at which data is taken is at distances of the order of laboratory dimensions. Thus we might adopt a description of a shielded Coulomb potential in which we take the limit as the shielding radius R goes to infinity but still assume R to be very much less than r_{source} and $r_{detector}$.

Let us assume a shielded Coulomb potential of the form

$$V(r) = \begin{cases} Z_1 Z_2 e^2/r, & r \leq R, \\ 0, & r > R. \end{cases} \qquad (5.29)$$

For $r \leq R$, then, the Schrödinger equation is given by Eq. (5.3) and has the solution given in Eq. (5.4), where $u_l(r)$ is the regular solution given by Eqs. (5.18) and (5.19). For $r > R$ the Schrödinger equation becomes the force-free equation. This is then the familiar case of a finite range potential and can be handled in the now familiar way.

Let us, for convenience, choose R so large that $kR >> \eta$. In the usual way, for $l >> kR$ the phase shift vanishes. For $l << kR$, $F_l(\eta, kr)$ takes on its asymptotic form at the shielding radius R,

$$F_l(\eta; kR) \sim \sin(kR - \tfrac{1}{2}\pi l + \sigma_l - \eta \log 2kR). \qquad (5.30)$$

Since this must join smoothly onto the free-particle solution $\sin(kr - \tfrac{1}{2}\pi l + \delta_l)$, the phase shift is, to order $1/R$,

$$\delta_l = \sigma_l - \eta \log 2kR. \qquad (5.31)$$

For angular momenta $l \approx kR$, the phase shifts are difficult to calculate. However, these phase shifts will not be of controlling importance in what follows. We know that they will decrease smoothly to zero with increasing l and that only a limited range of angular momenta will be involved in the scattering. In order to be completely explicit, however, we might imagine

5. Elastic Scattering by a Coulomb Potential

that there was a somewhat different shielded potential leading to a phase shift

$$\delta_l = \begin{cases} \sigma_l - \Lambda_L, & l \le L, \\ 0, & l > L, \end{cases} \quad (5.32)$$

where Λ_L is a constant of the order of $\eta \log 2L \sim \eta \log 2kR$. This form is consistent with Eq. (5.31) and could be produced by a suitably chosen shielding potential. Under these conditions the asymptotic wave function has the required form

$$\psi(\mathbf{r}) \xrightarrow[r \to \infty]{} e^{ikz} + \frac{e^{ikr}}{r} f(\theta) \quad (5.33)$$

with

$$f(\theta) = \sum_{l=0}^{L} (2l+1) \frac{\exp[2i(\sigma_l - \Lambda_L)] - 1}{2ik} P_l(\cos\theta)$$

$$= e^{-2i\Lambda_L} \sum_{l=0}^{L} (2l+1) \frac{e^{2i\sigma_l} - 1}{2ik} P_l(\cos\theta)$$

$$+ \frac{e^{-2i\Lambda_L} - 1}{2ik} \sum_{l=0}^{L} (2l+1) P_l(\cos\theta). \quad (5.34)$$

In the absence of the cutoff, the Coulomb scattering amplitude would be

$$\sum_{l=0}^{\infty} (2l+1) \frac{e^{2i\sigma_l} - 1}{2ik} P_l(\cos\theta) = \frac{-\eta \exp[2i\sigma_0 - i\eta \log(1 - \cos\theta)]}{k(1 - \cos\theta)}$$

$$= f_{\text{Coul}}(\theta). \quad (5.35)$$

Then we may write $f(\theta)$ as

$$f(\theta) = e^{-2i\Lambda_L} f_{\text{Coul}}(\theta) - e^{-2i\Lambda_L} \sum_{l=L+1}^{\infty} (2l+1) \frac{e^{2i\sigma_l} - 1}{2ik} P_l(\cos\theta)$$

$$+ \frac{e^{-2i\Lambda_L} - 1}{2ik} \sum_{l=0}^{L} (2l+1) P_l(\cos\theta). \quad (5.36)$$

The contribution of the last two terms may be estimated using the techniques of Section 4, Eq. (4.41) and following.

In the first term we use an estimate of σ_l that we obtain as follows: Since σ_l is the argument of $\Gamma(l+1+i\eta)$, it can be written

$$\sigma_l = \arg \Gamma(l+1+i\eta) = \arg (l+i\eta)(l-1+i\eta) \ldots (i\eta) \Gamma(i\eta)$$

$$= \tan^{-1}(\eta/l) + \tan^{-1}(\eta/(l-1)) + \cdots. \quad (5.37)$$

Then for large l, $\sigma_l \simeq (\eta/l) + \sigma_{l-1}$, or $\Delta\sigma_l = (\eta/l)\Delta l$. Integrating this, we have the approximate result $\sigma_l = \eta \log(l/L) + \sigma_L$. Replacing the sum in Eq. (5.36) by an integral and using the method of steepest descent, we find that the first of these two terms is appreciable only for $\theta \lesssim (2\eta/L)$. The second of these terms is proportional to

$$\sum_{l=0}^{L} (2l+1) P_l(\cos\theta) \simeq 2 \int_0^L l \, dl \, J_0(l\theta) = \frac{2L}{\theta} J_1(L\theta), \tag{5.38}$$

which is appreciable only for $\theta \lesssim (1/L)$.

Thus these correction terms, representing the effect of the shielding, are significant only in the forward direction. For $\theta > (2\eta/L)$ or $(1/L)$, whichever is larger, the cross section is identical with the Coulomb cross section since $|f(\theta)| = |f_{\text{Coul}}(\theta)|$. It is clear from the foregoing that this result is valid independent of the particular form of the shielding potential, subject only to the condition that η be small compared to kR. In general, L will be of the order of kR, so that we have for a shielded Coulomb potential

$$\sigma_{\text{shielded}}(\theta) = \sigma_{\text{Coul}}(\theta) \quad \text{for} \quad \theta > \max\left(\frac{2\eta}{kR}, \frac{1}{kR}\right). \tag{5.39}$$

For a 1 MeV proton incident on hydrogen, $kR \simeq 10^4$ and $\eta \simeq 0.2$. Thus deviations from the Coulomb cross section due to shielding will generally not affect most experimental observations. The shielding does, however, permit us to use our conventional mathematical machinery to compute the wave function and the phase shift, and it does lead to a finite cross section. As we can see from Eq. (5.34), so long as R (or L) is finite, the scattering amplitude will be finite at all angles.

The actual scattering amplitude differs from the Coulomb amplitude at all angles by the phase factor $e^{-2i\Lambda_L}$. This factor arises from the logarithmic terms in the Coulomb wave function, and its exact magnitude depends upon the form of the shielded potential. It might be supposed that this phase could be detected through the interference between the Coulomb scattered wave and, say, the nuclear scattered wave. That is, if we were to consider a potential of the form

$$V(r) = Z_1 Z_2(e^2/r) + V_1(r), \tag{5.40}$$

where $V_1(r)$ is an ordinary finite-range potential, we would expect the scattering amplitude to be of the form

$$f(\theta) = a(\theta) + b(\theta), \tag{5.41}$$

5. Elastic Scattering by a Coulomb Potential

where $a(\theta)$ is the scattering amplitude due to the Coulomb interaction and $b(\theta)$ is the scattering amplitude that arises because of the presence of the additional interaction $V_1(r)$, as was found in Eq. (2.52). Since the cross section is the absolute square of $f(\theta)$, the phase of $a(\theta)$ is relevant in the interference term. However, we can very easily see that the extra phase factor $e^{-2i\Lambda_L}$ is common to both $a(\theta)$ and $b(\theta)$ and is therefore irrelevant.

The presence of the finite-range potential would modify the treatment of the shielded potential in a very slight way. Let us assume that the shielding radius R of Eq. (5.29) is always very much larger than the range of $V_1(r)$. Then Eq. (5.31) would be modified only to the extent that there would be an additional phase shift $\delta_l^{(1)}$, i.e.,

$$\delta_l = \sigma_l - \Lambda_L + \delta_l^{(1)}, \tag{5.42}$$

as in the discussion following Eq. (2.38). Thus Eq. (5.32) would read

$$\delta_l = \begin{cases} \sigma_l + \delta_l^{(1)} - \Lambda_L & l \leq L \\ 0 & l > L, \end{cases} \tag{5.43}$$

so that, in all the results that follow that equation, we need simply replace σ_l by $\sigma_l + \delta_l^{(1)}$. For $\theta > \max\{(2\eta/kR), (1/kR)\}$, the scattering amplitude would be

$$f(\theta) = e^{-2i\Lambda_L}\{f_{\text{Coul}}(\theta) + f_1(\theta)\}, \tag{5.44}$$

where

$$f_1(\theta) = \frac{1}{k} \sum_l (2l+1) e^{2i\sigma_l} \frac{\exp(2i\delta_l^{(1)}) - 1}{2i} P_l(\cos\theta). \tag{5.45}$$

Thus the phase factor is again experimentally impossible to measure.

It is interesting to see what happens when we take the limit for the case of the shielded Coulomb potential as the cutoff (R or L) becomes infinite. Note that even in this limit we are maintaining the condition that $|r_{\text{source}}|$ and $|r_{\text{detector}}|$ obey the condition $r \gg L/k$ or $r \gg R$. In that case we maintain the condition

$$\psi(\mathbf{r}) \xrightarrow[r \to \infty]{} e^{ikz} + \frac{e^{ikr}}{r} f(\theta), \tag{5.46}$$

but we now have for the scattering amplitude

$$f(\theta) = e^{-2i\Lambda_\infty} \sum_{l=0}^{\infty} (2l+1) \frac{e^{2i\sigma_l}-1}{2i} P_l(\cos\theta)$$

$$+ \frac{e^{-2i\Lambda_\infty}-1}{2i} \sum_{l=0}^{\infty} (2l+1) P_l(\cos\theta). \tag{5.47}$$

If we make use of the identities

$$\sum_{l=0}^{\infty} (2l+1) \frac{e^{2i\sigma_l}-1}{2i} P_l(\cos\theta) = \frac{-\eta \exp[2i\sigma_0 - i\eta \log(1-\cos\theta)]}{k(1-\cos\theta)}$$

$$= f_{\text{Coul}}(\theta) \tag{5.48}$$

and

$$\sum_{l=0}^{\infty} (2l+1) P_l(\cos\theta) = 2\delta(1-\cos\theta), \tag{5.49}$$

we obtain the scattering amplitude

$$f(\theta) = e^{-2i\Lambda_\infty} f_{\text{Coul}}(\theta) - i(e^{-2i\Lambda_\infty}-1)\delta(1-\cos\theta). \tag{5.50}$$

We notice that, except in the forward direction, this scattering amplitude differs from the Coulomb amplitude $f_{\text{Coul}}(\theta)$ only by a phase factor. Thus the Coulomb scattering cross section will still be the familiar Rutherford formula. The addition to the scattering amplitude of the delta function in the forward direction is, of course, unobservable in any scattering experiment in the laboratory, and we have seen above that the phase is unmeasurable in the interference between the Coulomb scattered wave and the "nuclear" scattered. wave. Thus, so long as the source and detector distances are kept greater than the shielding radius, we can let the shielding radius increase to infinity without affecting the comparison with measured results.

Chapter

4

Green's Functions

The differential equation methods discussed in the previous chapter are capable of handling any problem involving two-body elastic scattering. However, when more particles are involved and reactions or inelastic processes are possible, integral equations become far more convenient. Even in the case of elastic scattering, the integral equation formulation provides additional insights that are not so readily apparent from the differential equation. In this and the succeeding chapter we shall study the integral formulas for two-body scattering. In later chapters these results will be generalized to more complex scattering processes using the operator formalism.

1. Integral Form of the Schrödinger Equation

We shall first derive the integral form of the radial Schrödinger equation. As in similar derivations in Chapter 3, we begin with the equation for the radial wave function $u_l(r)$,

$$u_l''(r) + \left\{ k^2 - \frac{l(l+1)}{r^2} - U(r) \right\} u_l(r) = 0, \qquad (1.1)$$

and the equation for a free-particle comparison function $w_l(r)$,

$$w_l''(r) + \left\{ k^2 - \frac{l(l+1)}{r^2} \right\} w_l(r) = 0. \qquad (1.2)$$

The two linearly-independent solutions of Eq. (1.2) will be denoted by $w_l^{(1)}(r)$ and $w_l^{(2)}(r)$, with $w_l^{(1)}(r)$ the solution that is regular at the origin.

In the now-familiar way, we multiply Eq. (1.1) by $w_l(r)$, multiply Eq. (1.2) by $u_l(r)$, subtract, and integrate to obtain the result

$$\{w_l(r)\,u_l'(r) - w_l'(r)\,u_l(r)\}\Big|_{r_a}^{r_b} = \int_{r_a}^{r_b} dr\, w_l(r)\, U(r)\, u_l(r). \tag{1.3}$$

If $w_l(r)$ is taken to be the regular solution $w_l^{(1)}(r)$, then r_a may be set equal to zero in Eq. (1.3). Replacing r_b simply by r, this gives

$$\{w_l^{(1)}(r)\,u_l'(r) - w_l^{(1)'}(r)\,u_l(r)\} = \int_0^r dr'\, w_l^{(1)}(r')\, U(r')\, u_l(r'). \tag{1.4}$$

Similarly, if in Eq. (1.3.) $w_l(r)$ is taken to be $w_l^{(2)}(r)$, the irregular solution of Eq. (1.2), and r_a is set equal to r, the resultant equation is

$$\{w_l^{(2)}(r_b)\,u_l'(r_b) - w_l^{(2)'}(r_b)\,u_l(r_b)\} - \{w_l^{(2)}(r)\,u_l'(r) - w_l^{(2)'}(r)\,u_l(r)\}$$
$$= \int_r^{r_b} dr'\, w_l^{(2)}(r')\, U(r')\, u_l(r'). \tag{1.5}$$

Suppose r_b is chosen greater than R, a radius such that the potential vanishes for $r > R$. Then the first bracket in Eq. (1.5) is easily evaluated, since beyond the range of the force the radial wave function $u_l(r)$ must be a linear combination of $w_l^{(1)}(r)$ and $w_l^{(2)}(r)$, viz.,

$$u_l(r) \underset{r>R}{=} N\{w_l^{(1)}(r) + D\, w_l^{(2)}(r)\}. \tag{1.6}$$

In this case the first bracket in Eq. (1.5) is

$$\{w_l^{(2)}(r_b)\,u_l'(r_b) - w_l^{(2)'}(r_b)\,u_l(r_b)\} = N\{w_l^{(2)}(r_b)\,w_l^{(1)'}(r_b) - w_l^{(2)'}(r_b)\,w_l^{(1)}(r_b)\}$$
$$\equiv NW, \tag{1.7}$$

where W is the Wronskian of $w_l^{(1)}(r)$ and $w_l^{(2)}(r)$ and is independent of r_b.

Let us now multiply Eq. (1.4) by $w_l^{(2)}(r)$ and multiply Eq. (1.5) by $w_l^{(1)}(r)$. Addition of the resultant equations gives the integral equation

$$u_l(r) = N\, w_l^{(1)}(r) - W^{-1}\, w_l^{(2)}(r) \int_0^r dr'\, w_l^{(1)}(r')\, U(r')\, u_l(r')$$
$$- W^{-1}\, w_l^{(1)}(r) \int_r^\infty dr'\, w_l^{(2)}(r')\, U(r')\, u_l(r'). \tag{1.8}$$

This result is linear in $u_l(r)$, so that N is an over-all normalization just as it is in Eq. (1.6). For convenience we will set it equal to unity. Similarly, it is convenient to normalize $w_l^{(1)}(r)$, the regular solution of Eq. (1.2), such that

$$w_l^{(1)}(r) = F_l(kr). \tag{1.9}$$

We normalize $w_l^{(2)}(r)$ so that

$$w_l^{(2)}(r) = G_l(kr) + \gamma F_l(kr), \tag{1.10}$$

where γ is an arbitrary constant. For the choices of normalization in Eqs. (1.9) and (1.10), the Wronskian is $W = k$. We shall see shortly that, with these choices, the normalization of $u_l(r)$ is determined by the value of the constant γ.

With the functions $u_l(r)$, $w_l^{(1)}(r)$, and $w_l^{(2)}(r)$ so normalized, Eq. (1.8) may be written as

$$u_l(r) = w_l^{(1)}(r) - k^{-1} \int_0^\infty dr'\, w_l^{(1)}(r_<)\, w_l^{(2)}(r_>)\, U(r')\, u_l(r'), \tag{1.11}$$

where $r_<$ is the smaller of (r, r') and $r_>$ is the greater. This equation has the familiar form of a linear inhomogeneous integral equation. It is completely equivalent to the differential form of the Schrödinger equation, under the boundary conditions of regularity at the origin and of asymptotic behavior given in Eq. (1.6), with $N = 1$.

Alternative Forms of the Radial Wave Function

Beyond the range of the potential the integral equation becomes

$$u_l(r) \underset{r>R}{=} w_l^{(1)}(r) - k^{-1} w_l^{(2)}(r) \int_0^\infty dr'\, w_l^{(1)}(r')\, U(r')\, u_l(r'). \tag{1.12}$$

In the previous chapter we sometimes imposed as a second boundary condition on the radial wave function $u_l(r)$ the asymptotic condition

$$u_l(r) \underset{r>R}{=} F_l(kr) + T_l\, H_l^{(+)}(kr) = F_l(kr) + e^{i\delta_l} \sin \delta_l\, \{G_l(kr) + i\, F_l(kr)\}$$

$$= e^{i\delta_l} \{\cos \delta_l\, F_l(kr) + \sin \delta_l\, G_l(kr)\}. \tag{1.13}$$

To achieve this behavior, the appropriate choice for $w_l^{(2)}(r)$ in Eq. (1.11) must be

$$w_l^{(2)}(r) = G_l(kr) + i\, F_l(kr) \equiv H_l^{(+)}(kr) \tag{1.14}$$

(i.e., $\gamma = i$). The function $H_l^{(+)}(kr)$ has the asymptotic behavior of an outgoing wave,

$$H_l^{(+)}(kr) \xrightarrow[kr \gg l]{} \exp[i(kr - \tfrac{1}{2}\pi l)]. \tag{1.15}$$

Comparing Eq. (1.12) with Eq. (1.13), we see that T_l, the scattering amplitude for the lth partial wave, is

$$T_l = e^{i\delta_l} \sin \delta_l = -k^{-1} \int_0^\infty dr' \, F_l(kr') \, U(r') \, u_l(r'). \tag{1.16}$$

This integral formula is identical with that derived in the previous chapter, when $u_l(r)$ was normalized as in Eq. (1.13).

The integral equation for $u_l(r)$ is, with this choice of normalization,

$$\begin{aligned} u_l(r) &= F_l(kr) - k^{-1} F_l(kr) \int_r^\infty dr' \, H_l^{(+)}(kr') \, U(r') \, u_l(r') \\ &\quad - k^{-1} H_l^{(+)}(kr) \int_0^r dr' \, F_l(kr') \, U(r') \, u_l(r') \\ &= F_l(kr) - k^{-1} \int_0^\infty dr' \, F_l(kr_<) \, H_l^{(+)}(kr_>) \, U(r') \, u_l(r'). \end{aligned} \tag{1.17}$$

The solution to this equation is the solution to the Schrödinger equation that satisfies the previously stated boundary conditions at the origin and at infinity; both the dynamics (or wave properties) and the boundary conditions are incorporated into the integral equation.

It sometimes is more convenient, particularly for computational purposes, for $u_l(r)$ to be a real function. This can be achieved for real potential by choosing $w_l^{(2)}(r) = G_l(kr)$ (i.e., $y = 0$). The integral equation for $u_l(r)$ then becomes

$$\begin{aligned} u_l(r) &= F_l(kr) - k^{-1} F_l(kr) \int_r^\infty dr' \, G_l(kr') \, U(r') \, u_l(r') \\ &\quad - k^{-1} G_l(kr) \int_0^r dr' \, F_l(kr') \, U(r') \, u_l(r'). \end{aligned} \tag{1.18}$$

Asymptotically, this equation takes the form

$$u_l(r) \underset{r > R}{=} F_l(kr) - k^{-1} G_l(kr) \int_0^\infty dr' \, F_l(kr') \, U(r') \, u_l(r'). \tag{1.19}$$

This amounts to defining a new function $u_l(r)$ that is $e^{-i\delta_l} \sec \delta_l$ times the solution of Eq. (1.17). As in Chapter 3, the coefficient of $G_l(kr)$ is just $\tan \delta_l$, so that

$$K_l = \tan \delta_l = -k^{-1} \int_0^\infty dr' \, F_l(kr') \, U(r') \, u_l(r'). \tag{1.20}$$

This result, which depends on the use of a wave function normalized as in Eq. (1.19), is identical with the result for $\tan \delta_l$ found in the last chapter.

1. Integral Form of the Schrödinger Equation

It is important to realize that the original integral equation, Eq. (1.11), will give the same shape for the wave function and the correct phase shift, regardless of the choice of $w_l^{(2)}(r)$. That is to say, the value of the phase shift is independent of the choice of the constant in Eq. (1.10). Thus, Eq. (1.11) can be rewritten as

$$
\begin{aligned}
u_l(r) = {} & \{1 - (\gamma/k) \int_0^\infty dr'\, F_l(kr')\, U(r')\, u_l(r')\}\, F_l(kr) \\
& - k^{-1}\, F_l(kr) \int_r^\infty dr'\, G_l(kr')\, U(r')\, u_l(r') \\
& - k^{-1}\, G_l(kr) \int_0^r dr'\, F_l(kr')\, U(r')\, u_l(r') \\
\underset{r>R}{=} {} & \{1 - (\gamma/k) \int_0^\infty dr'\, F_l(kr')\, U(r')\, u_l(r')\}\, F_l(kr) \\
& - k^{-1}\, G_l(kr) \int_0^\infty dr'\, F_l(kr')\, U(r')\, u_l(r'). \quad (1.21)
\end{aligned}
$$

In view of the fact that the asymptotic form of the wave function is in general

$$u_l(r) \underset{r>R}{=} N\,\{F_l(kr) + \tan \delta_l\, G_l(kr)\}, \quad (1.22)$$

this implies the general result

$$(\cot \delta_l - \gamma)^{-1} = -k^{-1} \int_0^\infty dr'\, F_l(kr')\, U(r')\, u_l(r'). \quad (1.23)$$

When $\gamma = i$, this reduces to Eq. (1.16). When $\gamma = 0$, it reduces to Eq. (1.20). The choice of γ, and thus of $w_l^{(2)}(r)$, is therefore solely a matter of convenience in any particular application.

The Two-Potential Formula

There is a far more significant variation that has application throughout the realm of scattering problems. Not only is the physical content of the integral equation independent of the choice of $w_l^{(2)}(r)$ that is used, but it is not even dependent on the use of the free-particle function as the comparison function. As we noted in the previous chapter, there are many problems in which the solution for the potential $U(r)$ may be difficult to find, but the solution for a related potential $U_0(r)$ is easily obtained. A common example is the separation of the calculable Coulomb scattering from the less easily determined nuclear potential scattering in the scattering of nucleons from nuclei. In such cases it is more convenient to define a comparison function that is the solution of the Schrödinger equation for the potential $U_0(r)$.

Then, let $v_l^{(1)}(r)$ and $v_l^{(2)}(r)$ be the regular and irregular solutions of the radial equation,

$$v_l''(r) + \{k^2 - l(l+1)r^{-2} - U_0(r)\} v_l(r) = 0. \tag{1.24}$$

We can then go through precisely the same derivation as before. We define the residual potential to be

$$U_1(r) = U(r) - U_0(r), \tag{1.25}$$

so that $u_l(r)$ satisfies the radial differential equation

$$u_l''(r) + \{k^2 - l(l+1)r^{-2} - U_0(r)\} u_l(r) = U_1(r) u_l(r). \tag{1.26}$$

Then manipulations identical with those of Eqs. (1.3)–(1.8) lead to the integral equation

$$u_l(r) = v_l^{(1)}(r) - W^{-1} v_l^{(1)}(r) \int_r^\infty dr' \, v_l^{(2)}(r') \, U_1(r') \, u_l(r')$$

$$- W^{-1} v_l^{(2)}(r) \int_0^r dr' \, v_l^{(1)}(r') \, U_1(r') \, u_l(r'), \tag{1.27}$$

where W is now the Wronskian of $v_l^{(1)}(r)$ and $v_l^{(2)}(r)$. This integral equation reduces to our previous result, Eq. (1.10), when $U_0(r)$ is set equal to zero so that $U_1(r) = U(r)$.

The relation of this general form to the scattering phase shift can be seen by making particular choices for $v_l^{(1)}(r)$ and $v_l^{(2)}(r)$. Suppose we use functions that behave asymptotically as

$$v_l^{(1)}(r) \xrightarrow[kr \gg l]{} e^{i\delta_{0l}} \sin(kr - \tfrac{1}{2}\pi l + \delta_{0l}) \tag{1.28}$$

and

$$v_l^{(2)}(r) \xrightarrow[kr \gg l]{} \exp[i(kr - \tfrac{1}{2}\pi l + 2\delta_{0l})]. \tag{1.29}$$

The functions $v_l^{(1)}(r)$ and $v_l^{(2)}(r)$ are analogous to $F_l(kr)$ and $H_l^{(+)}(kr)$, respectively, and δ_{0l} is the phase shift for scattering by $U_0(r)$.

With these choices, Eq. (1.27) becomes asymptotically

$$u_l(r) \xrightarrow[r \to \infty]{} e^{i\delta_{0l}} \sin(kr - \tfrac{1}{2}\pi l + \delta_{0l}) - k^{-1} \exp[i(kr - \tfrac{1}{2}\pi l)]$$

$$\times \int_0^\infty dr' \, \{v_l^{(1)}(r') \, U_1(r') \, u_l(r')\}. \tag{1.30}$$

2. Properties of Green's Functions

Since this choice also implies that $u_l(r)$ behaves as $e^{i\delta_l}\sin(kr - \frac{1}{2}\pi l + \delta_l)$ for large r, we find the identity

$$e^{2i\delta_{0l}}e^{i\delta_{1l}}\sin\delta_{1l} = -k^{-1}\int_0^\infty dr'\, v_l^{(1)}(r')\, U_1(r')\, u_l(r'), \qquad (1.31)$$

where $\delta_{1l} \equiv \delta_l - \delta_{0l}$. This result gives the additional phase shift δ_{1l} contributed by the residual potential $U_1(r)$. It is the partial-wave form of the general "two-potential formula."

2. Properties of Green's Functions

The structure of the integral equations developed in the previous section permits a useful separation of the scattering process into two distinct elements, namely, the interaction of the projectile with the target, and the propagation of the particle before and after this interaction. The interaction is, of course, governed by the potential energy $U(r)$, while the propagation of the particle is governed by the kinetic energy operator. The propagation is described in integral form by the Green's function, which relates the wave function at a field point **r** to the wave function at the point **r′**. In this section we shall examine some of the properties of the Green's function.

Definition

According to Eq. (1.11), the integral equation for the radial wave function can be written as

$$u_l(r) = w_l^{(1)}(r) - W^{-1}\int_0^\infty dr'\, w_l^{(1)}(r_<)\, w_l^{(2)}(r_>)\, U(r')\, u_l(r'), \qquad (2.1)$$

where $r_<$ is the lesser of (r, r'), $r_>$ is the greater of (r, r'), and $w_l^{(1)}(r)$ and $w_l^{(2)}(r)$ are regular and irregular solutions, respectively, of the free-particle Schrödinger equation. They are eigenfunctions of the kinetic energy operator satisfying appropriate boundary conditions. The radial Green's function is defined by

$$g_l(r, r') = -W^{-1}\, w_l^{(1)}(r_<)\, w_l^{(2)}(r_>) \qquad (2.2)$$

or, more explicitly, by

$$g_l(r, r') = -W^{-1}\begin{cases} w_l^{(1)}(r)\, w_l^{(2)}(r'), & r \leq r', \\ w_l^{(2)}(r)\, w_l^{(1)}(r'), & r \geq r'. \end{cases} \qquad (2.3)$$

In terms of this radial Green's function, the integral equation is

$$u_l(r) = w_l^{(1)}(r) + \int_0^\infty dr' \, g_l(r, r') \, U(r') \, u_l(r'). \tag{2.4}$$

The separation of effects referred to above is apparent here. The radial Green's function depends only on the total energy of the particle and the boundary conditions imposed on the radial wave function. It does not depend on the interaction potential, which appears as a separate element in the integral equation.

We shall see in the next chapter, using a time-dependent treatment, how the actual point-to-point propagation of the particle is described by the Green's function. Here we might consider the following special case: Suppose $U(r)$ is a radial delta-function potential

$$U(r) = \lambda \, \delta(r - R). \tag{2.5}$$

For this potential the wave function is, from Eq. (2.4),

$$u_l(r) = w_l^{(1)}(r) + g_l(r, R) \, \lambda \, u_l(R). \tag{2.6}$$

Since $w_l^{(1)}(r)$ is the initial unscattered wave, the Green's function $g_l(r, R)$ is just the scattered wave at r due to a unit source at R, if the source strength is measured by $\lambda \, u_l(R)$. For a continuous potential the scattered wave is a superposition of waves due to such thin-shell sources, weighted by the source strength $U(r') \, u_l(r')$. This is expressed in Eq. (2.4).

We derived in the last section another integral equation in which part of the interaction energy was included in the Hamiltonian for the comparison functions. Our result in that case, Eq. (1.27), may be written as

$$u_l(r) = v_l^{(1)}(r) - W^{-1} \int_0^\infty dr' \, v_l^{(1)}(r_<) \, v_l^{(2)}(r_>) \, U_1(r') \, u_l(r'), \tag{2.7}$$

where $v_l^{(1)}(r)$ and $v_l^{(2)}(r)$ are regular and irregular solutions of

$$v_l''(r) + \{k^2 - l(l+1)r^{-2} - U_0(r)\} \, v_l(r) = 0 \tag{2.8}$$

and $U_0(r) + U_1(r) = U(r)$. In this case also one may define a radial Green's function by

$$g_l(r, r') = -W^{-1} \begin{cases} v_l^{(1)}(r) \, v_l^{(2)}(r'), & r \leq r', \\ v_l^{(2)}(r) \, v_l^{(1)}(r'), & r \geq r'. \end{cases} \tag{2.9}$$

The structure of the integral equation is exactly as before, but the quantities that enter are differently defined. We can use a similar interpretation as before; that is, now $g_l(r, r')$ describes propagation of the particle in the presence of the potential $U_0(r)$. While this division between propagation and interaction may now seem somewhat artificial, it is extraordinarily useful in many applications. In the succeeding discussion we will for convenience restrict our attention to the free-particle Green's function, Eq. (2.3), but all the properties discussed are shared by the generalized Green's function as well.

DIFFERENTIAL EQUATION

Let us now reverse the previous derivation for the sake of additional insight into the properties of the Green's function. We will now seek a solution of the radial Schrödinger equation

$$u_l''(r) + \{k^2 - l(l + 1)r^{-2} - U(r)\} u_l(r) = 0 \tag{2.10}$$

having the form we found above, namely,

$$u_l(r) = w_l^{(1)}(r) + \int_0^\infty dr' \, g_l(r, r') \, U(r') \, u_l(r'). \tag{2.11}$$

Acting upon Eq. (2.11), the free-particle differential operator gives

$$\left\{\frac{d^2}{dr^2} + k^2 - \frac{l(l + 1)}{r^2}\right\} u_l(r) = U(r) u_l(r) = \int_0^\infty dr' \left\{\frac{d^2}{dr^2} + k^2 - \frac{l(l + 1)}{r^2}\right\}$$
$$\times g_l(r, r') \, U(r') \, u_l(r'). \tag{2.12}$$

The trivial relation

$$U(r) u_l(r) = \int_0^\infty dr' \, \delta(r - r') \, U(r') \, u_l(r') \tag{2.13}$$

brings Eq. (2.12) into the form

$$\int_0^\infty dr' \left\{\frac{d^2 g_l(r, r')}{dr^2} + \left(k^2 - \frac{l(l + 1)}{r^2}\right) g_l(r, r') - \delta(r - r')\right\} U(r') \, u_l(r') = 0.$$
$$\tag{2.14}$$

This must hold for all potentials $U(r)$, so that the radial Green's function must satisfy the inhomogeneous differential equation

$$\frac{d^2 g_l(r, r')}{dr^2} + \left(k^2 - \frac{l(l+1)}{r^2}\right) g_l(r, r') = \delta(r - r'). \quad (2.15)$$

We can see by integrating this equation from $r' - \varepsilon$ to $r' + \varepsilon$ that the solution of this equation is continuous, but its derivative suffers a discontinuity at $r = r'$. Direct integration of Eq. (2.15) yields the discontinuity

$$\lim_{\varepsilon \to 0} \frac{d g_l(r, r')}{dr} \bigg|_{r=r'-\varepsilon}^{r=r'+\varepsilon} = 1. \quad (2.16)$$

BOUNDARY CONDITIONS

In order to specify the radial Green's function completely, it is necessary to fix the boundary conditions that it satisfies. We will assign very general conditions, insisting only that the radial Green's function behave as

$$g_l(r, r') \sim w_l^{(1)}(r) \quad \text{near} \quad r = 0 \quad (2.17)$$

and

$$g_l(r, r') \sim w_l^{(2)}(r) \quad \text{for} \quad r \to \infty. \quad (2.18)$$

Here $w_l^{(1)}(r)$ is the regular solution of the free-particle wave equation and $w_l^{(2)}(r)$ is any irregular solution. These functions are linearly independent solutions of the homogeneous wave equation; if they were not, it would not be possible to reproduce the required discontinuity in the derivative of $g_l(r, r')$ at $r = r'$. As we know, the effect of a scattering center is to mix into the wave function some of the irregular free-particle solution and the Green's function is proportional to this irregular solution at large distances.

Except in the vicinity of $r = r'$, the Green's function must satisfy the homogeneous differential equation. Thus the boundary conditions, Eqs. (2.17) and (2.18), further imply the functional form

$$g_l(r, r') = \begin{cases} w_l^{(1)}(r) f(r'), & r \leq r', \\ w_l^{(2)}(r) g(r'), & r \geq r'. \end{cases} \quad (2.19)$$

The Green's function so defined is, as we shall soon see, a symmetric function of r and r'. Thus we must have

$$f(r') = C w_l^{(2)}(r') \quad (2.20)$$

and

$$g(r') = C w_l^{(1)}(r'). \quad (2.21)$$

The constant C is determined by the discontinuity in the derivative of $g_l(r, r')$ at $r = r'$:

$$\lim_{\varepsilon \to 0} \left. \frac{d\, g_l(r, r')}{dr} \right|_{r=r'-\varepsilon}^{r=r'+\varepsilon} = C\{w_l^{(2)'}(r')\, w_l^{(1)}(r')$$

$$- w_l^{(1)'}(r')\, w_l^{(2)}(r')\} = -CW = 1, \quad (2.22)$$

where W is the Wronskian of $w_l^{(1)}(r)$ and $w_l^{(2)}(r)$. This gives $C = -W^{-1}$ and we find

$$g_l(r, r') = -W^{-1} \begin{cases} w_l^{(1)}(r)\, w_l^{(2)}(r'), & r \leq r', \\ w_l^{(2)}(r)\, w_l^{(1)}(r'), & r \geq r', \end{cases} \quad (2.23)$$

which agrees with our earlier result. It may be noticed that this result is independent of the normalization of $w_l^{(1)}(r)$ and $w_l^{(2)}(r)$; the Green's function is sensitive only to their functional form.

SYMMETRY

We must now establish that the Green's function that satisfies the boundary condition, Eqs. (2.17) and (2.18), is symmetric. We begin with the differential equations satisfied by the Green's functions appropriate to unit singularities at $r = r'$ and $r = r''$, respectively. Then, in the familiar way, we multiply each equation by the solution of the other and subtract the resultant equations. This yields the relation

$$g_l(r, r'') \frac{d^2 g_l(r, r')}{dr^2} - g_l(r, r') \frac{d^2 g_l(r, r'')}{dr^2}$$

$$= \delta(r - r')\, g_l(r, r'') - \delta(r - r'')\, g_l(r, r'). \quad (2.24)$$

Integration of this result gives

$$\left[g_l(r, r'') \frac{dg_l(r, r')}{dr} - g_l(r, r') \frac{dg_l(r, r'')}{dr} \right] \bigg|_{r<r',r''}^{r>r',r''}$$

$$= g_l(r', r'') - g_l(r'', r'). \quad (2.25)$$

If the functions $g_l(r, r')$ obey homogeneous boundary conditions such as Eqs. (2.17) and (2.18) or Eq. (2.19), the quantity in square brackets vanishes at both its upper and lower limits. Thus we have established that

$$g_l(r', r'') = g_l(r'', r'), \quad (2.26)$$

and the Green's function is symmetric. This implies that a source of unit strength located at r'' produces the same effect (i.e., the same wave function) at r' as a source located at r' would produce at r''. This is the reciprocity relation familiar from the study of classical wave motion.

3. Eigenfunction Expansions of Green's Functions

In the previous section we obtained a closed-form expression for the radial Green's function. Another quite useful expression for the Green's function can be found in the form of an expansion in a complete set of energy eigenfunctions. In this section we shall investigate this expansion for the radial Schrödinger equation. In order to be specific, we will consider the free-particle Green's function and will use eigenfunctions of the free-particle radial Schrödinger equation

$$w_l''(k, r) + \{k^2 - l(l + 1)r^{-2}\} w_l(k, r) = 0. \tag{3.1}$$

Expansion in a Discrete Set of Functions

To avoid any ambiguity in the method of orthogonal expansions, it is convenient first to deal with the discrete spectrum; later we shall pass to the limit of the continuous spectrum. A discrete spectrum may be obtained by imposing the boundary condition that the wave function vanish at a finite radius R:

$$w_l(k, R) = 0. \tag{3.2}$$

Since we always require that the solution vanish at the origin,

$$w_l(k, 0) = 0. \tag{3.3}$$

The acceptable solutions of Eq. (3.1) are then the regular Bessel functions

$$w_l(k_n, r) = C F_l(k_n r), \tag{3.4}$$

where the eigenvalues k_n are defined by the condition

$$F_l(k_n R) = 0. \tag{3.5}$$

For large R, this gives $k_n = (n + l/2)\pi/R$.

Since these solutions of the Schrödinger equation fulfill homogeneous boun-

3. Eigenfunction Expansions of Green's Functions

dary conditions at $r = 0$ and $r = R$, they form an orthogonal set:

$$\int_0^R dr\, w_l^*(k_n, r)\, w_l(k_m, r) = 0, \quad \text{for } m \neq n. \tag{3.6}$$

They may be normalized to unity so that they form an orthonormal set:

$$\int_0^R dr\, w_l^*(k_n, r)\, w_l(k_m, r) = \delta_{nm}. \tag{3.7}$$

In particular, the normalized eigenfunctions of Eq. (3.1) are

$$w_l(k_n, r) = \left\{ \int_0^R dr\, F_l^2(k_n, r) \right\}^{-1/2} F_l(k_n r). \tag{3.8}$$

The functions $w_l(k_n, r)$ form a complete set, so that an arbitrary function satisfying the same boundary conditions can be expanded in terms of them. From this we may derive the important closure relation. Expansion of an arbitrary function $f(r)$ obeying the boundary conditions Eqs. (3.2) and (3.3) in terms of the set $w_l(k_n, r)$ gives

$$f(r) = \sum_n A_n\, w_l(k_n, r). \tag{3.9}$$

Using the orthonormality of the set $w_l(k_n, r)$, the coefficients A_n are given by

$$A_n = \int_0^R dr\, w_l^*(k_n, r)\, f(r). \tag{3.10}$$

Substituting this into Eq. (3.9) and exchanging the order of summation and integration, we obtain

$$f(r) = \int_0^R dr'\, f(r') \left\{ \sum_n w_l^*(k_n, r')\, w_l(k_n, r) \right\}, \tag{3.11}$$

which must hold for arbitrary functions $f(r)$. We must then have

$$\sum_n w_l^*(k_n, r')\, w_l(k_n, r) = \delta(r - r'), \tag{3.12}$$

which is the closure property. This provides a convenient series expansion for the delta function.

We are now in a position to derive an expansion for the radial Green's function obeying the boundary conditions, Eqs. (3.2) and (3.3). The Green's function $g_l(r, r')$ obeys the differential equation

$$g_l''(r, r') + \{k^2 - l(l+1)r^{-2}\}\, g_l(r, r') = \delta(r - r') \tag{3.13}$$

and is expanded in terms of normalized eigenfunctions as follows:

$$g_l(r, r') = \sum_n A_n(r') w_l(k_n, r). \tag{3.14}$$

Substitution of this into the differential equation yields

$$\sum_n (k^2 - k_n^2) A_n(r') w_l(k_n, r) = \delta(r - r'). \tag{3.15}$$

If we now multiply Eq. (3.15) by $w_l^*(k_n, r)$ and integrate, we find

$$(k^2 - k_n^2) A_n(r') = w_l^*(k_n, r), \tag{3.16}$$

or

$$A_n(r') = w_l^*(k_n, r') / (k^2 - k_n^2), \tag{3.17}$$

provided k is not one of the eigenvalues k_n. The radial Green's function is then

$$g_l(r, r') = \sum_n \frac{w_l^*(k_n, r') w_l(k_n, r)}{(k^2 - k_n^2)}. \tag{3.18}$$

The Green's function for the free-particle radial Schrödinger equation is explicitly

$$g_l(r, r') = \sum_n \frac{1}{\int_0^R dr\, F_l^2(k_n r)} \frac{F_l(k_n r) F_l(k_n r')}{k^2 - k_n^2}. \tag{3.19}$$

This result is manifestly symmetric in r and r'.

The reader should note that Eq. (3.18) could have been derived as readily in the more general case of the Green's function for the Schrödinger equation with interaction, Eq. (2.8). In that case the eigenfunctions would be the normalized eigenfunctions of Eq. (2.8), subject to the boundary conditions, Eqs. (3.2) and (3.3), and would, of course, no longer be given by Eq. (3.8).

STANDING-WAVE GREEN'S FUNCTION IN A BOUNDED DOMAIN

We have previously found a general prescription for obtaining a closed form of the Green's function. Let us apply that method to the problem at hand, of the free-particle Green's function subject to the boundary conditions that it vanish at $r = 0$ and $r = R$. The Green's function is in general

$$g_l(r, r') = -W^{-1} \begin{cases} w_l^{(1)}(k, r) w_l^{(2)}(k, r'), & r < r', \\ \\ w_l^{(2)}(k, r) w_l^{(1)}(k, r'), & r > r', \end{cases} \tag{3.20}$$

where $w_l^{(1)}(k, r)$ and $w_l^{(2)}(k, r)$ are linearly independent solutions of Eq. (3.1) appropriate to the boundary conditions and W is the Wronskian of $w_l^{(1)}(k, r)$ and $w_l^{(2)}(k, r)$. We choose $w_l^{(1)}(k, r)$ to be

$$w_l^{(1)}(k, r) = F_l(kr), \tag{3.21}$$

satisfying the boundary condition at the origin, and $w_l^{(2)}(k, r)$ to be the linear combination of regular and irregular Bessel functions,

$$w_l^{(2)}(k, r) = G_l(kr) + \gamma F_l(kr). \tag{3.22}$$

The Wronskian is $W = k$. According to Eq. (3.20) and the boundary condition at $r = R$, the wave function $w_l^{(2)}(k, R)$ must vanish,

$$w_l^{(2)}(k, R) = G_l(kR) + \gamma F_l(kR) = 0, \tag{3.23}$$

giving the constant γ as

$$\gamma = -\{G_l(kR) / F_l(kR)\}. \tag{3.24}$$

As a result, the Green's function for this case becomes

$$g_l(r, r') = \{G_l(kR) / k F_l(kR)\} F_l(kr) F_l(kr')$$
$$- \frac{1}{k} \begin{cases} F_l(kr) G_l(kr'), & r \le r' \le R, \\ G_l(kr) F_l(kr'), & r' \le r \le R. \end{cases} \tag{3.25}$$

CONNECTION BETWEEN EIGENFUNCTION EXPANSION AND CLOSED FORM

The connection between the Green's functions in this form and the eigenfunction expansion of Eq. (3.19) is by no means obvious. Yet, of course, they must be identical, and we can in fact obtain the eigenfunction expansion, Eq. (3.19), from Eq. (3.25). The connection is made by means of the Mittag–Leffler theorem, and Eq. (3.19) is the Mittag–Leffler expansion of the closed form given in Eq. (3.25).

The Mittag–Leffler theorem deals with meromorphic functions, that is, single-valued functions having no singularities other than poles anywhere in the complex plane (i.e., no branch points or essential singularities). If such a function has a finite number of poles it can be expanded as

$$f(z) = g(z) + \sum_{n=1}^{N} h_n(z), \tag{3.26}$$

where $g(z)$ is an *entire function*† and $h_n(z)$ may be written as

$$h_n(z) = \sum_{v=1}^{M_n} a^{(n)}_{-v}(z - z_n)^{-v}. \tag{3.27}$$

The function $f(z)$ so represented has N poles located at the points z_n, with corresponding *principal parts*‡ $h_n(z)$. If the number of poles becomes infinite, the expansion, Eq. (3.26), may not converge uniformly. In that case, however, we may obtain an equivalent expansion which is uniformly convergent, by writing

$$f(z) = F(z) + \sum_n [h_n(z) - \gamma_n(z)], \tag{3.28}$$

where $F(z)$ is an entire function and $\gamma_n(z)$ is a function defined as a finite polynomial coinciding with the first terms of the power series expansion of $h_n(z)$ about the origin. We assume, for the sake of convenience, that there is no pole at the origin. The Mittag-Leffler theorem shows that the functions $\gamma_n(z)$ in the expansion, Eq. (3.28), can be chosen so that the series, Eq. (3.28), is uniformly convergent. Thus even for a meromorphic function with an infinite number of poles, specification of the location of the poles and the principal parts of the function at those poles serves to define the function to within an entire function.

A familiar example is the expansion of the meromorphic function $f(z) = \pi \tan \pi z$. This function has no singularities other than first order poles at $z_n = (n + \tfrac{1}{2})$ and residues $a^{(n)}_{-1} = -1$ at these poles. The functions $h_n(z)$ are then

$$h_n(z) = (n + \tfrac{1}{2} - z)^{-1}. \tag{3.29}$$

However, the sum

$$\sum_n h_n(z) = \sum_{n=-\infty}^{\infty} (n + \tfrac{1}{2} - z)^{-1} \tag{3.30}$$

is not convergent. The power series expansion for $h_n(z)$ about the origin is

$$h_n(z) = \frac{1}{n + \tfrac{1}{2} - z} = \frac{1}{n + \tfrac{1}{2}} + \frac{z}{(n + \tfrac{1}{2})^2} + \frac{z^2}{(n + \tfrac{1}{2})^3} + \cdots . \tag{3.31}$$

†An *entire function* is one whose power series expansion, $F(z) = \sum_{m=0}^{\infty} a_m z^m$, converges in the entire complex plane.

‡A function with a pole at z_n may be expanded in the vicinity of that pole as a series in powers of $(z - z_n)$. That part of the expansion containing the negative powers is called the *principal part*.

3. Eigenfunction Expansions of Green's Functions

This suggests that we try the first term in this expansion for $\gamma_n(z)$, viz.,

$$\gamma_n(z) = \frac{1}{n + \frac{1}{2}}. \tag{3.32}$$

The sum

$$\sum_{n=-\infty}^{\infty} h_n(z) - \gamma_n(z) = \sum_{n=-\infty}^{\infty} \left(\frac{1}{n + \frac{1}{2} - z} - \frac{1}{n + \frac{1}{2}} \right)$$

$$= \sum_{n=-\infty}^{\infty} \frac{z}{(n + \frac{1}{2})(n + \frac{1}{2} - z)} \tag{3.33}$$

is uniformly convergent, so that we may write

$$\pi \tan \pi z = F(z) + \sum_{n=-\infty}^{\infty} \left\{ \frac{1}{(n + \frac{1}{2}) - z} - \frac{1}{(n + \frac{1}{2})} \right\}$$

$$= F(z) + \sum_{n=-\infty}^{\infty} \frac{4z}{(2n + 1)^2 - 2z(2n + 1)}, \tag{3.34}$$

where $F(z)$ is an entire function. By a further examination of the properties of the function $\pi \tan \pi z$, one can show that $F(z)$ is identically zero. Thus, we have

$$\pi \tan \pi z = \sum_{n=-\infty}^{\infty} \left\{ \frac{1}{(n + \frac{1}{2}) - z} - \frac{1}{(n + \frac{1}{2})} \right\}$$

$$= \sum_{n=-\infty}^{\infty} \frac{4z}{(2n + 1)^2 - 2z(2n + 1)}, \tag{3.35}$$

a result we shall find useful later in this section.

Let us now return to the Green's function in the eigenfunction representation, Eq. (3.19). It has first-order poles in k^2 at the eigenvalues of the energy. The Green's function given by the closed form, Eq. (3.25), has poles at the same energies, namely, at the zeros of the functions $F_l(kR)$. (The functions $F_l(kr)$ and $G_l(kr)$ are entire functions and have no singularities anywhere.) Since $F_l(kR)$ is either an odd or an even function of k, depending on whether l is even or odd, the positions of the singularities depend only on k^2. Thus we see immediately that the Green's function of Eq. (3.25) has its only singularities at the points $k^2 = k_n^2$, where k_n is defined by Eq. (3.5).

The character of the singularity may be ascertained by expanding the denominator function $F_l(kR)$ in the neighborhood of $k^2 = k_n^2$. This is most easily done by returning to the differential equations for $F_l(kr)$ and $F_l(k_n r)$.

Multiplying the former by $F_l(k_n r)$, multiplying the latter by $F_l(kr)$, subtracting, and integrating, we obtain

$$F_l(kr) = \frac{k^2 - k_n^2}{\left[\dfrac{d}{dr} F_l(k_n r)\right]_{r=R}} \int_0^R dr\, F_l(kr)\, F_l(k_n r). \tag{3.36}$$

We may use the Wronskian relation to remove the derivative. Then, for k^2 in the neighborhood of k_n^2, we have

$$F_l(kR) = \frac{k^2 - k_n^2}{k_n} G_l(k_n R) \left\{ \int_0^R dr\, F_l^2(k_n r) + \mathcal{O}((k^2 - k_n^2)^2) \right\}. \tag{3.37}$$

As a result, the singularity of the Green's function in Eq. (3.25) is of order $1/(k^2 - k_n^2)$, as expected, and the residue is

$$F_l(k_n r)\, F_l(k_n r') \bigg/ \int_0^R dr\, F_l^2(k_n r). \tag{3.38}$$

This is identical with the residue in the series expansion, Eq. (3.19).

This is not quite enough to allow us to conclude that Eqs. (3.19) and (3.25) are identical. Two meromorphic functions with identical first order poles and identical residues at these poles can still differ by an entire function. However, in the present case this entire function must necessarily be a solution of the homogeneous equation for wave number k, subject to the boundary conditions at $r = 0$ and $r = R$. The only function that can meet these requirements is the null function. Thus we have explicitly demonstrated the equality of these two expressions. Of course, since they were each derived as the Green's function for Eq. (3.1) under the boundary conditions, Eqs. (3.2) and (3.3), they must necessarily be identical.

General Boundary Conditions

In exactly the same way we could obtain the two forms of the Green's function for any other boundary conditions. In general, the boundary condition at $r = R$ will specify the logarithmic derivative

$$\left[\frac{g_l'(r, r')}{g_l(r, r')}\right]_{r=R} = L. \tag{3.39}$$

3. Eigenfunction Expansions of Green's Functions

Two linearly independent solutions of the homogeneous equation are

$$w_l^{(1)}(k, r) = F_l(kr) \tag{3.40}$$

and

$$w_l^{(2)}(k, r) = G_l(kr) + \gamma F_l(kr), \tag{3.41}$$

where γ is determined by the boundary condition and is given by

$$\gamma = -\frac{G_l'(kR) - LG_l(kR)}{F_l'(kR) - L F_l(kR)} \tag{3.42}$$

The Green's function is

$$g_l(r, r') = -\frac{1}{k} \begin{cases} F_l(kr)[G_l(kr') + \gamma F_l(kr')], & r \leq r' \leq R, \\ F_l(kr')[G_l(kr) + \gamma F_l(kr)], & r' \leq r \leq R. \end{cases} \tag{3.43}$$

The eigenvalues of k^2 are those values for which $\gamma \to \infty$, as previously. In fact, one can easily show that, for $k^2 \to k_n^2$,

$$\gamma \xrightarrow[k^2 \to k_n^2]{} -\frac{k_n \bigg/ \int_0^R dr\, F_l^2(k_n r)}{k^2 - k_n^2}, \tag{3.44}$$

independent of the choice of L. Then Eq. (3.37) reproduces the series expansion for any boundary condition.

It is worthwhile to note that L must be real if we are to obtain real eigenvalues, as can be seen from the eigenvalue equation

$$\frac{F_l'(k_n R)}{F_l(k_n R)} = L. \tag{3.45}$$

However, we are not constrained to consider only real values of L, since these correspond only to standing waves. That is, if the function $w_l^{(2)}(k, r)$ is to be a standing wave, then γ and L must be real. We may, however, wish to consider traveling waves, in which case we must use complex values of L.

Let us calculate the total current emerging from a sphere of radius R. We take our wave function to be

$$\psi_l(k, \mathbf{r}) = (C/r)\{G_l(kr) + \gamma F_l(kr)\} Y_{l0}(\theta), \tag{3.46}$$

where C is any constant of normalization. The total outward current is then found to be

$$\int_{\text{surf}} \mathbf{j} \cdot d\mathbf{s} = |C|^2 (\hbar k/m) \, \text{Im} \, \gamma, \tag{3.47}$$

so that the outward current is proportional to the imaginary part of γ. Introducing the logarithmic derivative L at the radius R, we may rewrite Eq. (3.47) as

$$\int_{\text{surf}} \mathbf{j} \cdot \mathbf{ds} = \frac{|C|^2 \, (\hbar k/m) \, \text{Im} \, L}{|F_l'(kR) - L F_l(kR)|^2}, \tag{3.48}$$

which shows the need for a complex value of L if a net current is to be obtained.
For any boundary conditions the normalized free-particle eigenfunctions are

$$w_l(k_n, r) = F_l(k_n r) \left/ \left(\int_0^R dr \, F_l^2(k_n r) \right)^{1/2} \right. . \tag{3.49}$$

If the boundary condition is complex, the functions $w_l(k_n, r)$ satisfy only the unusual orthonormality relation

$$\int_0^R dr \, w_l(k_n, r) \, w_l(k_m, r) = \delta_{nm}, \tag{3.50}$$

and the normalization integral is

$$\int_0^R dr \, F_l^2(k_n r) = \tfrac{1}{2} R \{ F_l^2(k_n R) - F_{l-1}(k_n R) F_{l+1}(k_n R) \} \xrightarrow[k_n R \gg l]{} \tfrac{1}{2} R. \tag{3.51}$$

Because Eq. (3.50) may be unfamiliar, we shall derive this orthogonality relation. The regular function $w_l(k_n, r)$ obeys the free-particle Schrödinger equation, Eq. (3.1), subject to the boundary condition $\{w_l'(k_n R)/w_l(k_n, R)\} = L$ which determines the eigenvalues k_n. Subtracting and integrating the Schrödinger equations for the eigenvalues k_n and k_m in the familiar way, we obtain

$$w_l(k_n, R) \, w_l'(k_m, R) - w_l(k_m, R) \, w_l'(k_n, R)$$
$$= (k_n^2 - k_m^2) \int_0^R w_l(k_m, r) \, w_l(k_n r) \, dr. \tag{3.52}$$

Assuming L to be finite, we may rewrite the left-hand side of Eq. (3.46) as

$$w_l(k_n, R) \, w_l(k_m, R) \left\{ \frac{w_l'(k_m, R)}{w_l(k_m, R)} - \frac{w_l'(k_n, R)}{w_l(k_n, R)} \right\} = 0, \tag{3.53}$$

which vanishes by the boundary condition at $r = R$. This gives Eq. (3.50).

In order to derive the more familiar orthogonality relation, we would proceed as above, except that we would use the complex conjugate of one of the radial equations. We would then obtain

$$\{w_l^*(k_n^*, R) \, w_l'(k_m, R) - w_l(k_m, R) \, w_l^{*\prime}(k_n^*, R)\} = w_l^*(k_n^*, R) \, w_l(k_m, R)[L - L^*]$$
$$= (k_n^{2*} - k_m^2) \int_0^R w_l^*(k_n^*, R) \, w_l(k_n, R). \tag{3.54}$$

If L is real, then k_n is also real and the familiar orthogonality relation is established; otherwise, only Eq. (3.44) holds.

This orthogonality discussion need not be restricted to solutions of the free-particle Schrödinger equation. Clearly Eqs. (3.52) and (3.53) hold as well for any potential function. Similarly Eq. (3.54) holds provided only that $V(r)$ is real. The usual orthogonality relation,

$$\int u_l^*(k_n, r) \, u_l(k_m, r) \, dr = 0, \qquad n \neq m, \tag{3.55}$$

is a well-known consequence of the hermiticity of the Hamiltonian. This is a property not of the Hamiltonian or potential function alone, but depends on the boundary conditions as well. In particular, the reality of $V(r)$ and L is sufficient to establish hermiticity.

These considerations allow us to rewrite Eq. (3.18) in the more general form

$$g_l(r, r') = \sum_n \frac{w_l(k_n, r') \, w_l(k_n, r)}{(k^2 - k_n^2)}, \tag{3.56}$$

where the functions $w_l(k, r)$ obey the orthonormality relation, Eq. (3.50). In this form the radial Green's function is manifestly symmetric in r and r'.

OUTGOING-WAVE GREEN'S FUNCTION IN A BOUNDED DOMAIN

An interesting and important example of a complex logarithmic derivative is obtained from the logarithmic derivative of an outgoing wave,

$$\frac{d}{dr}(e^{ikr}) \,/\, e^{ikr} = ik.$$

For large R this boundary condition, $L = ik$, implies that Eq. (3.42)

$$\gamma = -\frac{-k \sin(kR - \tfrac{1}{2}\pi l) - ik \cos(kR - \tfrac{1}{2}\pi l)}{k \cos(kR - \tfrac{1}{2}\pi l) - ik \sin(kR - \tfrac{1}{2}\pi l)} = i. \tag{3.57}$$

Unlike the previous case, Eq. (3.24), where γ was a function of R, we find here that γ is independent of R. If we impose this boundary condition

$$L = \left[\frac{g_l'(r, r')}{g_l(r, r')}\right]_{r=R} = ik, \tag{3.58}$$

and assume that $kR \gg l$, the Green's function becomes

$$g_l(r, r') = -\frac{1}{k}\begin{cases} F_l(kr)\, H_l^{(+)}(kr'), & r < r' < R, \\ H_l^{(+)}(kr)\, F_l(kr'), & r' < r < R. \end{cases} \qquad (3.59)$$

The outgoing-wave function $H_l^{(+)}(kr)$ is defined, as we recall, by

$$H_l^{(+)}(kr) \equiv G_l(kr) + i\, F_l(kr) \xrightarrow[kr \gg l]{} \exp[i(kr - \tfrac{1}{2}\pi l)]. \qquad (3.60)$$

As we expect, this radial Green's function behaves like an outgoing wave for large r or r'.

In a similar fashion we could obtain an ingoing-wave Green's function by imposing the boundary condition $L = -ik$; since the results are completely analogous to those for the outgoing-wave boundary condition, we will continue to direct our attention to the case of outgoing waves.

We now want to consider the relation between this closed form of the outgoing-wave radial Green's function, Eq. (3.59), and the Green's function in the form of a sum over eigenfunctions. We must therefore find the expansion when the boundary condition is $L = ik$.

The eigenvalue equation is

$$F_l'(k_n R) / F_l(k_n R) = ik, \qquad (3.61)$$

which, for large R, becomes

$$\cot(k_n R - \tfrac{1}{2}\pi l) = i(k/k_n), \qquad k_n R \gg l. \qquad (3.62)$$

The set of eigenvalues k_n determined by this equation will be functions of the wave number k, which for this purpose is treated as a fixed parameter. They will be complex and may be written as

$$k_n = \kappa_n + i\lambda_n = k\, \frac{-\sinh 2\lambda_n R + i \sin(2\kappa_n R - \pi l)}{\cosh 2\lambda_n R + \cos(2\kappa_n R - \pi l)}. \qquad (3.63)$$

The positions of these eigenvalues in k-space are shown in Fig. 4.1.

The Green's function may be written immediately as

$$g_l(r, r') = \sum_n \frac{1}{\int_0^R dr\, F^2(k_n r)} \frac{F_l(k_n r)\, F_l(k_n r')}{(k^2 - k_n^2)}, \qquad (r, r') < R, \qquad (3.64)$$

with k_n defined by Eq. (3.61). The reader should experience no difficulty in showing that this is the Mittag-Leffler expansion of the Green's function, Eq. (3.59). However, it must be noted that the wave number k plays two roles: it determines the boundary condition as well as the energy. These roles must

3. Eigenfunction Expansions of Green's Functions

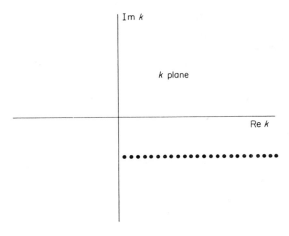

Fig. 4.1

be distinguished in finding the singularities of the Green's function. If the boundary condition is kept constant by letting $L = ik_a$, where k_a is a fixed wave number, then the proof of the equivalence of these two forms of the Green's function parallels the previous demonstration.

Standing-Wave Green's Function for the Unbounded Domain

We now want to take the limit of these Green's functions as R becomes very large. A finite radius was originally used simply to obtain a discrete set of eigenfunctions, in order that conventional orthogonal expansions could be used. In a realistic scattering problem there is no such well-defined radius, and we must examine the eigenfunction expansions in the limit that R goes to infinity.

Let us first deal with the general Green's function of Eq. (3.43). If we were to increase R continuously, always adjusting γ so that the logarithmic derivative at R was equal to L, then we see from Eq. (3.42) that γ would in general oscillate without bound as R increases, and there would be no unique, uniform limit. If, on the other hand, we have a certain value of γ at the radius R_v, we can obtain the same value of γ at a larger radius by increasing R_v through a discrete amount to R_{v+1}, such that

$$\gamma = -\frac{G_l'(kR_v) - L\, G_l(kR_v)}{F_l'(kR_v) - L\, F_l(kR_v)} = -\frac{G_l'(kR_{v+1}) - L\, G_l(kR_{v+1})}{F_l'(kR_{v+1}) - L\, F_l(kR_{v+1})}. \quad (3.65)$$

Such a procedure provides one means of passing to the limit $R \to \infty$. However, in a scattering problem the radius R_v is not specified, and we can obtain any value for γ that we wish. In fact, the only restriction on γ is that it will be real if L is real and complex if L is complex, with the imaginary parts of L and γ having the same signs. (The proof of this follows easily from Eq. (3.39).) Thus, in the limit as $R \to \infty$, the only part of the boundary condition that survives uniquely is the specification of whether we have standing waves, outgoing waves, or incoming waves, i.e., whether Im L is zero, positive, or negative.

The Green's function of Eq. (3.25) satisfies the boundary condition $L^{-1} = 0$. As one example of the limiting process, if we choose as our sequence of radii values of R_v corresponding to the zeros of $G_l(kr)$, then the Green's function for the infinite domain becomes simply

$$g_l(r, r') = -\frac{1}{k} \begin{cases} F_l(kr)\, G_l(kr'), & r \le r', \\ G_l(kr)\, F_l(kr'), & r \ge r'. \end{cases} \quad (3.66)$$

As Eq. (3.43) shows, the most general Green's function is the sum of a term identical with Eq. (3.66) and a term proportional to the product of the regular functions, $F_l(kr)\, F_l(kr')$. The former term is a solution of an inhomogeneous differential equation, having a discontinuous first derivative and an amplitude fixed by the condition that this discontinuity be unity, as required by Eq. (2.22). The amplitude of the latter term, which is a solution of the homogeneous differential equation having continuous derivatives, is chosen so that the boundary condition is satisfied. As we have observed, γ, the amplitude of this homogeneous term, is in general not uniquely determined in the limit $R \to \infty$. On the other hand, in the special case when $L = ik$, we have seen that $\gamma = i$, independent of R. Because of this, the limit of the outgoing-wave Green's function is independent of the sequence R_v and is uniquely

$$g_l(r, r') = -\frac{1}{k} \begin{cases} F_l(kr)\, H_l^{(+)}(kr'), & r \le r', \\ H_l^{(+)}(kr)\, F_l(kr'), & r \ge r'. \end{cases} \quad (3.67)$$

Now let us return to the radial Green's function in the form of a sum over energy eigenfunctions, Eq. (3.57). We recall that this is identical with the closed form, Eq. (3.43), if the same boundary conditions are satisfied. Thus every step followed in obtaining the closed-form Green's function for the infinite domain should have its analog in the derivation of the Green's function for the infinite domain as an expansion in energy eigenfunctions. Further-

3. Eigenfunction Expansions of Green's Functions

more, the results may differ in form, but they must necessarily yield identical functions.

Consider first the standing-wave Green's function. At first glance it would appear that we can take the limit of Eq. (3.19) very easily. We note that the energy eigenvalues k_n^2 are given by the zeros of $F_l(k_n R)$. In the limit as R increases, these are solutions of the asymptotic form

$$F_l(k_n R) = \sin(k_n R - \tfrac{1}{2}\pi l) = 0, \qquad k_n R \gg l. \tag{3.68}$$

Then the eigenvalues are given by

$$k_n = (n + \tfrac{1}{2}l)\pi/R, \tag{3.69}$$

and the spacing of the energy eigenvalues decreases as R increases, since

$$\Delta k_n = (\pi/R)\,\Delta n. \tag{3.70}$$

It follows from this that the sum over n in Eq. (3.19) is replaced by $(R/\pi)\int_0^\infty dk'$, and, in passing to the limit of large R, the expansion becomes

$$g_l(r, r') = \frac{2}{\pi}\int_0^\infty dk'\,\frac{F_l(k'r)\,F_l(k'r')}{k^2 - k'^2}. \tag{3.71}$$

However, this integral is undefined because of the vanishing denominator at $k' = k$, so that this is not as easy as we might have assumed.

We must specify how we will perform the integration in the neighborhood of the singularity at $k' = k$. The integration path is fixed and is along the real axis since k_n and k' are real, so that only a prescription for avoiding the singular point is needed to make Eq. (3.71) well defined. The reader will note that we have assumed throughout that k is not an eigenvalue, although it may be arbitrarily close to an eigenvalue k_n. Thus the point $k = k'$ is excluded from the integral.

In taking the limit of the closed form, we were not able to let R increase continuously. Rather, we found that the limit would have different values depending on the way we defined the sequence R_v of increasing values of R. In arriving at Eq. (3.71) we attempted to let R pass continuously to infinity. Clearly we may not do this, since if R were increased continuously k would coincide with an eigenvalue k_n for some values of R. The same difficulty would appear also in taking the limit of the closed form, Eq. (3.25); $F_l(k_n R)$ would go to zero for those same values of R [cf. Eq. (3.30)].

If k is not an eigenvalue, it must lie between two eigenvalues. Let us label these two neighboring eigenvalues by k_p and k_{p+1}:

$$k_p < k < k_{p+1}. \tag{3.72}$$

Now let us impose the same constraints as led to the derivation of the limit of the closed form, Eq. (3.66). There the values of R_v were so chosen that kR_v always coincided with a zero of $G_l(kR_v)$; the eigenvalues k_n will occur at the zeros of $F_l(k_nR_v)$. For large values of R_v one can easily see that this gives eigenvalues positioned such that k lies midway between two eigenvalues:

$$\tfrac{1}{2}(k_p + k_{p+1}) = k. \qquad (3.73)$$

In general, if we choose R_v such that k is always equidistant from the nearest eigenvalues, γ will vanish for large R and the radial Green's function will be given by Eq. (3.66), for *any* real boundary condition.

If we now impose this requirement on R_v, we find from Eq. (3.19) that the Green's function for the infinite domain is

$$\begin{aligned}
g_l(r, r') &= \lim_{v \to \infty}(2/R_v) \sum_{k_n=0}^{\infty} \frac{F_l(k_n r) F_l(k_n r')}{k^2 - k_n^2} \\
&= \lim_{v \to \infty}(2/R_v)\left\{ \sum_{k_n=0}^{k_p} \frac{F_l(k_n r) F_l(k_n r')}{k^2 - k_n^2} + \sum_{k_n=k_{p+1}}^{\infty} \frac{F_l(k_n r) F_l(k_n r')}{k^2 - k_n^2} \right\} \\
&= \frac{2}{\pi} \lim_{\varepsilon \to 0}\left\{ \int_0^{k-\varepsilon} dk' \frac{F_l(k'r) F_l(k'r')}{k^2 - k'^2} + \int_{k+\varepsilon}^{\infty} dk' \frac{F_l(k'r) F_l(k'r')}{k^2 - k'^2} \right\} \\
&\equiv \frac{2}{\pi} \mathscr{P} \int_0^{\infty} dk' \frac{F_l(k'r) F_l(k'r')}{k^2 - k'^2}, \qquad (3.74)
\end{aligned}$$

where we have introduced $\varepsilon = k - k_p = k_{p+1} - k = \tfrac{1}{2}\pi/R_v$, and the symbol $\mathscr{P}$ stands for the principal value of the integral.

Since we have imposed the same constraints on the sequence R_v in each case, we have now shown that for the infinite domain the Green functions of Eq. (3.74) and Eq. (3.66) are identical. That is, we have shown that

$$\begin{aligned}
g_l(r, r') &= \frac{2}{\pi} \mathscr{P} \int_0^{\infty} dk' \frac{F_l(k'r) F_l(k'r')}{k^2 - k'^2} \\
&= -\frac{1}{k} \begin{cases} F_l(kr) G_l(kr'), & r \leq r', \\ G_l(kr) F_l(kr'), & r \geq r'. \end{cases}
\end{aligned} \qquad (3.75)$$

The inhomogeneous part of the Green's function may thus be represented by a principal-value integral.

Let us now examine the result of choosing a limiting process that does not give eigenvalues satisfying the condition of Eq. (3.63). We may choose

3. Eigenfunction Expansions of Green's Functions

the sequence R_v to given eigenvalues such that

$$\bar{k} \equiv \tfrac{1}{2}(k_p + k_{p+1}) = k - \pi x/R_v, \tag{3.76}$$

with x a fixed number between $-\tfrac{1}{2}$ and $+\tfrac{1}{2}$. Under these circumstances Eq. (3.65) becomes $y = \tan \pi x$ and the closed form of the standing-wave Green's function for the infinite domain is

$$g_l(r, r') = -\frac{1}{k} \tan \pi x \, F_l(kr) \, F_l(kr')$$

$$-\frac{1}{k} \begin{cases} F_l(kr) \, G_l(kr'), & r \le r', \\ G_l(kr) \, F_l(kr'), & r \ge r'. \end{cases} \tag{3.77}$$

Let us verify that under these same circumstances the limit of the eigenfunction expansion is

$$g_l(r, r') = -\frac{1}{k} \tan \pi x \, F_l(kr) \, F_l(kr') + \frac{2}{\pi} \mathscr{P} \int_0^\infty dk' \, \frac{F_l(k'r) \, F_l(k'r')}{k^2 - k'^2}. \tag{3.78}$$

From Eq. (3.19) we have

$$g_l(r, r') = \lim_{R_v \to \infty} \frac{2}{R_v} \sum_{k_n=0}^{\infty} \frac{F_l(k_n r) \, F_l(k_n r')}{k^2 - k_n^2}. \tag{3.79}$$

If we define $\bar{k}$ as the average of the two eigenvalues closest to k, as in Eq. (3.76), the denominator in Eq. (3.73) may be written

$$\frac{1}{k^2 - k_n^2} = \frac{1}{\bar{k}^2 - k_n^2} + \frac{\bar{k}^2 - k^2}{(k^2 - k_n^2)(\bar{k}^2 - k_n^2)}. \tag{3.80}$$

Introducing the first term into Eq. (3.73) gives

$$\lim_{R_v \to \infty} \frac{2}{R_v} \sum_{k_n=0}^{\infty} \frac{F_l(k_n r) \, F_l(k_n r')}{\bar{k}^2 - k_n^2} = \frac{2}{\pi} \mathscr{P} \int_0^\infty dk' \, \frac{F_l(k'r) \, F_l(k'r')}{k^2 - k'^2}. \tag{3.81}$$

since the sum is identical with Eq. (3.74).

We can see that the contribution of the second term in Eq. (3.74) is negligible except for values of k_n in the immediate neighborhood of k. Using the definition of $\bar{k}$, we can express the eigenvalues as

$$k_n = \bar{k} + (m + \tfrac{1}{2})\pi/R_v = k + (m + \tfrac{1}{2} - x)\pi/R_v. \tag{3.82}$$

We then find for large R_v

$$\frac{(\bar{k}^2 - k^2)}{(k^2 - k_n^2)(\bar{k}^2 - k_n^2)} = \frac{R_v}{\pi k} \left\{ \frac{2x}{(2m+1)^2 - 2x(2m+1)} + \mathcal{O}(1/kR_v) \right\}. \tag{3.83}$$

The leading term decreases rapidly for increasing m so that, according to Eq. (3.76), this expression is negligible unless $|k_n - k| \sim 1/R_v$.

If Eq. (3.77) is inserted into Eq. (3.79), we obtain

$$\lim_{R_v \to \infty} \frac{2}{R_v} \sum_{k_n=0}^{\infty} \frac{(\bar{k}^2 - k^2) F_l(k_n r) F_l(k_n r')}{(k^2 - k_n^2)(\bar{k}^2 - k_n^2)}$$

$$= -\frac{1}{\pi k} F_l(kr) F_l(kr') \sum_{m=-\infty}^{\infty} \frac{4x}{(2m+1)^2 - 2x(2m+1)}$$

$$= -k^{-1} \tan \pi x \, F_l(kr) F_l(kr'), \qquad (3.84)$$

where we have identified the expansion of $\tan \pi x$ from Eq. (3.35). Thus we have obtained the expected result, Eq. (3.79).

These results can be summarized by the symbolic expression

$$\frac{1}{k^2 - k_n^2} \xrightarrow[R \to \infty]{} \mathscr{P} \frac{1}{k^2 - k'^2} - \pi \tan \pi x \, \delta(k^2 - k'^2). \qquad (3.85)$$

Inserted into the sum, Eq. (3.73), this yields Eq. (3.72). More will be said about such expressions in our later work. We note here only that the amplitude of the principal-value term is fixed by the condition on the discontinuity in the derivative of $g_l(r, r')$, while the amplitude of the delta-function term depends upon the boundary condition and the way in which we take the limit $R \to \infty$.

Outgoing-Wave Green's Function for the Unbounded Domain

Now let us turn our attention to the situation in which the eigenvalues of k may be complex. We treat the case where the boundary condition at $r = R$ requires that the logarithmic derivative be $L = ik$; for large values of R this implies pure outgoing waves, or $\gamma = i$. The radial Green's function for the infinite domain is given in closed form by Eq. (3.67), and we wish to examine the relation between the Green's function in this form and the Green's function in the form of an integral over eigenfunctions.

We start with the expansion Eq. (3.56), with the eigenvalues k_n defined by Eq. (3.63), and pass to the limit $R \to \infty$. As noted earlier, in this case we need not concern ourselves with choosing a sequence R_v, but can take the

3. Eigenfunction Expansions of Green's Functions

limit directly. In the limit of large R, Eq. (3.63) gives for the eigenvalues $k_n = \kappa_n + i\lambda_n$ with

$$\kappa_n = \begin{cases} (n + \tfrac{1}{2}l)(\pi/R), & \kappa_n < k, \\ (n + \tfrac{1}{2}l + \tfrac{1}{2})(\pi/R), & \kappa_n > k, \end{cases} \tag{3.86}$$

and

$$\lambda_n = (2R)^{-1} \log\left| (k + \kappa_n)/(k - \kappa_n) \right|. \tag{3.87}$$

The eigenvalue spacing is $\Delta k_n = (\pi/R) \Delta n$, approaching zero in the limit of large R. The imaginary part of k_n also approaches zero in this limit. For κ_n far from k, λ_n goes as $(1/R)$; for κ_n in the neighborhood of k, λ_n varies as $R^{-1} \log R$, which vanishes as $R \to \infty$. Thus the summation passes over into an integral with the path of integration infinitesimally below the real axis. If the line of eigenvalues is followed, the path has the form shown in Fig. 4.2(a). As $R \to \infty$, this path approaches the real axis.

The resulting integral is

$$g_l(r, r') = \frac{2}{\pi} \int_C dk' \, \frac{F_l(k'r) \, F_l(k'r')}{k^2 - k'^2}, \tag{3.88}$$

where the contour C may be chosen in various ways. It may be the path shown in Fig. 4.2(a). However, the only singularity is at $k' = k$, and the contour may be distorted in any way that proves convenient, provided only that it continues to pass by this singularity in the counterclockwise direction. Thus the contour shown in Fig. 4.2(b), corresponding to a constant imaginary part for k', or the contour shown in Fig. 4.2(c) could be used equally well. In each case, of course, we must take the limit of the integral as the contour approaches the real axis, *after* performing the integration.

The reader may note that the integral would vary smoothly if the wave number k were given a variable but positive imaginary part. However, the singularity would cross over the integration contours in Fig. 4.2 if k were given a negative imaginary part, and the integral would then be converted into quite a different function. Thus, the outgoing-wave Green's function is an analytic function of k for values of k in the upper right-hand quadrant of the k plane (Im $k > 0$), but it is not analytic in the lower right-hand quadrant (Im $k < 0$). This behavior is closely related to the causality condition imposed on the time-dependent wave function. We saw in Chapter 2 that this condition, which requires that the wave function reduce to the initial free-particle wave function for very early times, leads to the outgoing-wave boundary condition on the time-independent wave function. This outgoing-wave

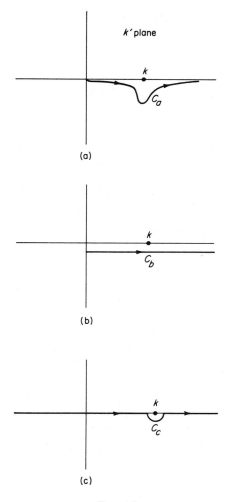

Fig. 4.2

boundary condition leads, in turn, to analyticity of the Green's function in the upper right-hand k plane. This observation is the starting point in developing the analytic properties of the scattering amplitude. However, in this book we shall not discuss this particular aspect of scattering theory, which is the subject of much current research in high-energy physics.

The integral in Eq. (3.88) can be calculated rather easily using the contours in Fig. 4.2(b) and (c). For C_b it is convenient to replace the variable k' in Eq. (3.88) by $k'-i\lambda$, with $\lambda > 0$ and k' real. The integration is then along the

3. Eigenfunction Expansions of Green's Functions

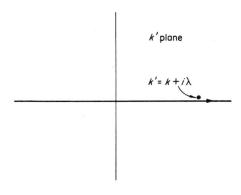

FIG. 4.3

real axis rather than along the line Im $k' = -\lambda$; in effect, this shifts the singularity to the point $k' = k + i\lambda$ (see Fig. 4.3). The Green's function is then given by the integral

$$g_l(r, r') = \lim_{\lambda \to 0} \frac{2}{\pi} \int_0^\infty dk' \, \frac{F_l(k'r) F_l(k'r')}{k^2 - (k' - i\lambda)^2}, \tag{3.89}$$

or, more conventionally but completely equivalently,

$$g_l(r, r') = \lim_{\varepsilon \to 0} \frac{2}{\pi} \int_0^\infty dk' \, \frac{F_l(k'r) F_l(k'r')}{k^2 - k'^2 + i\varepsilon}. \tag{3.90}$$

In either form the denominator has a *positive* imaginary part, and both give identical results.

In order to verify explicitly that these are equivalent to the closed form, Eq. (3.67), we must perform these integrals. Extending the integration over the entire real axis, the radial Green's function for outgoing waves becomes

$$g_l(r, r') = \lim_{\varepsilon \to 0} \frac{1}{\pi} \int_{-\infty}^\infty dk' \, \frac{F_l(k'r) F_l(k'r')}{k^2 - k'^2 + i\varepsilon}. \tag{3.91}$$

If we express the function $F_l(z)$ as

$$\begin{aligned} F_l(z) &= (1/2i)\{G_l(z) + i F_l(z)\} - (1/2i)\{G_l(z) - i F_l(z)\} \\ &= (1/2i)\{H_l^{(+)}(z) - H_l^{(-)}(z)\}, \end{aligned} \tag{3.92}$$

then Eq. (3.91) becomes a sum of four terms,

$$g_l(r, r') = \sum_{n=1}^4 I_n, \tag{3.93}$$

where

$$I_1 = -\lim_{\varepsilon \to 0} \frac{1}{4\pi} \int_{-\infty}^{\infty} dk' \, \frac{H_l^{(+)}(k'r) H_l^{(+)}(k'r')}{k^2 - k'^2 + i\varepsilon}, \quad (3.94)$$

$$I_2 = \lim_{\varepsilon \to 0} \frac{1}{4\pi} \int_{-\infty}^{\infty} dk' \, \frac{H_l^{(+)}(k'r) H_l^{(-)}(k'r')}{k^2 - k'^2 + i\varepsilon}, \quad (3.95)$$

$$I_3 = \lim_{\varepsilon \to 0} \frac{1}{4\pi} \int_{-\infty}^{\infty} dk' \, \frac{H_l^{(-)}(k'r) H_l^{(+)}(k'r')}{k^2 - k'^2 + i\varepsilon}, \quad (3.96)$$

and

$$I_4 = -\lim_{\varepsilon \to 0} \frac{1}{4\pi} \int_{-\infty}^{\infty} dk' \, \frac{H_l^{(-)}(k'r) H_l^{(-)}(k'r')}{k^2 - k'^2 + i\varepsilon}. \quad (3.97)$$

The integrals I_n may be evaluated by the technique of contour integration. Let us first consider I_1. Since $H_l^{(+)}(z) \to \exp[i(z - \tfrac{1}{2}\pi l)]$ for large z, we may complete the contour for I_1 in the upper half plane as indicated in Fig. 4.4,

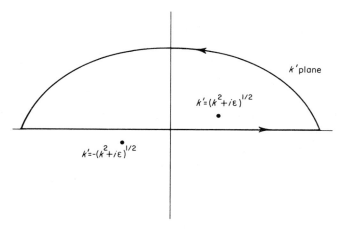

FIG. 4.4

since there will be no contribution from the arc. Similarly, to evaluate I_2 we may complete the contour in the upper half plane if $r > r'$; if $r < r'$, the contour is completed in the lower half plane. In the latter case we can use the relation $H_l^{(\pm)}(-z) = (-1)^l H_l^{(\pm)}(z)$ to regain positive arguments in the functions evaluated at $k' = -(k^2 + i\varepsilon)^{1/2}$. We now have merely to apply

3. Eigenfunction Expansions of Green's Functions

Cauchy's integral formula to evaluate the integrals. The results are:

$$I_1 = (i/4k) H_l^{(+)}(kr) H_l^{(+)}(kr'), \tag{3.98}$$

$$I_2 = -(i/4k) \begin{cases} H_l^{(-)}(kr) H_l^{(+)}(kr'), & r < r', \\ H_l^{(+)}(kr) H_l^{(-)}(kr'), & r > r', \end{cases} \tag{3.99}$$

$$I_3 = -(i/4k) \begin{cases} H_l^{(-)}(kr) H_l^{(+)}(kr'), & r < r', \\ H_l^{(+)}(kr) H_l^{(-)}(kr'), & r > r', \end{cases} \tag{3.100}$$

and

$$I_4 = (i/4k) H_l^{(+)}(kr) H_l^{(+)}(kr'). \tag{3.101}$$

The Green's function is then

$$g_l(r, r') = I_1 + I_2 + I_3 + I_4 = -\frac{1}{k} \begin{cases} F_l(kr) H_l^{(+)}(kr'), & r < r', \\ H_l^{(+)}(kr) F_l(kr'), & r > r', \end{cases} \tag{3.102}$$

which agrees with Eq. (3.67).

The integration path given in Fig. 4.2(c) may be used equally well. The integration path consists of two straight-line portions and a semicircle:

$$g_l(r, r') = \lim_{\varepsilon \to 0} \frac{2}{\pi} \Big\{ \int_0^{k-\varepsilon} dk' \frac{F_l(k'r) F_l(k'r')}{k^2 - k'^2} + \int_{\text{semi-circle}} dk' \frac{F_l(k'r) F_l(k'r')}{k^2 - k'^2} + \int_{k+\varepsilon}^{\infty} dk' \frac{F_l(k'r) F_l(k'r')}{k^2 - k'^2} \Big\}. \tag{3.103}$$

The first and third integrals combine to give a principal-value integral. A semicircle need not be used to pass around the singularity in the second integral; any other shape would do equally well, but the semicircle is the most convenient. The integral around the semicircle is, with $k' = k + \varepsilon e^{i\theta}$ and ε small,

$$\int_{\text{semi-circle}} dk' \frac{F_l(k'r) F_l(k'r')}{k^2 - k'^2} \xrightarrow[\varepsilon \to 0]{} \frac{2}{\pi} F_l(kr) F_l(kr') \int_{\text{semi-circle}} \frac{dk'}{k^2 - k'^2}$$

$$= \frac{2}{\pi} F_l(kr) F_l(kr') \int_\pi^{2\pi} \frac{d\theta}{2ik} = -\frac{i}{k} F_l(kr) F_l(kr'). \tag{3.104}$$

Combining this with the principal-value integral, Eq. (3.75), we obtain

$$g_l(r, r') = -\frac{1}{k} \begin{cases} F_l(kr) H_l^{(+)}(kr'), & r < r', \\ H_l^{(+)}(kr) F_l(kr'), & r > r', \end{cases} \qquad (3.105)$$

as before.

Completely analogous results may be obtained for the ingoing-wave Green's function. The imaginary part of the eigenvalue changes sign, and the integration contours change correspondingly. In particular, the semicircle in Fig. 4.2(c) lies in the upper half plane, and its contribution has the opposite sign.

We thus find here, as in the standing-wave case, that the contribution from the neighborhood of the singularity [Eq. (3.104)] is a solution of the homogeneous differential equation with an amplitude that depends upon the boundary conditions imposed on the Green's function. The inhomogeneous part of the Green's function is in all cases described by the principal-value integral.

We also note that the result of the integrations in Eqs. (3.94)–(3.104) can be summarized in the symbolic expression

$$\frac{1}{k^2 - k'^2 + i\varepsilon} \xrightarrow[\varepsilon \to 0]{} \mathscr{P} \frac{1}{k^2 - k'^2} - \pi i\, \delta(k^2 - k'^2). \qquad (3.106)$$

Inserted into Eq. (3.90), this yields Eq. (3.105). Its significance can be understood by examining Fig. 4.5, which shows how each part selects out the proper

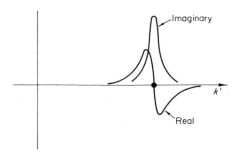

FIG. 4.5

segment of the integration path. The effect of the boundary condition can be seen by comparing this with the standing-wave case, Eq. (3.85). As in that case, the amplitude of the delta-function term is dependent on the boundary conditions imposed on the Green's function.

Another feature of these results may be noted. For both the bounded and unbounded domains, only the eigenvalues depend on the boundary condition. The eigenfunctions in both the sum and integral forms are simply the regular solutions $F_l(k_n r)$. When one passes to the limit of the infinite domain, the dependence on the boundary condition appears only in the integration path. In fact, as we have seen, in that limit the boundary condition affects only the way in which the singularity of the integral is treated.

We have seen also that the inhomogeneous part of the Green's function is represented by a principal-value integral and thus contains contributions only from eigenvalues $k' \neq k$. The homogeneous term, dependent on the boundary condition at infinity, arises only from eigenvalues in the immediate neighborhood of k. This will be discussed further in Section 3 of Chapter 5.

4. Green's Function in Three Dimensions

In this section we shall investigate the three-dimensional Green's function. Our procedure will parallel that of the previous section, where we studied the radial Green's function appearing in the solution of the radial Schrödinger equation. By summing over angular momentum states, we could recover the solution of the three-dimensional Schrödinger equation from the solutions of the radial Schrödinger equations. However, that will not be our program in this section. Instead, we shall look at the problem as one three-dimensional problem, rather than as a superposition of one-dimensional problems.

Properties of Three-Dimensional Green's Functions

The three-dimensional Green's function, which we shall denote by $G(\mathbf{r}, \mathbf{r}')$, satisfies a differential equation analogous to that satisfied by the one-dimensional Green's function. For free particles it is a solution of the equation

$$(\nabla^2 + k^2)\, G(\mathbf{r}, \mathbf{r}') = \delta(\mathbf{r} - \mathbf{r}'). \tag{4.1}$$

It satisfies the homogeneous differential equation for $\mathbf{r} \neq \mathbf{r}'$, but is singular when the field point $\mathbf{r}$ coincides with the source point $\mathbf{r}'$.

If we integrate Eq. (4.1) throughout a sphere of radius R centered about $\mathbf{r}'$, we find

$$\int_{|\mathbf{r} - \mathbf{r}'| \leq R} d^3 r\, (\nabla^2 + k^2)\, G(\mathbf{r}, \mathbf{r}') = 1. \tag{4.2}$$

This result must hold even in the limit $R \to 0$. Assuming that the Laplacian of the Green's function is more singular than the Green's function itself, this gives

$$\lim_{R \to 0} \int_{|\mathbf{r}-\mathbf{r}'|<R} d^3r \, \nabla^2 G(\mathbf{r}, \mathbf{r}') = \lim_{R \to 0} \int_{|\mathbf{r}-\mathbf{r}'|=R} d\mathbf{S} \cdot \nabla G(\mathbf{r}, \mathbf{r}')$$

$$= \lim_{R \to 0} 4\pi R^2 \frac{d}{dR} G(R) = 1. \quad (4.3)$$

The second equality results from the fact that Eq. (4.1) is unchanged under a translation or rotation of the coordinate system, so that in the neighborhood of the singularity the Green's function $G(\mathbf{r}, \mathbf{r}')$ can depend only on the distance $|\mathbf{r}-\mathbf{r}'|$. Eq. (4.3) implies in turn that

$$G(\mathbf{r}, \mathbf{r}') \xrightarrow[\mathbf{r} \to \mathbf{r}']{} -\frac{1}{4\pi} \frac{1}{|\mathbf{r}-\mathbf{r}'|}. \quad (4.4)$$

Thus the Green's function is indeed singular when $\mathbf{r} = \mathbf{r}'$.

Let us solve Eq. (4.1) by using the eigenfunction expansion technique. The eigenfunctions will be denoted by $\phi_i(\mathbf{r})$, where the index i encompasses all the eigenvalues such as the energy, angular momentum, and magnetic quantum number, or the linear momentum $\mathbf{k}$. These functions satisfy the homogeneous equation

$$(\nabla^2 + k_i^2) \phi_i(\mathbf{r}) = 0. \quad (4.5)$$

Eigenfunctions corresponding to different energies will be orthogonal if they satisfy homogeneous boundary conditions on some bounding surface Σ. The full set of functions $\phi_i(\mathbf{r})$ can be chosen to form an orthogonal set and, with proper normalization, will satisfy the orthonormality relation

$$\int d^3r \, \phi_i^*(\mathbf{r}) \phi_j(\mathbf{r}) = \delta_{ij}, \quad (4.6)$$

where the integral is over the volume enclosed by the surface Σ. As in the one-dimensional case, these functions also form a complete set and satisfy the closure relation

$$\sum_i \phi_i(\mathbf{r}) \phi_i^*(\mathbf{r}') = \delta(\mathbf{r} - \mathbf{r}'). \quad (4.7)$$

To be completely explicit, let us choose Σ to be the surface of a cube with sides of length L. By imposing periodic boundary conditions at the walls, we obtain the normalized momentum eigenfunctions

$$\phi_i(\mathbf{r}) = L^{-3/2} e^{i\mathbf{k} \cdot \mathbf{r}}, \quad (4.8)$$

where the eigenvalues $\mathbf{k} = (k_x, k_y, k_z)$ are given by

$$k_x = (2\pi/L)\lambda, \qquad k_y = (2\pi/L)\mu, \qquad k_z = (2\pi/L)\nu, \tag{4.9}$$

and λ, μ, ν are integers.

In the limit as the bounding surface goes to infinity, the discrete spectrum will become a continuous spectrum. In this limit the sum over eigenfunctions in Eq. (4.7) becomes

$$L^{-3} \sum_\lambda \sum_\mu \sum_\nu \exp[i\mathbf{k}_{\lambda\mu\nu} \cdot (\mathbf{r} - \mathbf{r}')] \xrightarrow[L \to \infty]{} (2\pi)^{-3} \int d^3k \, \exp[i\mathbf{k} \cdot (\mathbf{r} - \mathbf{r}')] = \delta(\mathbf{r} - \mathbf{r}'), \tag{4.10}$$

since $\Delta\lambda = (L/2\pi) \Delta k_x$, etc. Through an obvious relabeling of variables this also implies the relation

$$(2\pi)^{-3} \int d^3r \, \exp[i(\mathbf{k} - \mathbf{k}') \cdot \mathbf{r}] = \delta(\mathbf{k} - \mathbf{k}'), \tag{4.11}$$

which is just the orthogonality relation for the continuous spectrum,

$$\int d^3r \, \phi_{\mathbf{k}'}^*(\mathbf{r}) \phi_{\mathbf{k}}(\mathbf{r}) = \delta(\mathbf{k} - \mathbf{k}'). \tag{4.12}$$

In the continuous limit the closure relation is

$$\int d^3k \, \phi_{\mathbf{k}}^*(\mathbf{r}) \phi_{\mathbf{k}}(\mathbf{r}') = \delta(\mathbf{r} - \mathbf{r}') \tag{4.13}$$

and the eigenfunctions $\phi_{\mathbf{k}}(\mathbf{r})$ are

$$\phi_{\mathbf{k}}(\mathbf{r}) = (2\pi)^{-3/2} e^{i\mathbf{k} \cdot \mathbf{r}}. \tag{4.14}$$

Now we treat $\mathbf{r}'$ as a fixed parameter and write $G(\mathbf{r}, \mathbf{r}')$ as an expansion in the functions $\phi_i(\mathbf{r})$ for the bounded domain:

$$G(\mathbf{r}, \mathbf{r}') = \sum_i A_i(\mathbf{r}') \phi_i(\mathbf{r}). \tag{4.15}$$

Inserting this into the differential equation, Eq. (4.1), and using the orthonormality of the set of eigenfunctions, we have

$$(k^2 - k_i^2) A_i(\mathbf{r}') = \phi_i^*(\mathbf{r}'). \tag{4.16}$$

Provided the energy $\hbar^2 k^2/2m$ does not coincide with one of the eigenvalues, $E_i = \hbar^2 k_i^2/2m$, this gives

$$A_i(\mathbf{r}') = \frac{\phi_i^*(\mathbf{r}')}{k^2 - k_i^2}, \tag{4.17}$$

or

$$G(\mathbf{r}, \mathbf{r}') = \sum_i \frac{\phi_i(\mathbf{r}) \phi_i^*(\mathbf{r}')}{k^2 - k_i^2}. \tag{4.18}$$

Just as in the case of the radial Green's function, we can pass to the limit of the unbounded domain if, in the resulting integral, the behavior of the integral in the neighborhood of the singularity is specified. As in Eq. (4.10), this limit is

$$G(\mathbf{r}, \mathbf{r}') = (2\pi)^{-3} \int_C d^3k \frac{\exp[i\mathbf{k}' \cdot (\mathbf{r} - \mathbf{r}')]}{(k^2 - k'^2)}, \tag{4.19}$$

where the contour depends on the boundary condition satisfied by $G(\mathbf{r}, \mathbf{r}')$. The angular integration can be performed without specifying the contour by using the integral

$$\int d\Omega \, e^{i\mathbf{k} \cdot \mathbf{r}} = (4\pi/kr) F_0(kr) = 4\pi(\sin kr)/kr, \tag{4.20}$$

with the result

$$G(\mathbf{r}, \mathbf{r}') = \frac{1}{2\pi^2 |\mathbf{r} - \mathbf{r}'|} \int_C k' \, dk' \frac{\sin k' |\mathbf{r} - \mathbf{r}'|}{(k^2 - k'^2)}. \tag{4.21}$$

The Green's function for the infinite domain therefore depends only on the source-to-field-point distance $|\mathbf{r} - \mathbf{r}'|$ and is symmetric with respect to an interchange $\mathbf{r}$ and $\mathbf{r}'$. As we noted earlier, the dependence on $|\mathbf{r} - \mathbf{r}'|$ follows in general from the invariance of the defining differential equation, Eq. (4.1), under translations and rotations, while the symmetry property will hold whenever the Green's function satisfies homogeneous boundary conditions on the bounding surface Σ.

OUTGOING-WAVE GREEN'S FUNCTION

To proceed further we must define the integration contour. The structure of the integral is identical with that of the radial Green's function discussed in Section 3 so that, for instance, we can obtain the outgoing-wave Green's function by using the contours of Fig. 4.2. We saw there that integration along these contours, which pass beneath the singularity at k, is equivalent to integration along the real axis but with the denominator in Eq. (4.21) replaced by $k^2 - (k' - i\lambda)^2$ or $(k^2 - k'^2 + i\varepsilon)$. Extending the integral over the entire

4. Green's Function in Three Dimensions

real axis, we find using Cauchy's integral formula

$$G(\mathbf{r}, \mathbf{r}') = \lim_{\varepsilon \to 0} \frac{1}{4\pi^2 |\mathbf{r} - \mathbf{r}'|} \int_{-\infty}^{\infty} k' \, dk' \, \frac{\sin k' |\mathbf{r} - \mathbf{r}'|}{k^2 - k'^2 + i\varepsilon}$$

$$= \frac{1}{4\pi} \frac{\exp(ik|\mathbf{r} - \mathbf{r}'|)}{|\mathbf{r} - \mathbf{r}'|}. \qquad (4.22)$$

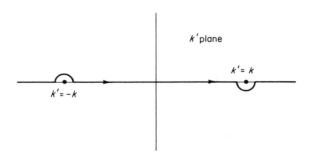

FIG. 4.6

The outgoing-wave Green's function may also be evaluated by integrating along the contour of Fig. 4.2(c) or, when extended over the entire real axis, that of Fig. 4.6. The integral along this contour may be written as

$$\begin{aligned} G(\mathbf{r}, \mathbf{r}') &= \frac{1}{4\pi^2} \frac{1}{|\mathbf{r} - \mathbf{r}'|} \mathscr{P} \int_{-\infty}^{\infty} k' \, dk' \, \frac{\sin k' |\mathbf{r} - \mathbf{r}'|}{(k^2 - k'^2)} \\ &+ \lim_{\varepsilon \to 0} \left\{ \frac{1}{4\pi^2} \frac{1}{|\mathbf{r} - \mathbf{r}'|} \int_{\substack{\text{semicircle} \\ \text{at } k' = -k}} k' \, dk' \, \frac{\sin k' |\mathbf{r} - \mathbf{r}'|}{(k^2 - k'^2)} \right. \\ &+ \left. \frac{1}{4\pi^2} \frac{1}{|\mathbf{r} - \mathbf{r}'|} \int_{\substack{\text{semicircle} \\ \text{at } k' = k}} k' \, dk' \, \frac{\sin k' |\mathbf{r} - \mathbf{r}'|}{(k^2 - k'^2)} \right\}. \qquad (4.23) \end{aligned}$$

Using $k' = \pm k + \varepsilon e^{i\theta}$, we obtain

$$\int_{\substack{\text{semicircle} \\ \text{at } k' = -k}} k' \, dk' \, \frac{\sin k' |\mathbf{r} - \mathbf{r}'|}{k^2 - k'^2} = \int_{\substack{\text{semicircle} \\ \text{at } k' = k}} k' \, dk' \, \frac{\sin k' |\mathbf{r} - \mathbf{r}'|}{k^2 - k'^2}$$

$$= \frac{i}{2} \sin k |\mathbf{r} - \mathbf{r}'| \int_{\pi}^{2\pi} d\theta = \tfrac{1}{2} \pi i \sin k |\mathbf{r} - \mathbf{r}'|. \qquad (4.24)$$

Together with Eq. (4.22), this implies that the principal-value integral is

$$\frac{1}{4\pi^2} \frac{1}{|\mathbf{r} - \mathbf{r}'|} \mathcal{P} \int_{-\infty}^{\infty} k' \, dk' \frac{\sin k' |\mathbf{r} - \mathbf{r}'|}{k^2 - k'^2}$$

$$= -\frac{1}{4\pi} \frac{1}{|\mathbf{r} - \mathbf{r}'|} \{\exp(ik |\mathbf{r} - \mathbf{r}'|) - i \sin k |\mathbf{r} - \mathbf{r}'|\}$$

$$= -\frac{1}{4\pi} \frac{\cos k |\mathbf{r} - \mathbf{r}'|}{|\mathbf{r} - \mathbf{r}'|}. \tag{4.25}$$

Combining these results, the free-particle outgoing-wave Green's function is

$$G(\mathbf{r}, \mathbf{r}') = \lim_{\varepsilon \to 0} (2\pi)^{-3} \int d^3k' \frac{\exp[i\mathbf{k}' \cdot (\mathbf{r} - \mathbf{r}')]}{k^2 - k'^2 + i\varepsilon} = -\frac{1}{4\pi} \frac{\exp(ik |\mathbf{r} - \mathbf{r}'|)}{|\mathbf{r} - \mathbf{r}'|}$$

$$= (2\pi)^{-3} \mathcal{P} \int d^3k \frac{\exp[i\mathbf{k}' \cdot (\mathbf{r} - \mathbf{r}')]}{k^2 - k'^2} - \frac{i}{4\pi} \frac{\sin k |\mathbf{r} - \mathbf{r}'|}{|\mathbf{r} - \mathbf{r}'|}. \tag{4.26}$$

From this the symbolic expression

$$\frac{1}{k^2 - k'^2 + i\varepsilon} \xrightarrow[\varepsilon \to 0]{} \mathcal{P} \frac{1}{k^2 - k'^2} - \pi i \, \delta(k^2 - k'^2) \tag{4.27}$$

follows immediately. This is the same relation we found in the previous section in connection with the radial Green's function.

GENERAL FORM OF THE GREEN'S FUNCTION

We can see now that the most general form for the Green's function is

$$G(\mathbf{r}, \mathbf{r}') = -\frac{1}{4\pi} \left\{ \frac{\cos k |\mathbf{r} - \mathbf{r}'|}{|\mathbf{r} - \mathbf{r}'|} + A(k) \frac{\sin k |\mathbf{r} - \mathbf{r}'|}{|\mathbf{r} - \mathbf{r}'|} \right\}, \tag{4.28}$$

where $A(k)$ depends on the boundary conditions. For the outgoing-wave boundary condition, we have seen that $A = i$. For ingoing-wave boundary conditions, $A = -i$. For standing-wave boundary conditions, A will be a real number which, as we saw in the last section, depends on the way the limit $L \to \infty$ is defined.

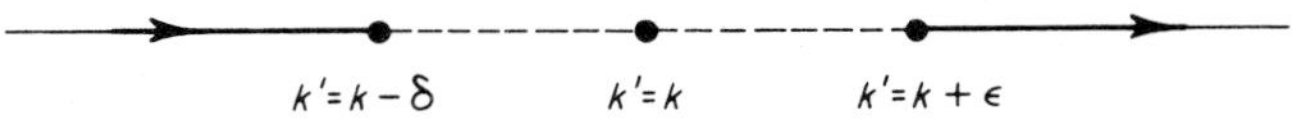

$k' = k - \delta \qquad k' = k \qquad k' = k + \epsilon$

FIG. 4.7

This last result is obtained by using the contour of Fig. (4.7). The contour of integration is along the real axis from $k' = -\infty$ to $k' = k - \delta$ and

4. Green's Function in Three Dimensions

from $k' = k + \varepsilon$ to $k' = \infty$, and the singularity is avoided since the region along the real axis $k - \delta < k' < k + \varepsilon$ is omitted. The quantities δ and ε are real numbers, of course. With this contour the Green's function can be expressed as the sum of a principal-value integral and a homogeneous term:

$$G(\mathbf{r}, \mathbf{r}') = \frac{1}{2\pi^2} \frac{1}{|\mathbf{r} - \mathbf{r}'|} \left\{ \mathscr{P} \int_0^\infty k' \, dk' \, \frac{\sin k' |\mathbf{r} - \mathbf{r}'|}{k^2 - k'^2} \right.$$

$$\left. + \int_{k + \varepsilon}^{k + \delta} k' \, dk' \, \frac{\sin k' |\mathbf{r} - \mathbf{r}'|}{k^2 - k'^2} \right\}$$

$$= -\frac{1}{4\pi} \left\{ \frac{\cos k |\mathbf{r} - \mathbf{r}'|}{|\mathbf{r} - \mathbf{r}'|} - \frac{1}{\pi} \log(\varepsilon/\delta) \frac{\sin k |\mathbf{r} - \mathbf{r}'|}{|\mathbf{r} - \mathbf{r}'|} \right\}. \quad (4.29)$$

In the limit as δ and ε approach zero, the value of the integral will depend on the ratio (ε/δ). If (ε/δ) is unity, the omitted region is symmetric about $k' = k$ and Eq. (4.29) reduces simply to the principal-value integral, as expected.

The integration contour of Fig. (4.7) is appropriate for the general case of standing waves, since proper choice of the ratio (ε/δ) can yield any real value for $A(k)$ in Eq. (4.28). Complex values of $A(k)$ will result from integration

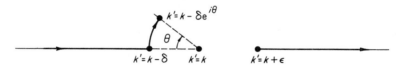

FIG. 4.8

contours such as in Fig. (4.8). That is to say, the integration may be taken along the real axis to the point $k' = k - \delta e^{i\theta}$, where it is interrupted and is resumed at the point $k' = k + \varepsilon$. (As usual, ε, δ, and θ are real.) In the limit $(\varepsilon, \delta) \to 0$, we can see that this contour yields for $A(k)$

$$A(k) = -(1/\pi)\{\log(\varepsilon/\delta) + i\theta\}. \quad (4.30)$$

As Eq. (4.28) shows, the three-dimensional Green's function is a linear combination of a singular part, $-\cos(k |\mathbf{r} - \mathbf{r}'|)/4\pi |\mathbf{r} - \mathbf{r}'|$, and a nonsingular part proportional to $\sin(k |\mathbf{r} - \mathbf{r}'|)/|\mathbf{r} - \mathbf{r}'|$. The amplitude of the singular part is uniquely determined by the amplitude of the source function in Eq. (4.1), and thus satisfies Eq. (4.4). The nonsingular part satisfies the homogeneous differential equation and has an amplitude determined by the boundary conditions imposed on $G(\mathbf{r}, \mathbf{r}')$. This is completely analogous to the results we found in the previous section for the radial Green's function.

RELATION TO THE RADIAL GREEN'S FUNCTION

The relation of $G(\mathbf{r}, \mathbf{r}')$ to the radial Green's function can be derived by introducing the Legendre expansion for the plane wave into the integral representation of $G(\mathbf{r}, \mathbf{r}')$, Eq. (4.19). Using Eq. (1.24) of Chapter 3 and letting $\mathbf{r} \cdot \mathbf{r}' = rr' \cos \theta$, we have

$$\int d\Omega_k \exp[i\mathbf{k} \cdot (\mathbf{r} - \mathbf{r}')] = 4\pi (k^2 rr')^{-1} \sum_{l=0}^{\infty} (2l + 1) F_l(kr) F_l(kr') P_l(\cos \theta). \quad (4.31)$$

Inserted into Eq. (4.19), this yields

$$G(\mathbf{r}, \mathbf{r}') = \frac{1}{2\pi^2 rr'} \sum_{l=0}^{\infty} (2l + 1) P_l(\cos \theta) \int_C dk' \frac{F_l(k'r) F_l(k'r')}{k^2 - k'^2}. \quad (4.32)$$

The integral may be recognized from Eq. (3.82) as $\tfrac{1}{2}\pi g_l(r, r')$, so that the three-dimensional Green's function has the Legendre expansion

$$G(\mathbf{r}, \mathbf{r}') = (4\pi rr')^{-1} \sum_{l=0}^{\infty} (2l + 1) P_l(\cos \theta) g_l(r, r'). \quad (4.33)$$

For instance, the outgoing-wave Green's function is

$$\begin{aligned} G(\mathbf{r}, \mathbf{r}') &= -\frac{1}{4\pi} \frac{\exp(ik|\mathbf{r} - \mathbf{r}'|)}{|\mathbf{r} - \mathbf{r}'|} \\ &= -\frac{1}{4\pi} \frac{1}{krr'} \sum_{l=0}^{\infty} (2l + 1) P_l(\cos \theta) \\ &\quad \times \begin{cases} F_l(kr) H_l^{(+)}(kr'), & r < r', \\ H_l^{(+)}(kr) F_l(kr'), & r > r'. \end{cases} \end{aligned} \quad (4.34)$$

One implication of this result is that the integration contours in the one- and three-dimensional cases are identical for any particular boundary condition.

We have seen that for the infinite domain $G(\mathbf{r}, \mathbf{r}')$ depends only on the distance $|\mathbf{r} - \mathbf{r}'|$. If we choose the coordinate origin to be at $\mathbf{r}'$, only the $l = 0$ term contributes to Eq. (4.34), which is then

$$G(\mathbf{r}, 0) = -\frac{1}{4\pi} \left\{ \lim_{r' \to 0} \frac{\sin kr'}{kr'} \right\} \frac{e^{ikr}}{r} = -\frac{1}{4\pi} \frac{e^{ikr}}{r}. \quad (4.35)$$

4. Green's Function in Three Dimensions

Thus $G(\mathbf{r}, \mathbf{r}')$ produces a spherically symmetric wave emerging from the point $\mathbf{r}'$; this of course could be seen from the closed form in Eq. (4.34) as well.

GREEN'S FUNCTION FOR MOTION IN A POTENTIAL

Let us generalize these results to the case of a particle propagating in a potential $V(\mathbf{r})$, rather than moving as a free particle. The Green's function in this case will satisfy the differential equation

$$\{\nabla^2 + k^2 - U(\mathbf{r})\} G(\mathbf{r}, \mathbf{r}') = \lambda \delta(\mathbf{r} - \mathbf{r}'), \qquad (4.36)$$

We have previously set $\lambda = 1$. In describing motion in a general potential, and especially in passing to the operator formalism, it is convenient instead to choose $\lambda = 2m/\hbar^2$, so that Eq. (4.36) becomes

$$\{E - H(\mathbf{r})\} G(\mathbf{r}, \mathbf{r}') = \delta(\mathbf{r} - \mathbf{r}'), \qquad (4.37)$$

with

$$H(\mathbf{r}) = -(\hbar^2/2m)\nabla^2 + V(\mathbf{r}). \qquad (4.38)$$

The Green's function may again be expressed in terms of an eigenfunction expansion, using solutions of the homogeneous equation

$$(E_i - H(\mathbf{r}))\psi_i(\mathbf{r}) = 0 \qquad (4.39)$$

The set $\psi_i(\mathbf{r})$ is chosen to satisfy homogeneous boundary conditions on the surface Σ. Using the orthonormality and completeness of this set, the solution of Eq. (4.37) is found to be

$$G(\mathbf{r}, \mathbf{r}') = \sum_i \frac{\psi_i(\mathbf{r}) \psi_i^*(\mathbf{r}')}{E - E_i}. \qquad (4.40)$$

It is instructive to obtain Eq. (4.40) from Eq. (4.37) in a slightly different manner. We write Eq. (4.37) as

$$G(\mathbf{r}, \mathbf{r}') = [E - H(\mathbf{r})]^{-1} \delta(\mathbf{r} - \mathbf{r}'). \qquad (4.41)$$

Since $\{E - H(\mathbf{r})\}$ is a differential operator, the inverse operator $\{E - H(\mathbf{r})\}^{-1}$ is an integral operator and has the boundary conditions incorporated into it. The closure relation

$$\delta(\mathbf{r} - \mathbf{r}') = \sum_i \psi_i^*(\mathbf{r}') \psi_i(\mathbf{r}), \qquad (4.42)$$

together with the fact that $\psi_i(\mathbf{r})$ is an eigenfunction of $\{E - H(\mathbf{r})\}^{-1}$, or

$$[E - H(\mathbf{r})]^{-1} \psi_i(\mathbf{r}) = (E - E_i)^{-1} \psi_i(\mathbf{r}), \tag{4.43}$$

implies that Eq. (4.41) is equivalent to

$$G(\mathbf{r}, \mathbf{r}') = [E - H(\mathbf{r})]^{-1} \delta(\mathbf{r} - \mathbf{r}') = \sum_i \frac{\psi_i(\mathbf{r}) \psi_i^*(\mathbf{r}')}{(E - E_i)}. \tag{4.44}$$

This is identical with Eq. (4.40).

The same proof as used for the free-particle Green's function shows that, in general,

$$G(\mathbf{r}, \mathbf{r}') \xrightarrow[\mathbf{r} \to \mathbf{r}']{} -\frac{\lambda}{4\pi} \frac{1}{|\mathbf{r} - \mathbf{r}'|}. \tag{4.45}$$

Likewise, the general Green's function is symmetric under interchange of $\mathbf{r}$ and $\mathbf{r}'$. To see this, we return to the differential equation satisfied by the Green's function, Eq. (4.37). If we consider the equations for $G(\mathbf{r}, \mathbf{r}')$ and $G(\mathbf{r}, \mathbf{r}'')$, we find, using Green's theorem,

$$\int d^3r \, \{G(\mathbf{r}, \mathbf{r}'') (E - H(\mathbf{r})) G(\mathbf{r}, \mathbf{r}') - G(\mathbf{r}, \mathbf{r}') (E - H(\mathbf{r})) G(\mathbf{r}, \mathbf{r}'')\}$$
$$= (\hbar^2/2m) \int d^3r \, \{G(\mathbf{r}, \mathbf{r}'') \nabla^2 G(\mathbf{r}, \mathbf{r}') - G(\mathbf{r}, \mathbf{r}') \nabla^2 G(\mathbf{r}, \mathbf{r}'')\}$$
$$= (\hbar^2/2m) \int_\Sigma d\mathbf{S} \cdot \{G(\mathbf{r}, \mathbf{r}'') \nabla G(\mathbf{r}, \mathbf{r}') - G(\mathbf{r}, \mathbf{r}') \nabla G(\mathbf{r}, \mathbf{r}'')\}$$
$$= \int d^3r \, \{G(\mathbf{r}, \mathbf{r}'') \delta(\mathbf{r} - \mathbf{r}') - G(\mathbf{r}, \mathbf{r}') \delta(\mathbf{r} - \mathbf{r}'')\}$$
$$= G(\mathbf{r}', \mathbf{r}'') - G(\mathbf{r}'', \mathbf{r}'). \tag{4.46}$$

If the boundary condition is such that the integral on the surface Σ vanishes, the Green's function is symmetric in $\mathbf{r}$ and $\mathbf{r}'$. Since the boundary conditions for scattering problems are invariably homogeneous, symmetry is guaranteed. Of course, $\mathbf{r}'$ and $\mathbf{r}''$ must be enclosed within the surface Σ in order for the proof to be applicable.

Let us now demonstrate explicitly that the eigenfunction expansion, Eq. (4.40), is symmetric in $\mathbf{r}$ and $\mathbf{r}'$. We recall that index i includes eigenvalues for variables other than the energy. If these other variables are denoted collectively by β, the Green's function becomes

$$G(\mathbf{r}, \mathbf{r}') = \sum_{E_i, \beta} \frac{\psi_{E_i\beta}(\mathbf{r}) \psi_{E_i\beta}^*(\mathbf{r}')}{E - E_i}. \tag{4.47}$$

Any solution of the homogeneous equation, Eq. (4.39), having the eigenvalue E_i must be a linear combination of the eigenfunctions $\psi_{E_i\beta}(\mathbf{r})$. In particular, the hermiticity of H tells us that $\psi_{E_i\beta}^*(\mathbf{r}')$ is a solution and can be written as

$$\psi_{E_i\beta}^*(\mathbf{r}') = \sum_\gamma C_{\beta\gamma} \psi_{E_i\gamma}(\mathbf{r}') = \sum_\gamma \psi_{E_i\gamma}(\mathbf{r}') \int d^3r'' \, \psi_{E_i\gamma}^*(\mathbf{r}'') \psi_{E_i\beta}^*(\mathbf{r}''). \tag{4.48}$$

Inserted into Eq. (4.47), this yields a completely symmetric numerator.

If the bounding surface Σ that defines the eigenfunctions is allowed to recede to infinity, the sum in Eq. (4.40) will become an integral. When the bounding surface is a finite distance from the origin, the functions $\psi_i(\mathbf{r})$ will be regular solutions of Eq. (4.39), with the eigenvalues E_i so chosen that these functions satisfy the same boundary conditions as does the Green's function. When the surface Σ recedes to infinity, the spectrum becomes continuous, and the boundary condition is reflected in the treatment of the singularity in the denominator. The eigenfunctions $\psi_i(\mathbf{r})$ can then be *any* regular solution of the Schrödinger equation. This is identical with the free-particle case, in which we used the plane wave functions $e^{i\mathbf{k}\cdot\mathbf{r}}$ but could have used any other complete set of solutions of the free-particle equation.

Let us see how this works in practice. If the eigenfunction is characterized by the momentum $\mathbf{k}$, then $\psi_\mathbf{k}(\mathbf{r})$ must be of the general form

$$\psi_\mathbf{k}(\mathbf{r}) = (kr)^{-1} \sum_{l=0}^{\infty} (2l+1) b_l u_l(k,r) P_l(\cos\theta). \tag{4.49}$$

Here $u_l(r)$ is the regular solution of the radial Schrödinger equation, normalized so that its asymptotic form is $e^{i\delta_l} \sin(kr - \tfrac{1}{2}\pi l + \delta_l)$, and θ is the angle between $\mathbf{k}$ and $\mathbf{r}$. The coefficient b_l may be complex, but if $\psi_\mathbf{k}(\mathbf{r})$ is normalized according to $\int d^3r\, \psi_{\mathbf{k}'}^*(\mathbf{r}) \psi_\mathbf{k}(\mathbf{r}) = \delta(\mathbf{k} - \mathbf{k}')$, then $|b_l| = 1$. The phase of this coefficient is determined by the boundary condition that $\psi_\mathbf{k}(\mathbf{r})$ satisfies at infinity. Thus, if $b_l = i^l$, $\psi_\mathbf{k}(\mathbf{r})$ has the form of a plane wave plus an outgoing wave, as in Chapter 3; if $b_l = i^l e^{-2i\delta_l}$, $\psi_\mathbf{k}(\mathbf{r})$ will behave as a plane wave plus an ingoing wave.

If we pass to the continuum limit in Eq. (4.40) and insert Eq. (4.49), we find

$$G(\mathbf{r},\mathbf{r}') = (2\pi)^{-3} \int_C d^3k' \frac{\psi_{\mathbf{k}'}(\mathbf{r}) \psi_{\mathbf{k}'}^*(\mathbf{r}')}{E - E'}$$

$$= \frac{1}{2\pi^2 rr'} \sum_{l=0}^{\infty} (2l+1) P_l(\cos\theta) \int_C dk' \frac{u_l(k',r) u_l(k',r')}{E - E'}, \tag{4.50}$$

where θ is now the angle between r and r' and there is no longer any dependence on b_l. This result is completely analogous to the free-particle result, Eq. (4.32). As there, the integral can be identified with $\tfrac{1}{2}\pi g_l(r,r')$, where $g_l(r,r')$ is now the radial Green's function for propagation in the field of the potential $V(\mathbf{r})$. We know from our previous work that the behavior of $g_l(r,r')$, as either r or r' tends to infinity, is determined by the choice of integration contour; the asymptotic behavior of $G(\mathbf{r},\mathbf{r}')$ also depends only on this choice.

Asymptotic Behavior

Using the radial Green's functions developed in Section 2, the three-dimensional Green's function has the Legendre expansion

$$G(\mathbf{r}, \mathbf{r}') = -\left(\frac{2m}{4\pi\hbar^2}\right)\left(\frac{1}{rr'}\right)\sum_{l=0}^{\infty}(2l+1)P_l(\cos\theta)$$

$$\times \frac{1}{W_l}\begin{cases} u_l^{(1)}(r)\,u_l^{(2)}(r'), & r < r', \\ u_l^{(2)}(r)\,u_l^{(1)}(r'), & r > r', \end{cases} \quad (4.51)$$

where $u_l^{(1)}(r)$ and $u_l^{(2)}(r)$ are regular and irregular solutions, respectively, of the radial Schrödinger equation. In the particular case of outgoing waves, the irregular function must behave asymptotically as e^{ikr}. We can then normalize the radial functions to have the asymptotic behavior

$$u_l^{(1)}(r) \xrightarrow[r\to\infty]{} e^{i\delta_l}\sin(kr - \tfrac{1}{2}\pi l + \delta_l) \quad (4.52)$$

and

$$u_l^{(2)}(r) \xrightarrow[r\to\infty]{} \exp[i(kr - \tfrac{1}{2}\pi l + 2\delta_l)], \quad (4.53)$$

as in Eqs. (1.28) and (1.29).

When we take the limit of Eq. (4.51) as the source distance $r' \to \infty$, we find with this normalization

$$G(\mathbf{r}, \mathbf{r}') \xrightarrow[r'\to\infty]{} -\frac{2m}{4\pi\hbar^2}\left(\frac{e^{ikr'}}{r'}\right)\left\{\frac{1}{kr}\sum_{l=0}^{\infty}(-i)^l(2l+1)P_l(\cos\theta)\,u_l^{(1)}(r)\right\}. \quad (4.54)$$

This result has a clear interpretation. The Green's function $G(\mathbf{r}, \mathbf{r}')$ gives the wave amplitude at $\mathbf{r}$ due to a source at $\mathbf{r}'$. We have let the source point recede to infinity and have previously defined θ as the angle between $\mathbf{r}$ and $\mathbf{r}'$. If we let $\theta' = \pi - \theta$, then $(-1)^l P_l(\cos\theta) = P_l(\cos\theta')$ and Eq. (4.54) becomes simply

$$G(\mathbf{r}, \mathbf{r}') \xrightarrow[r'\to\infty]{} -\frac{2m}{4\pi\hbar^2}\left(\frac{e^{ikr'}}{r'}\right)\psi(\mathbf{r}), \quad (4.55)$$

and $\psi(\mathbf{r})$ is the wave function that describes scattering of a beam initially traveling from the source along the vector $-\mathbf{r}'$, as discussed in Section 1 of Chapter 3.

There is an alternative way of characterizing this wave function which is

4. Green's Function in Three Dimensions

more general and which will appear in our discussion of the operator formalism. If we were to take the complex conjugate of the expression in the bracket in Eq. (4.54), we would recover the usual factor of i^l appearing in the Legendre expansions of the plane wave and of scattering wave functions. However, we would also encounter the complex conjugate function $u_l^{(1)*}(r)$. The asymptotic behavior of this radial wave function is

$$u_l^{(1)*}(r) \xrightarrow[r \to \infty]{} e^{-i\delta_l} \sin(kr - \tfrac{1}{2}\pi l + \delta_l)$$

$$= \sin(kr - \tfrac{1}{2}\pi l) + e^{-i\delta_l} \sin \delta_l \exp[-i(kr - \tfrac{1}{2}\pi l)]. \quad (4.56)$$

Thus the complex conjugate function is the usual incident wave plus an ingoing wave, and the latter has an amplitude equal to the complex conjugate of T_l. The asymptotic behavior of the complex conjugate of $\psi(\mathbf{r})$ is then

$$\psi^*(\mathbf{r}) \xrightarrow[r \to \infty]{} e^{ikr \cos \theta} + \frac{e^{-ikr}}{r} f^*(\theta). \quad (4.57)$$

This is an "ingoing-wave function" generated by an initial wave in the direction of $\mathbf{r}'$ (since $\cos \theta = \hat{\mathbf{r}}' \cdot \hat{\mathbf{r}}$). For real potentials it satisfies the same Schrödinger equation as $\psi(\mathbf{r})$ and is just that solution having the asymptotic form given in Eq. (4.57). If we use a subscript to indicate the direction of the incident beam and a superscript to indicate the nature of the scattered wave, we have then the identity

$$\psi(\mathbf{r}) \equiv \psi_{-\mathbf{k}'}^{(+)}(\mathbf{r}) = \psi_{\mathbf{k}'}^{(-)*}(\mathbf{r}), \quad (4.58)$$

where $\mathbf{k}' = k\hat{\mathbf{r}}'$. While this identity holds only for real central potentials, the identification of the wave function in Eq. (4.54) with the complex conjugate of an ingoing-wave function is completely general. One may easily see that for any potential $V(\mathbf{r})$ the asymptotic form of the Green's function is

$$G(\mathbf{r}, \mathbf{r}') \xrightarrow[r \to \infty]{} -\left(\frac{2m}{4\pi\hbar^2}\right)\left(\frac{e^{ikr'}}{r'}\right) \psi_{\mathbf{k}'}^{(-)*}(\mathbf{r}). \quad (4.59)$$

If we now let the observation point $r \to \infty$, but keep r less than r', the Green's function becomes

$$G(\mathbf{r}, \mathbf{r}') \xrightarrow[\substack{r, r' \to \infty \\ r' > r}]{} -\left(\frac{2m}{4\pi\hbar^2}\right) \frac{e^{ikr'}}{r'} \left[e^{-ikr \cos \theta} + f(\theta) \frac{e^{ikr}}{r} \right]. \quad (4.60)$$

Thus, if desired, complete information on scattering from the potential $V(\mathbf{r})$ can be obtained from this Green's function.

Green's Function for Motion in a Non-Hermitian Potential

In the development to this point we have assumed hermiticity of the Hamiltonian. Let us relax this condition to obtain a somewhat more general treatment of the Green's function. Hermiticity ensures the reality of the eigenvalues of H and the mutual orthogonality of the eigenfunctions of H. The eigenfunctions of a hermitian Hamiltonian satisfy a closure relation like Eq. (4.7). If, however, the Hamiltonian is not hermitian, then the eigenfunctions of H need not be eigenfunctions of H^*. Thus, consider the eigenequations

$$H\psi_j = E_j \psi_j \tag{4.61}$$

and

$$H^* \Psi_i = \varepsilon_i \Psi_i. \tag{4.62}$$

The solutions of Eqs. (4.61) and (4.62) are not completely defined until the boundary conditions are specified. Let us take a general homogeneous boundary condition,

$$\left(a\psi + b \frac{\partial \psi}{\partial n}\right)_\Sigma = 0 \tag{4.63}$$

on the bounding surface Σ, where $\partial/\partial n$ is the directional derivative in the direction of the outward normal to the surface. We assume that Ψ obeys the complex conjugate boundary condition

$$(a^* + b^* \partial \Psi/\partial n) = 0. \tag{4.64}$$

From Eqs. (4.61) and (4.64) we see that Ψ_j^* obeys the same equation as ψ_j subject to the same boundary conditions; hence, the eigenvalues E_j and ε_j must be complex conjugates of each other:

$$E_j = \varepsilon_j^*. \tag{4.56}$$

From Eq. (4.61) and the complex conjugate of Eq. (4.62), we obtain, using Green's theorem,

$$-(\hbar^2/2m) \int_\Sigma \left\{ \Psi_i^*(\mathbf{r}) \frac{\partial \psi_j(\mathbf{r})}{\partial n} - \psi_j(\mathbf{r}) \frac{\partial \Psi_i^*(\mathbf{r})}{\partial n} \right\} dS$$
$$= (E_j - E_i) \int d^3r \, \Psi_i^*(\mathbf{r}) \psi_j(\mathbf{r}). \tag{4.66}$$

The left-hand side of Eq. (4.66) vanishes under the boundary conditions,

Eqs. (4.63) and (4.64), so that we have the orthonormality condition

$$\int d^3r \, \Psi_i^*(\mathbf{r}) \, \psi_j(\mathbf{r}) = \delta_{ij}, \tag{4.67}$$

and the closure relation

$$\sum_i \Psi_i(\mathbf{r}) \, \psi_i^*(\mathbf{r}') = \delta(\mathbf{r} - \mathbf{r}'). \tag{4.68}$$

The functions $\Psi_j(r)$ and $\psi_j(r)$ satisfying these relations are sometimes referred to as biorthonormal. If the Green's function satisfies the differential equation, Eq. (4.37), subject to the boundary condition, Eq. (4.63), then it becomes

$$G(\mathbf{r}, \mathbf{r}') = \sum_i \frac{\Psi_i(\mathbf{r}) \, \psi_i^*(\mathbf{r}')}{E_i - E}, \tag{4.69}$$

which is the analog of Eq. (4.40).

5. Many-Particle Green's Functions

For completeness, and because some of these results will be useful in later chapters (especially Chapter 7), we will give here a brief discussion of many-particle Green's functions. Very few rigorous results have been proved concerning general many-particle Green's functions. We shall content ourselves with a discussion of the Green's function for a system of free particles, and shall then indicate some properties that Green's functions for more complicated systems are expected to have.

THE FREE-PARTICLE GREEN'S FUNCTION

The Green's function for n free particles having masses m_i satisfies the differential equation

$$\{(\hbar^2/2m_1) \nabla_1^2 + \cdots + (\hbar^2/2m_n) \nabla_n^2 + E\} \, G^{(n)}(\mathbf{r}_1, \ldots, \mathbf{r}_n; \mathbf{r}_1', \ldots, \mathbf{r}_n')$$
$$= \lambda \, \delta(\mathbf{r}_1 - \mathbf{r}_1') \cdots \delta(\mathbf{r}_n - \mathbf{r}_n'). \tag{5.1}$$

The set of $3n$ coordinates $(\mathbf{r}_1 \ldots \mathbf{r}_n)$ can be considered as the components of a $3n$-dimensional vector. If we define the $3n$-dimensional vector $\boldsymbol{\rho} = (\boldsymbol{\rho}_1, \ldots, \boldsymbol{\rho}_n)$ with $\boldsymbol{\rho}_i = (2m_i/\hbar^2)^{1/2} \mathbf{r}_i$, the differential operator in Eq. (5.1) becomes simply

$$\nabla_\rho^2 = \sum_{i=1}^n \nabla_{\rho_i} \cdot \nabla_{\rho_i}.$$

Also, the delta function is

$$\delta(\mathbf{r}_i - \mathbf{r}'_i) = (2m_i/\hbar^2)^{3/2}\,\delta(\boldsymbol{\rho}_i - \boldsymbol{\rho}'_i).$$

Choosing $\lambda = \prod_{i=1}^{n}(\hbar^2/2m_i)^{3/2}$, the differential equation becomes

$$(\nabla_\rho^2 + E)\,G^{(n)}(\boldsymbol{\rho};\boldsymbol{\rho}') = \delta(\boldsymbol{\rho} - \boldsymbol{\rho}'), \tag{5.2}$$

with the $3n$-dimensional delta function defined as the product of n 3-dimensional delta functions.

This equation is identical in form with that satisfied by the one-particle Green's function. As in that simpler case, it can be solved by eigenfunction expansion or Fourier transformation. Taking the $3n$-dimensional Fourier transform of Eq. (5.2) and treating the singularity so as to obtain outgoing waves, we have

$$G^{(n)}(\boldsymbol{\rho};\boldsymbol{\rho}') = (2\pi)^{-3n}\int d^3\kappa_1 \cdots d^3\kappa_n\,\frac{\exp[i\boldsymbol{\kappa}\cdot(\boldsymbol{\rho}-\boldsymbol{\rho}')]}{k^2 - \kappa'^2 + i\varepsilon}. \tag{5.3}$$

Here $\boldsymbol{\kappa}$ is a $3n$-dimensional vector of length $k = E^{1/2}$, the scalar product is defined by $\boldsymbol{\kappa}\cdot\boldsymbol{\rho} = \sum_{i=1}^{n}\boldsymbol{\kappa}_i\cdot\boldsymbol{\rho}_i$, and $\kappa^2 = \sum_{i=1}^{n}\kappa_i^2$.

This integral can be performed if $3n$-dimensional spherical coordinates are introduced, by analogy with ordinary 3-dimensional spherical coordinates:

$$\begin{aligned}
\kappa_{1z} &= \kappa\cos\theta_1 \\
\kappa_{1y} &= \kappa\sin\theta_1\cos\theta_2 \\
\kappa_{1x} &= \kappa\sin\theta_1\sin\theta_2\cos\theta_3 \\
\kappa_{2z} &= \kappa\sin\theta_1\sin\theta_2\sin\theta_3\cos\theta_4 \\
&\;\;\vdots \\
\kappa_{ny} &= \kappa\sin\theta_1\cdots\sin\theta_{3n-2}\cos\varphi \\
\kappa_{nx} &= \kappa\sin\theta_1\cdots\sin\theta_{3n-2}\sin\varphi.
\end{aligned} \tag{5.4}$$

The spherical variables span the ranges $0 \leq \kappa < \infty$, $0 \leq \theta_i \leq \pi$, and $-\pi \leq \varphi \leq \pi$. The $3n$-dimensional volume element is

$$d^3\kappa_1 \cdots d^3\kappa_n = \kappa^{3n-1}\,d\kappa\,d\Omega_n, \tag{5.5}$$

with the element of solid angle given by

$$d\Omega_n = \sin^{3n-2}\theta_1\,\sin^{3n-3}\theta_2\cdots\sin\theta_{3n-2}\,d\theta_1\cdots d\theta_{3n-2}\,d\varphi. \tag{5.6}$$

5. Many-Particle Green's Functions

The solid angle can be evaluated using the integral

$$\int_0^\pi \sin^m \theta \, d\theta = \pi^{1/2} \left\{ \Gamma\left(\frac{m+1}{2}\right) \Big/ \Gamma\left(\frac{m+2}{2}\right) \right\}, \tag{5.7}$$

where the Γ function has the recurrence relation $\Gamma(z+1) = z\,\Gamma(z)$ and is easy to evaluate since we know that $\Gamma(1) = 1$ and $\Gamma(\tfrac{1}{2}) = \pi^{1/2}$. With this we find the total solid angle, for instance, to be

$$\Omega_n = \int d\Omega_n = 2\pi^{3n/2}[\Gamma(3n/2)]^{-1}. \tag{5.8}$$

If we choose to define the polar axis along $(\rho - \rho')$, then $\kappa \cdot (\rho - \rho') = \kappa\,|\rho - \rho'|\cos\theta_1$ and the Green's function becomes

$$G^{(n)}(\rho; \rho') = \frac{1}{(4\pi)^\beta \pi\, \Gamma(\beta)} \int_0^\infty \frac{\kappa'^{2\beta}\, d\kappa'}{\kappa^2 - \kappa'^2 + i\varepsilon}$$

$$\times \int_0^\pi \sin^{2\beta-1}\theta_1 \exp(i\kappa'\,|\rho - \rho'|\cos\theta_1)\, d\theta_1, \tag{5.9}$$

where $\beta \equiv \tfrac{1}{2}(3n - 1)$. The angular integral is a well-known integral representation of the Bessel function:

$$\int_0^\pi \sin^{2\beta-1}\exp(i\kappa'\,|\rho - \rho'|\cos\theta_1)\, d\theta_1$$

$$= \int_0^\pi \sin^{2\beta-1}\theta_1 \cos(\kappa'\,|\rho - \rho'|\cos\theta_1)\, d\theta_1$$

$$= \tfrac{1}{2}\pi^{\frac{1}{2}}\Gamma(\beta)\,(\tfrac{1}{2}\kappa'\,|\rho - \rho'|)^{-\beta+\frac{1}{2}}\, J_{\beta-\frac{1}{2}}(\kappa'\,|\rho - \rho'|). \tag{5.10}$$

If this is inserted into Eq. (5.9), integration methods similar to those used in earlier sections of this chapter and in Section 4 of Chapter 5 give the result

$$G^{(n)}(\rho; \rho') = -\tfrac{1}{4}i(E^{1/2}/2\pi)^{\beta-\frac{1}{2}}\,|\rho - \rho'|^{-\beta+\frac{1}{2}}\,H^{(1)}_{\beta-\frac{1}{2}}(E^{1/2}\,|\rho - \rho'|). \tag{5.11}$$

In Eq. (5.11), $H^{(1)}_\nu = J_\nu + iN_\nu$ is the Hankel function of the first kind.

Asymptotic Behavior

The function defined by Eq. (5.11) has only outgoing waves for large ρ or ρ'. When $\rho \to \infty$ with ρ' held fixed,

$$G^{(n)}(\rho; \rho') \xrightarrow[\rho\to\infty]{} \tfrac{1}{2}E^{-1/2}\exp[-\tfrac{1}{2}\pi i(\beta + 1)]\,(E/4\pi^2)^{\frac{1}{2}\beta}\,\rho^{-\beta}$$

$$\times \exp(iE^{1/2}\rho)\exp[-iE^{1/2}(\rho \cdot \rho'/\rho)]. \tag{5.12}$$

This is indeed an outgoing wave, having a positive exponential $\exp(iE^{1/2}\rho)$ for large ρ (for $n=1$, this is just e^{ikr}). The outgoing wave boundary condition can be formulated in general by requiring that for large ρ

$$\nabla_\rho G^{(n)}(\rho;\rho') \xrightarrow[\rho\to\infty]{} i(\rho/\rho) E^{1/2} G^{(n)}(\rho,\rho'). \tag{5.13}$$

By using the last exponential in Eq. (5.12), we can obtain the wave number of each of the outgoing particles. The exponent is

$$-iE^{1/2}\frac{\boldsymbol{\rho}\cdot\boldsymbol{\rho}'}{\rho} = -i\frac{E^{1/2}}{\rho}\sum_{i=1}^{n}\left(\frac{2m_i}{\hbar^2}\right)\mathbf{r}_i\cdot\mathbf{r}'_i. \tag{5.14}$$

Since each particle is free, we expect this to have the form $-i\sum_{i=1}^{n}\mathbf{k}_i\cdot\mathbf{r}'_i$, which implies

$$\mathbf{k}_i = \frac{2m_i}{\hbar^2} E^{1/2} \frac{\mathbf{r}_i}{\rho}. \tag{5.15}$$

This definition of the wave number k_i is satisfactory, since the relation $\rho^2 = \sum_{i=1}^{n}(2m_i/\hbar^2)r_i^2$ implies $\sum_{i=1}^{n}(\hbar^2 k_i^2/2m_i) = E$, and the sum of the kinetic energies is just the total energy E. We can use this definition of k_i to show that $G^{(n)}(\rho,\rho')$ is outgoing in each single-particle coordinate; this is expressed in the relation

$$\nabla_i G^{(n)}(\rho,\rho') \xrightarrow[r_i\to\infty]{} i\frac{\mathbf{r}_i}{r_i}\left(\frac{2m_i E}{\hbar^2}\right)^{1/2} G^{(n)}(\rho,\rho'). \tag{5.16}$$

All of these results have the conventional form for $n=1$.

The free-particle Green's function goes to zero as $\rho^{-(3n-1)/2}$ for large ρ, and the probability density goes as $\rho^{-(3n-1)}$. Since the surface area of a "sphere" of radius ρ in $3n$-dimensional space is proportional to ρ^{3n-1}, this shows that there is a nonvanishing current crossing such a surface at infinity. This in turn is merely an expression of the fact that for positive energy all of the particles can escape to infinity. When E is negative, the square root of the energy is defined such that $\exp(iE^{1/2}\rho) = \exp(-|E|^{1/2}\rho)$, which vanishes when $\rho \to \infty$. Hence for negative energy there can be no outgoing wave at infinity.

GREEN'S FUNCTION FOR MOTION IN A POTENTIAL

We are also interested in Green's functions describing propagation in the presence of interactions. Let us, as a particular example, discuss the case of two particles moving in the presence of a third. Suppose, for illustrative purposes, that particle 2, but not particle 1, interacts with the third particle. Then if $\mathbf{r}_1$ and $\mathbf{r}_2$ are the positions of particles 1 and 2, respectively, relative to the position of the third particle, the relevant Green's function will satisfy the equation

$$\left\{ E + \frac{\hbar}{2m_1} \nabla_1^2 + \frac{\hbar^2}{2m_2} \nabla_2^2 - V(\mathbf{r}_2) \right\} G^{(2)}(\mathbf{r}_1, \mathbf{r}_2; \mathbf{r}_1', \mathbf{r}_2')$$
$$= \delta(\mathbf{r}_1 - \mathbf{r}_1') \delta(\mathbf{r}_2 - \mathbf{r}_2'). \quad (5.17)$$

In general, it is impossible explicitly to write down the solution to this equation. We can, however, infer some properties of $G^{(2)}(\mathbf{r}_1, \mathbf{r}_2; \mathbf{r}_1', \mathbf{r}_2')$ from our discussion of the free-particle Green's function. For instance, if $G^{(2)}(\mathbf{r}_1, \mathbf{r}_2; \mathbf{r}_1', \mathbf{r}_2')$ satisfies outgoing-wave boundary conditions, it will behave asymptotically as

$$\nabla_i G^{(2)}(\mathbf{r}_1, \mathbf{r}_2; \mathbf{r}_1', \mathbf{r}_2') \xrightarrow[r_i \to \infty]{} i \frac{\mathbf{r}_i}{r_i} \left(\frac{2m_i E}{\hbar^2} \right)^{1/2} G^{(2)}(\mathbf{r}_1, \mathbf{r}_2; \mathbf{r}_1', \mathbf{r}_2'). \quad (5.18)$$

If $V(\mathbf{r}_2)$ is an attractive potential having bound states, then for negative values of the total energy the Green's function describes the motion of one bound particle and one free particle. For positive values of the energy, particle 2 can be unbound or bound, depending upon what fraction of the total energy is available to it. If particle 2 is unbound, it eventually passes outside the range of $V(\mathbf{r}_2)$, and Eq. (5.17) becomes identical with the free-particle equation. In that circumstance the Green's function goes to zero for large values of either r_1 or r_2 as $r_i^{-5/2}$. On the other hand, if particle 2 is bound, the Green's function goes to zero for large r_1 as r_1^{-1}.

In general, the outgoing-wave solution to Eq. (5.17) may be written as an eigenfunction expansion

$$G^{(2)}(\mathbf{r}_1, \mathbf{r}_2; \mathbf{r}_1', \mathbf{r}_2') = \sum_\lambda \int \frac{d^3k}{(2\pi)^3} \frac{\exp[i\mathbf{k} \cdot (\mathbf{r}_1 - \mathbf{r}_1')] \eta_\lambda(\mathbf{r}_2) \eta_\lambda^*(\mathbf{r}_2')}{E - (\hbar^2 k^2/2m_1) - E_\lambda + i\varepsilon}, \quad (5.19)$$

where the functions $\eta_\lambda(r_2)$ are regular solutions of the eigenequation

$$\left\{ E_\lambda + \frac{\hbar^2}{2m_2} \nabla_2^2 - V(\mathbf{r}_2) \right\} \eta_\lambda(\mathbf{r}_2) = 0, \quad (5.20)$$

and the sum over λ includes a sum over the discrete bound-state solutions and an integral over the continuum unbound solutions of this equation. The integral over **k** yields

$$G^{(2)}(\mathbf{r}_1,\mathbf{r}_2;\mathbf{r}_1',\mathbf{r}_2') = \frac{2m_2}{4\pi\hbar^2} \sum_\lambda \frac{\exp\{i[(2m_1/\hbar^2)(E-E_\lambda)]^{1/2}|\mathbf{r}_1-\mathbf{r}_1'|\}}{|\mathbf{r}_1-\mathbf{r}_1'|}$$
$$\times \eta_\lambda(\mathbf{r}_2)\eta_\lambda^*(\mathbf{r}_2'). \qquad (5.21)$$

The bound-state terms in the sum over λ vanish exponentially as $r_2 \to \infty$. When $r_1 \to \infty$, these terms describe a single particle escaping to infinity and go to zero as r_1^{-1}. The unbound functions ($E_\lambda > 0$) form a continuum and give the contribution discussed in the previous paragraph.

Chapter

5

Integral Equations of Scattering Theory and Their Solutions

In this chapter we continue our discussion of the integral formulation of scattering theory, begun in the previous chapter. In Chapter 4 we derived the radial and three-dimensional Green's functions and discussed their properties at some length. We also derived the integral equation for the radial wave function. Now we shall use the Green's function to construct the integral equations that represent the solution of the three-dimensional Schrödinger equation. We shall then study the properties of the resulting integral equations and methods for explicitly obtaining solutions.

In what follows we will use $\psi(\mathbf{r})$ to denote the solution of the Schrödinger equation for the potential $V(\mathbf{r})$, whereas $\phi(\mathbf{r})$ will represent the free-particle solution and $\chi(\mathbf{r})$ the solution for the "intermediate" potential $V_0(\mathbf{r})$. The use of such a consistent convention helps make the structure of the theory somewhat more transparent.

1. Integral Equations for the Wave Function

We wish to solve a differential equation of the general form

$$(E - H_0(\mathbf{r}))\,\psi(\mathbf{r}) = \rho(\mathbf{r}). \tag{1.1}$$

In the case of greatest interest, the Hamiltonian operator $H_0(\mathbf{r})$ is

$$H_0(\mathbf{r}) \equiv -(\hbar^2/2m)\,\nabla^2 + V_0(\mathbf{r}) \tag{1.2}$$

and the source density is

$$\rho(\mathbf{r}) = V_1(\mathbf{r})\,\psi(\mathbf{r}). \tag{1.3}$$

We may solve Eq. (1.1) with the aid of the solution for a delta-function source:

$$(E - H_0(\mathbf{r})) G(\mathbf{r}, \mathbf{r}') = \delta(\mathbf{r} - \mathbf{r}'). \tag{1.4}$$

This, of course, is the equation for the Green's function appropriate to Eq. (1.1).

With the aid of this Green's function we may write the solution of Eq. (1.1) as

$$\psi(\mathbf{r}) = \int d^3r' \, G(\mathbf{r}, \mathbf{r}') \rho(\mathbf{r}'). \tag{1.5}$$

To this we may add any solution of the homogeneous equation,

$$(E - H_0(\mathbf{r})) \chi(\mathbf{r}) = 0, \tag{1.6}$$

so that the general solution of Eq. (1.1) is

$$\psi(\mathbf{r}) = \chi(\mathbf{r}) + \int d^3r' \, G(\mathbf{r}, \mathbf{r}') \rho(\mathbf{r}'). \tag{1.7}$$

The function of primary interest to us here is a special case of this, viz.,

$$\psi(\mathbf{r}) = \chi(\mathbf{r}) + \int d^3r' \, G(\mathbf{r}, \mathbf{r}') V_1(\mathbf{r}') \psi(\mathbf{r}'). \tag{1.8}$$

It is to this equation that we shall direct our attention. We emphasize that the boundary conditions are included in Eq. (1.8); that is, a given choice for the Green's function and for the function $\chi(\mathbf{r})$ implies a definite boundary condition on $\psi(\mathbf{r})$.

Integral Equation with a Free-Particle Incident Wave

We first treat the simplest case, and then generalize. Thus we choose $H_0(\mathbf{r})$ to be simply the kinetic energy operator, so that $G(\mathbf{r}, \mathbf{r}')$ describes the propagation of a free particle. Under boundary conditions appropriate to the scattering problem, the wave function must have the form of a free-particle incident wave plus an outgoing spherical scattered wave, for large values of r. We choose the Green's function to have this outgoing-wave boundary condition and indicate this by the superscript "(+)." From the previous chapter we take the result that for free particles the outgoing-wave Green's function is

$$G^{(+)}(\mathbf{r}, \mathbf{r}') = -\left(\frac{2m}{4\pi\hbar^2}\right) \frac{\exp(ik|\mathbf{r} - \mathbf{r}'|)}{|\mathbf{r} - \mathbf{r}'|}. \tag{1.9}$$

We see immediately that the integral in Eq. (1.8) then contributes an outgoing

1. Integral Equations for the Wave Function

spherical wave for large values of r. In order to satisfy the boundary condition, we must take the function $\chi(\mathbf{r})$ to be a plane wave solution of Eq. (1.6). We write $\chi(\mathbf{r})$ as

$$\chi_{\mathbf{k}}(\mathbf{r}) = \phi_{\mathbf{k}}(\mathbf{r}) = A\, e^{i\mathbf{k}\cdot\mathbf{r}}. \tag{1.10}$$

The integral equation with the boundary conditions made explicit becomes

$$\psi_{\mathbf{k}}^{(+)}(\mathbf{r}) = A\, e^{i\mathbf{k}\cdot\mathbf{r}} + \int d^3r'\, G^{(+)}(\mathbf{r}, \mathbf{r}')\, V(\mathbf{r}')\, \psi_{\mathbf{k}}^{(+)}(\mathbf{r}'). \tag{1.11}$$

The superscript "(+)" on the wave function will serve to remind us of the boundary conditions inherent in this equation. The constant A is a normalization constant and is conventionally set equal to unity. For any other value of A we can simply multiply the result so obtained by A to find the solution to Eq. (1.11).

For the wave function normalized so that $A = 1$, we obtain the integral equation

$$\psi_{\mathbf{k}}^{(+)}(\mathbf{r}) = e^{i\mathbf{k}\cdot\mathbf{r}} + \int d^3r'\, G^{(+)}(\mathbf{r}, \mathbf{r}')\, V(\mathbf{r}')\, \psi_{\mathbf{k}}^{(+)}(\mathbf{r}')$$

$$= e^{i\mathbf{k}\cdot\mathbf{r}} - \left(\frac{1}{4\pi}\right) \int d^3r'\, \frac{\exp(ik|\mathbf{r} - \mathbf{r}'|)}{|\mathbf{r} - \mathbf{r}'|}\, U(\mathbf{r}')\, \psi_{\mathbf{k}}^{(+)}(\mathbf{r}'), \tag{1.12}$$

where

$$U(\mathbf{r}) \equiv (2m/\hbar^2)\, V(\mathbf{r}). \tag{1.13}$$

Since $|\mathbf{r} - \mathbf{r}'| \to r[1 - (\mathbf{r}\cdot\mathbf{r}')/r^2 + \mathcal{O}(r'^2/r^2)]$ for large r, the free-particle Green's function behaves asymptotically as

$$G^{(+)}(\mathbf{r}, \mathbf{r}') \xrightarrow[r\to\infty]{} -\left(\frac{2m}{4\pi\hbar^2}\right) \frac{e^{ikr}}{r}\, e^{-i\mathbf{k}'\cdot\mathbf{r}'}, \tag{1.14}$$

where the vector $\mathbf{k}'$ is directed along $\mathbf{r}$ and is defined by $\mathbf{k}' = k\hat{\mathbf{r}}$. Thus, provided $V(\mathbf{r})$ is negligible beyond some finite radius, the wave function defined by Eq. (1.12) has the asymptotic form

$$\psi_{\mathbf{k}}^{(+)}(\mathbf{r}) \xrightarrow[r\to\infty]{} e^{i\mathbf{k}\cdot\mathbf{r}} + \frac{e^{ikr}}{r}\left(-\frac{1}{4\pi}\right) \int d^3r'\, e^{-i\mathbf{k}'\cdot\mathbf{r}'}\, U(\mathbf{r}')\, \psi_{\mathbf{k}}^{(+)}(\mathbf{r}'). \tag{1.15}$$

Identifying $f(\theta)$ as the coefficient of the outgoing wave, we find the integral expression for the scattering amplitude

$$f(\theta) = (-1/4\pi) \int d^3r'\, e^{-i\mathbf{k}'\cdot\mathbf{r}'}\, U(\mathbf{r}')\, \psi_{\mathbf{k}}^{(+)}(\mathbf{r}'). \tag{1.16}$$

This basic formula was derived in Section 1 of Chapter 3 and will be re-examined in greater detail later in this chapter.

Integral Equation with a Distorted Incident Wave: The Two-Potential Formula

In Eq. (1.11) we chose $H_0(\mathbf{r})$ to be the kinetic energy operator, so that the free-particle Green's function appears in the integral equation. More generally, we can let $H_0(\mathbf{r}) = -(\hbar^2/2m)\nabla^2 + V_0(\mathbf{r})$, where $V_0(\mathbf{r})$ is an arbitrary potential selected to facilitate solution of the scattering problem. To satisfy the boundary conditions, we must use the outgoing-wave Green's function and must also require that $\chi(\mathbf{r})$ have the asymptotic form of a plane wave plus an outgoing wave. Thus we have the general solution

$$\psi_\mathbf{k}^{(+)}(\mathbf{r}) = \chi_\mathbf{k}^{(+)}(\mathbf{r}) + \int d^3r'\, G_0^{(+)}(\mathbf{r}, \mathbf{r}')\, V_1(\mathbf{r}')\, \psi_\mathbf{k}^{(+)}(\mathbf{r}'), \qquad (1.17)$$

where $G_0^{(+)}(\mathbf{r}, \mathbf{r}')$ is the Green's function for motion in the potential $V_0(\mathbf{r})$.

In the asymptotic limit the "distorted wave" function $\chi_\mathbf{k}^{(+)}(\mathbf{r})$ behaves as

$$\chi_\mathbf{k}^{(+)}(\mathbf{r}) \xrightarrow[r\to\infty]{} e^{i\mathbf{k}\cdot\mathbf{r}} + (1/r)\, e^{ikr} f_0(\theta), \qquad (1.18)$$

where, as in Eqs. (1.15) and (1.16),

$$f_0(\theta) = -(1/4\pi)\int d^3r'\, e^{-i\mathbf{k}'\cdot\mathbf{r}'}\, U_0(\mathbf{r}')\, \chi_\mathbf{k}^{(+)}(\mathbf{r}'). \qquad (1.19)$$

This is simply the scattering amplitude for the potential $V_0(\mathbf{r})$, as if it were the only potential present.

The asymptotic form of the Green's function in Eq. (1.17) can be obtained from Eq. (4.55) *et seq.* of Chapter 4 and is

$$G_0^{(+)}(\mathbf{r}, \mathbf{r}') \xrightarrow[r\to\infty]{} -(2m/4\pi\hbar^2)(e^{ikr}/r)\, \chi_{\mathbf{k}'}^{(-)*}(\mathbf{r}'), \qquad (1.20)$$

with $\mathbf{k}' \equiv k\hat{\mathbf{r}}$.

From this result the second term of Eq. (1.17) has the asymptotic form

$$\int d^3r'\, G_0^+(\mathbf{r}, \mathbf{r}')\, U_1(\mathbf{r}')\, \psi_\mathbf{k}^{(+)}(\mathbf{r}')$$

$$\xrightarrow[r\to\infty]{} r^{-1} e^{ikr} (-1/4\pi) \int d^3r'\, \chi_{\mathbf{k}'}^{(-)*}(\mathbf{r}')\, U_1(\mathbf{r}')\, \psi_\mathbf{k}^{(+)}(\mathbf{r}'). \qquad (1.21)$$

1. Integral Equations for the Wave Function

The structure of this integral is analogous to that of Eq. (1.16) or Eq. (1.19); it is the overlap in the region of $U_1(\mathbf{r})$ of the full scattering wave function with a wave function computed by ignoring the effect of $U_1(\mathbf{r})$. The asymptotic form of Eq. (1.17) gives the three-dimensional form of the "two-potential" formula derived earlier, namely, the scattering amplitude $f(\theta)$ is

$$f(\theta) \equiv f_0(\theta) + f_1(\theta), \qquad (1.22)$$

where $f_0(\theta)$ is given by Eq. (1.19) and $f_1(\theta)$ is

$$f_1(\theta) = -(1/4\pi)\int d^3r'\, \chi_{\mathbf{k}'}^{(-)*}(\mathbf{r}')\, U_1(\mathbf{r}')\, \psi_{\mathbf{k}}^{(+)}(\mathbf{r}'). \qquad (1.23)$$

To see that this is equivalent to Eq. (1.16) and to the expressions in earlier chapters, we introduce the Legendre expansions of the wave functions. The use of

$$\chi_{\mathbf{k}'}^{(-)}(\mathbf{r}') = (kr')^{-1} \sum_{l=0}^{\infty} (2l+1) i^l\, v_l^*(r')\, P_l(\cos\theta_{rr'}) \qquad (1.24)$$

and of the analogous expansions for the plane wave and $\psi_{\mathbf{k}}^{(+)}(\mathbf{r})$ in Eq. (1.22) yields

$$f(\theta) = -k^{-2}\sum_{l=0}^{\infty}(2l+1)\left\{\int_0^{\infty} dr'\, F_l(kr')\, U_0(r')\, v_l(r') \right.$$
$$\left. + \int_0^{\infty} dr'\, v_l(r')\, U_1(r')\, u_l(r')\right\} P_l(\cos\theta). \qquad (1.25)$$

Comparing this with Eqs. (2.24) and (2.46) of Chapter 3, we see that the scattering amplitude $f(\theta)$ is simply

$$f(\theta) = k^{-1}\sum_{l=0}^{\infty}(2l+1)\{e^{i\delta_{l0}}\sin\delta_{l0} + e^{2i\delta_{l0}} e^{i\delta_{l1}}\sin\delta_{l1}\} P_l(\cos\theta)$$

$$= k^{-1}\sum_{l=0}^{\infty}(2l+1)\, e^{i\delta_l}\sin\delta_l\, P_l(\cos\theta), \qquad (1.26)$$

with $\delta_l = \delta_{l0} + \delta_{l1}$. This is the familiar partial-wave expansion for the scattering amplitude. Equations (1.22) and (1.23) thus provide a general expression for the scattering amplitude due to the sum of two potentials, $V(r) = V_0(r) + V_1(r)$.

Explicit Solution for the Wave Function

In the preceding we have considered integral equations involving the free-particle Green's function and the Green's function for an intermediate potential $V_0(\mathbf{r})$. The final case is that involving the Green's function for the full potential $V(\mathbf{r})$. The use of Eq. (1.8) in that case would lead only to an uninteresting identity, since the inhomogeneous term is identically zero, but by using a different procedure it is possible to obtain a new and useful result, an "explicit solution" for the wave function.

Let us consider the equation for the wave function $\chi(\mathbf{r})$. We rewrite Eq. (1.6) as

$$\{E - H(\mathbf{r})\} \chi(\mathbf{r}) = -V_1(\mathbf{r}) \chi(\mathbf{r}). \tag{1.27}$$

Clearly, Eq. (1.27) has the same structure as Eqs. (1.1)–(1.3), and we can in a similar fashion express the solution in the form of an integral equation. The solution satisfying outgoing-wave boundary conditions is

$$\chi_\mathbf{k}^{(+)}(\mathbf{r}) = \psi_\mathbf{k}^{(+)}(\mathbf{r}) - \int d^3r' \; \bar{G}^{(+)}(\mathbf{r}, \mathbf{r}') V_1(\mathbf{r}') \chi_\mathbf{k}^{(+)}(\mathbf{r}'), \tag{1.28}$$

where the Green's function satisfies the differential equation

$$\{E - H(\mathbf{r})\} \bar{G}(\mathbf{r}, \mathbf{r}') = \{E + (\hbar^2/2m) \nabla^2 - V(\mathbf{r})\} \bar{G}(\mathbf{r}, \mathbf{r}') = \delta(\mathbf{r} - \mathbf{r}'), \tag{1.29}$$

and is subject to the outgoing-wave boundary condition. The bar above the G in Eqs. (1.27) and (1.28) is a reminder that the Green's function under consideration is not the free-particle Green's function, but rather is the Green's function for the full potential $V(\mathbf{r})$.

If we now insert into Eq. (1.28) the integral equation for the wave function, Eq. (1.17), we obtain

$$\chi_\mathbf{k}^{(+)}(\mathbf{r}) = \chi_\mathbf{k}^{(+)}(\mathbf{r}) + \int d^3r' \; G_0^{(+)}(\mathbf{r}, \mathbf{r}') V_1(\mathbf{r}') \psi_\mathbf{k}^{(+)}(\mathbf{r}')$$

$$- \int d^3r' \; \bar{G}^{(+)}(\mathbf{r}, \mathbf{r}') V_1(\mathbf{r}') \chi_\mathbf{k}^{(+)}(\mathbf{r}'). \tag{1.30}$$

Thus we see that Eq. (1.28) implies the identity

$$\int d^3r' \; G_0^{(+)}(\mathbf{r}, \mathbf{r}') V_1(\mathbf{r}') \psi_\mathbf{k}^{(+)}(\mathbf{r}') = \int d^3r' \; \bar{G}^{(+)}(\mathbf{r}, \mathbf{r}') V_1(\mathbf{r}') \chi_\mathbf{k}^{(+)}(\mathbf{r}'). \tag{1.31}$$

1. Integral Equations for the Wave Function

It is apparent that Eq. (1.28) can be rewritten as

$$\psi_{\mathbf{k}}^{(+)}(\mathbf{r}) = \chi_{\mathbf{k}}^{(+)}(\mathbf{r}) + \int d^3 r'\, \bar{G}^{(+)}(\mathbf{r},\mathbf{r}')\, V_1(\mathbf{r}')\, \chi_{\mathbf{k}}^{(+)}(\mathbf{r}'), \tag{1.32}$$

which is an explicit solution for the scattering wave function. It must be remembered, however, that it is as difficult to calculate the Green's function $\bar{G}^{(+)}(\mathbf{r},\mathbf{r}')$ as it is to solve the "implicit" integral equation, Eq. (1.17), so that this does not bypass the real difficulties of solving the Schrödinger equation. This result is of use, though, in performing manipulations to obtain approximate expressions in more difficult cases, such as those involving many particles, and in deriving general theoretical results.

In the limit as $r \to \infty$ the Green's function $\bar{G}^{(+)}(\mathbf{r},\mathbf{r}')$ becomes

$$\bar{G}^{(+)}(\mathbf{r},\mathbf{r}') \xrightarrow[r\to\infty]{} -(2m/4\pi\hbar^2)(e^{ikr}/r)\, \psi_{\mathbf{k}'}^{(-)*}(\mathbf{r}'), \tag{1.33}$$

so that in that limit Eq. (1.31) implies that the contribution $f_1(\theta)$ to the scattering amplitude is given by the two equivalent expressions,

$$f_1(\theta) = -(2m/4\pi\hbar^2) \int d^3 r'\, \chi_{\mathbf{k}'}^{(-)*}(\mathbf{r}')\, V_1(\mathbf{r}')\, \psi_{\mathbf{k}}^{(+)}(\mathbf{r}')$$

$$= -(2m/4\pi\hbar^2) \int d^3 r'\, \psi_{\mathbf{k}'}^{(-)*}(\mathbf{r}')\, V_1(\mathbf{r}')\, \chi_{\mathbf{k}}^{(+)}(\mathbf{r}'). \tag{1.34}$$

For H_0 taken to be the kinetic energy operator, $\chi_{\mathbf{k}}^{(+)}(\mathbf{r})$ becomes a plane wave and we have the identity

$$f(\theta) = -(2m/4\pi\hbar^2) \int d^3 r'\, \phi_{\mathbf{k}'}^{*}(\mathbf{r}')\, V(\mathbf{r}')\, \psi_{\mathbf{k}}^{(+)}(\mathbf{r}')$$

$$= -(2m/4\pi\hbar^2) \int d^3 r'\, \psi_{\mathbf{k}'}^{(-)*}(\mathbf{r}')\, V(\mathbf{r}')\, \phi_{\mathbf{k}}(\mathbf{r}'), \tag{1.35}$$

where, we remember, $\phi_{\mathbf{k}}(\mathbf{r}) = e^{i\mathbf{k}\cdot\mathbf{r}}$. The two equivalent expressions for the scattering amplitudes given in Eq. (1.34) and in Eq. (1.35) will be discussed at length in later chapters.

THE SOURCE AT A FINITE DISTANCE

The method of integral equations applies to a wider class of problems than the problem to which we have applied it. As an illustration, we will apply this method to one problem of a slightly different character.

We have heretofore considered the case where the source was, in effect, infinitely distant, so that the incident radiation was in the form of a plane wave. Let us now consider a point source of radiation of wave number k located at the finite distance $r' = r_s$ from the center of the potential. The reader will note that, in the language of physical optics, we are now considering Fresnel instead of Fraunhofer diffraction by the potential.

The Hamiltonian of the system is $H(\mathbf{r})$ as before. The time-independent wave equation for this situation is then

$$\{E - H(\mathbf{r})\}\,\psi(\mathbf{r}, \mathbf{r}_s) = Q\,\delta(\mathbf{r} - \mathbf{r}_s), \tag{1.36}$$

where Q is the source strength. Now, we note that Eq. (1.36) is precisely Eq. (1.29), if $Q = 1$. Since Q is a normalization constant, we have the relation

$$\psi(\mathbf{r}, \mathbf{r}_s) = Q\,\bar{G}(\mathbf{r}, \mathbf{r}_s). \tag{1.37}$$

We may subject the solution to the boundary condition that we have only an outgoing spherical wave as $r \to \infty$, i.e.,

$$\lim_{r \to \infty} \hat{\mathbf{r}} \cdot \nabla G(\mathbf{r}, \mathbf{r}_s)/G(\mathbf{r}, \mathbf{r}_s) = \lim_{r \to \infty} \hat{\mathbf{r}} \cdot \nabla \psi^{(+)}(\mathbf{r}, \mathbf{r}_s)/\psi^{(+)}(\mathbf{r}, \mathbf{r}_s) = ik. \tag{1.38}$$

The Green's function has been shown in Eq. (4.46) of Chapter 4 to be symmetric, so that $\psi^{(+)}(\mathbf{r}, \mathbf{r}_s)$ is also symmetric and

$$\psi^{(+)}(\mathbf{r}, \mathbf{r}_s) = \psi^{(+)}(\mathbf{r}_s, \mathbf{r}). \tag{1.39}$$

The solution of Eq. (1.36) has been written as an eigenfunction expansion in Eq. (4.40) of Chapter 4. In coordinate space the solution is easy to write as an integral equation. We use $H(\mathbf{r}) = -(\hbar^2/2m)\nabla^2 + V(r)$ to rewrite Eq. (1.36) as

$$\{E + (\hbar^2/2m)\nabla^2\}\,\psi(\mathbf{r}, \mathbf{r}_s) = Q\,\delta(\mathbf{r} - \mathbf{r}_s) + V(r)\,\psi(\mathbf{r}, \mathbf{r}_s). \tag{1.40}$$

Assuming that there is no wave in the absence of the source, and therefore that there is no homogeneous term, the solution of this equation is

$$\begin{aligned}\psi^{(+)}(\mathbf{r}, \mathbf{r}_s) = Q\,\bar{G}^{(+)}(\mathbf{r}, \mathbf{r}_s) &= \int d^3 r'\, G^{(+)}(\mathbf{r}, \mathbf{r}')\{Q\,\delta(\mathbf{r}' - \mathbf{r}_s) \\ &\quad + V(r')\,\psi^{(+)}(\mathbf{r}', \mathbf{r}_s)\} \\ &= Q\,G^{(+)}(\mathbf{r}, \mathbf{r}_s) + \int d^3 r'\, G^{(+)}(\mathbf{r}, \mathbf{r}')\,V(r')\,\psi^{(+)}(\mathbf{r}', \mathbf{r}_s) \\ &= -Q\left(\frac{2m}{4\pi\hbar^2}\right)\left\{\frac{\exp(ik|\mathbf{r} - \mathbf{r}_s|)}{|\mathbf{r} - \mathbf{r}_s|}\right. \\ &\quad \left. + \int d^3 r'\, \frac{\exp(ik|\mathbf{r} - \mathbf{r}'|)}{|\mathbf{r} - \mathbf{r}'|}\,V(r')\,\bar{G}^{(+)}(\mathbf{r}', \mathbf{r}_s)\right\}. \end{aligned} \tag{1.41}$$

1. Integral Equations for the Wave Function

This is an integral equation for the wave function; because of Eq. (1.37), it is also an integral equation for the Green's function for the potential $V(r)$.

THE RECIPROCITY CONDITION

In the limit as the source is moved to infinity, $\psi^{(+)}(\mathbf{r}, \mathbf{r}_s)$ should reduce to the wave function of ordinary scattering. For r_s very large, the outgoing spherical wave becomes

$$\frac{\exp(ik|\mathbf{r} - \mathbf{r}_s|)}{|\mathbf{r} - \mathbf{r}_s|} \xrightarrow[r_s \to \infty]{} \frac{e^{ikr_s}}{r_s} e^{i\mathbf{k}\cdot\mathbf{r}}, \tag{1.42}$$

where $\mathbf{k}$ is a vector of magnitude k in the $-\mathbf{r}_s$ direction, i.e., *from* the source toward the origin. If we were to choose the source strength Q such that

$$-Q(1/4\pi)(2m/\hbar^2)(e^{ikr_s}/r_s) = 1, \tag{1.43}$$

then in the limit as $r_s \to \infty$ Eq. (1.41) would become

$$\psi^{(+)}(\mathbf{r}) = e^{i\mathbf{k}\cdot\mathbf{r}} - \frac{2m}{4\pi\hbar^2} \int d^3r' \frac{\exp(ik|\mathbf{r} - \mathbf{r}'|)}{|\mathbf{r} - \mathbf{r}'|} V(\mathbf{r}') \psi^{(+)}(\mathbf{r}'), \tag{1.44}$$

which we recognize as the integral equation for the wave function subject to the usual outgoing wave scattering boundary conditions.

In the limit as both r and r_s become infinite, that is, as we go from the "Fresnel" situation to the "Fraunhofer" case, with the source strength chosen in accordance with Eq. (1.43), the "Fresnel" function $\psi^{(+)}(\mathbf{r}, \mathbf{r}_s)$ approaches

$$\psi^{(+)}(\mathbf{r}, \mathbf{r}_s) \xrightarrow[r, r_s \to \infty]{} e^{i\mathbf{k}_i \cdot \mathbf{r}} + f(\mathbf{k}_f, \mathbf{k}_i)(e^{ikr}/r), \tag{1.45}$$

where $\mathbf{k}_i$ is a vector of magnitude k in the $-\mathbf{r}_s$ direction, $\mathbf{k}_f$ is a vector of magnitude k in the $\mathbf{r}$ direction, and

$$f(\mathbf{k}_f, \mathbf{k}_i) = -(2m/4\pi\hbar^2) \lim_{r_s \to \infty} \int d^3r' \, e^{-i\mathbf{k}_f \cdot \mathbf{r}'} V(\mathbf{r}') \psi^{(+)}(\mathbf{r}', \mathbf{r}_s). \tag{1.46}$$

Similarly, $\psi^{(+)}(\mathbf{r}_s, \mathbf{r})$ becomes in this same limit, with Q properly chosen

$$\psi^{(+)}(\mathbf{r}_s, \mathbf{r}) \xrightarrow[r, r_s \to \infty]{} e^{i\mathbf{k}_i' \cdot \mathbf{r}_s} + f(\mathbf{k}_f', \mathbf{k}_i')(e^{ikr_s}/r_s)$$

$$= e^{-i\mathbf{k}_f \cdot \mathbf{r}_s} + f(-\mathbf{k}_i, -\mathbf{k}_f)(e^{ikr_s}/r_s), \tag{1.47}$$

where $\mathbf{k}_i'$ is a vector of magnitude k in the $-\mathbf{r}$ direction, $\mathbf{k}_f'$ is a vector of magnitude k in the $\mathbf{r}_s$ direction, and

$$\begin{aligned} f(\mathbf{k}_f', \mathbf{k}_i') &= f(-\mathbf{k}_i, -\mathbf{k}_f) \\ &= -(2m/4\pi\hbar^2) \lim_{r \to \infty} \int d^3r' \, e^{-i\mathbf{k}_f' \cdot \mathbf{r}'} V(\mathbf{r}') \psi^{(+)}(\mathbf{r}', \mathbf{r}) \\ &= -(2m/4\pi\hbar^2) \lim_{r \to \infty} \int d^3r' \, e^{i\mathbf{k}_i \cdot \mathbf{r}'} V(\mathbf{r}') \psi^{(+)}(\mathbf{r}', \mathbf{r}). \end{aligned} \quad (1.48)$$

The reciprocity relation, Eq. (1.39), holds here if Q is the same for both functions. If we are to obtain the normalization we have chosen in Eq. (1.45) and Eq. (1.47), with Q fixed by Eq. (1.43), then we require $r = r_s$. Then the reciprocity relation, Eq. (1.39), implies that the scattering amplitude satisfies the reciprocity condition

$$f(\mathbf{k}_f, \mathbf{k}_i) = f(-\mathbf{k}_i, -\mathbf{k}_f). \quad (1.49)$$

This condition relates scattering processes in which initial and final states are interchanged and the direction of motion is reversed. For these two "time-reversed" situations, the scattering amplitudes are equal.

2. Time-Dependent Derivation of the Integral Equation

We have derived the integral equation for the time-independent wave function by requiring that it have the proper form at large distances from the scattering center. As we recall from Chapter 2, this form was required in turn by the condition that the time-dependent solution for the scattering problem, a superposition of stationary-state functions, must contain no scattered wave at very early times. Let us here deduce the integral equation for the time-independent wave function directly, using time-dependent Green's function techniques. This will show in a somewhat different way how the causality condition on the time-dependent function leads to the outgoing-wave boundary condition on the stationary-state solution.

The Time-Dependent Green's Function

The time-dependent Schrödinger equation for which a solution must be found is

$$\left\{ -\frac{\hbar^2}{2m} \nabla^2 - i\hbar \frac{\partial}{\partial t} \right\} \psi(\mathbf{r}, t) = -V(\mathbf{r}) \psi(\mathbf{r}, t). \quad (2.1)$$

In order to solve this we use the free-particle Green's function satisfying the

2. Time-Dependent Derivation of the Integral Equation

differential equation

$$\left\{-\frac{\hbar^2}{2m}\nabla^2 - i\hbar\frac{\partial}{\partial t}\right\} G(\mathbf{r},t;\mathbf{r}',t') = -i\hbar\,\delta(\mathbf{r}-\mathbf{r}')\,\delta(t-t'). \quad (2.2)$$

As in the time-independent case, $G(\mathbf{r},t;\mathbf{r}',t')$ depends only on the coordinate differences $|\mathbf{r}-\mathbf{r}'|$ and $(t-t')$, so that the Green's function $G(\mathbf{r},t;\mathbf{r}',t')$ also satisfies the differential equation

$$\left\{-\frac{\hbar^2}{2m}\nabla'^2 + i\hbar\frac{\partial}{\partial t'}\right\} G(\mathbf{r},t;\mathbf{r}',t') = i\hbar\,\delta(\mathbf{r}-\mathbf{r}')\,\delta(t-t'). \quad (2.3)$$

In our case the Green's function will satisfy the "retarded" boundary condition

$$G(\mathbf{r},t;\mathbf{r}',t') = 0 \quad \text{for} \quad t \le t'. \quad (2.4)$$

This is the causality requirement, stating that the effects of a disturbance can be felt only *after* the disturbance has occurred.

Following our standard procedure, we can obtain an expression for this Green's function by expanding it in terms of solutions of the corresponding homogeneous equation. The basis functions $\phi_{\mathbf{k}}(\mathbf{r},t)$ satisfy the homogeneous differential equation

$$\left\{-\frac{\hbar^2}{2m}\nabla^2 - i\hbar\frac{\partial}{\partial t}\right\} \phi_{\mathbf{k}}(\mathbf{r},t) = 0 \quad (2.5)$$

and may be taken to be plane waves. With this choice the properly normalized eigenfunctions are

$$\phi_{\mathbf{k}}(\mathbf{r},t) = (2\pi)^{-3} e^{i\mathbf{k}\cdot\mathbf{r}} \exp[-i(\hbar k^2/2m)t]. \quad (2.6)$$

Thus the time-dependent Green's function may be expressed as

$$G(\mathbf{r},t;\mathbf{r}',t') = \begin{cases} (2\pi)^{-3}\int d^3k\, g(\mathbf{k})\exp[i\mathbf{k}\cdot(\mathbf{r}-\mathbf{r}')] \exp[-i(\hbar k^2/2m)(t-t')], & t > t', \\ 0, & t \le t'. \end{cases} \quad (2.7)$$

This solution automatically satisfies the differential equation, Eq. (2.2), when $t \ne t'$, and it has been defined in accordance with the causality condition.

The remaining condition to be satisfied by this function follows from the differential equation, which requires a discontinuity in the time derivative of the Green's function. We integrate Eq. (2.2) over time from $t = t' - \eta$ to $t = t' + \eta$ and allow η to go to zero. Assuming that G is continuous and that the integral over $\nabla^2 G$ vanishes as $\eta \to 0$, we have

$$\lim_{\eta \to 0} \int_{t'-\eta}^{t'+\eta} dt \frac{\partial}{\partial t} G(\mathbf{r}, t; \mathbf{r}', t') = \lim_{\eta \to 0} G(\mathbf{r}, t' + \eta; \mathbf{r}', t') = \delta(\mathbf{r} - \mathbf{r}'). \quad (2.8)$$

From this result and the well-known relation

$$\delta(\mathbf{r} - \mathbf{r}') = (2\pi)^{-3} \int d^3k \exp[i\mathbf{k} \cdot (\mathbf{r} - \mathbf{r}')], \quad (2.9)$$

we recognize that the factor $g(\mathbf{k})$ in Eq. (2.7) must be unity for all $\mathbf{k}$.

If we define the step function by

$$u(t - t') \equiv \begin{cases} 1, & t > t', \\ 0, & t \leq t', \end{cases} \quad (2.10)$$

then the Green's function can be expressed as

$$G(\mathbf{r}, t; \mathbf{r}', t')$$
$$= u(t - t')(2\pi)^{-3} \int d^3k \exp[i\mathbf{k} \cdot (\mathbf{r} - \mathbf{r}')] \exp[-i(\hbar k^2/2m)(t - t')]. \quad (2.11)$$

Use of the Fourier expansion for the step function

$$u(t - t') = -(2\pi i)^{-1} \int_{-\infty}^{\infty} d\omega \frac{\exp[-i\omega(t - t')]}{\omega + i\varepsilon} \quad (2.12)$$

permits Eq. (2.11) to be rewritten as

$$G(\mathbf{r}, t; \mathbf{r}', t') = -(2\pi)^{-3}(2\pi i)^{-1} \int d^3k \int_{-\infty}^{\infty} d\omega \frac{\exp[i\mathbf{k} \cdot (\mathbf{r} - \mathbf{r}') - i\omega(t - t')]}{\{\omega - (\hbar k^2/2m) + i\varepsilon\}}. \quad (2.13)$$

The integration in Eq. (2.11) or, equivalently, in Eq. (2.13) may be performed analytically. The result yields the time-dependent Green's function in closed form,

$$G(\mathbf{r}, t; \mathbf{r}'\, t') = u(t - t') \{-im/2\pi\hbar(t - t')\}^{3/2} \exp[im(\mathbf{r} - \mathbf{r}')^2/2\hbar(t - t')]. \quad (2.14)$$

We shall use this form of the Green's function later in this section.

INTEGRAL EQUATION FOR THE WAVE FUNCTION

Let us now use the differential equations for $\psi(\mathbf{r}, t)$ and $G(\mathbf{r}, t; \mathbf{r}', t')$ to obtain an integral expression for the time-dependent wave function $\psi(\mathbf{r}, t)$.

2. Time-Dependent Derivation of the Integral Equation

In the familiar way we multiply Eq. (2.1) by the Green's function, multiply Eq. (2.3) by $\psi(\mathbf{r}, t)$, and subtract. Integrating the result over all space and over t' between $-T$ and $t+$, with T large and $t+$ any time greater than t, we obtain

$$(\hbar^2/2m) \int_{-T}^{t} dt' \int d^3r' \{\psi(\mathbf{r}', t') \nabla'^2 G(\mathbf{r}, t; \mathbf{r}', t') - G(\mathbf{r}, t; \mathbf{r}', t') \nabla'^2 \psi(\mathbf{r}', t)\}$$

$$- i\hbar \int_{-T}^{t} dt' \int d^3r' \left\{ \psi(\mathbf{r}', t') \frac{\partial}{\partial t'} G(\mathbf{r}, t; \mathbf{r}', t') + G(\mathbf{r}, t; \mathbf{r}', t') \frac{\partial}{\partial t'} \psi(\mathbf{r}', t') \right\}$$

$$= i\hbar \, \psi(\mathbf{r}, t) - \int_{-T}^{t} dt' \int d^3r' \, G(\mathbf{r}, t; \mathbf{r}', t') V(\mathbf{r}') \psi(\mathbf{r}', t'). \quad (2.15)$$

In this we have let $t+ \to t$, since the contribution to the integral for $t' > t$ is identically zero.

According to Green's theorem, the first term gives a surface integral which is zero if $\psi(\mathbf{r}, t)$ and $\nabla \psi(\mathbf{r}, t)$ vanish at infinity. The integrand in the second term is a total time derivative which gives no contribution at the upper limit of the time integral, because of the retardation condition on the Green's function. Thus this result can be written

$$\psi(\mathbf{r}, t) = \int d^3r' \, G(\mathbf{r}, t; \mathbf{r}', -T) \psi(\mathbf{r}', -T)$$

$$+ (i\hbar)^{-1} \int_{-T}^{t} dt' \int d^3r' \, G(\mathbf{r}, t; \mathbf{r}', t') V(\mathbf{r}') \psi(\mathbf{r}', t'). \quad (2.16)$$

This has the general structure that was used in Chapter 2 and earlier in this chapter. The first term depends only on the value of the wave function at an arbitrary early time $-T$ and does not depend on the potential. The second term involves the potential and gives the scattered wave. Using the Fourier representation of the Green's function, Eq. (2.11), the first term of Eq. (2.16) is

$$\int d^3r' \, G(\mathbf{r}, t; \mathbf{r}', -T) \psi(\mathbf{r}', -T)$$

$$= (2\pi)^{-3} \int d^3r' \int d^3k \, \exp[i\mathbf{k} \cdot (\mathbf{r} - \mathbf{r}')] \exp[-i(\hbar k^2/2m)(t + T)] \psi(\mathbf{r}', -T)$$

$$= (2\pi)^{-3} \int d^3k \, \exp[i\{\mathbf{k} \cdot \mathbf{r} - i(\hbar k^2/2m)t\}]$$

$$\times \int d^3r' \, \exp[-i\{\mathbf{k} \cdot \mathbf{r}' + (\hbar k^2/2m)T\}] \psi(\mathbf{r}', -T). \quad (2.17)$$

The **r'**—integral is just the Fourier amplitude of the incident wave, so that this term is simply

$$\int d^3r'\, G(\mathbf{r}, t; \mathbf{r}', -T)\, \psi(\mathbf{r}', -T)$$

$$= (2\pi)^{-3} \int d^3k\, A(\mathbf{k})\, \exp[i\{\mathbf{k}\cdot\mathbf{r} - (\hbar k^2/2m)t\}] = \phi(\mathbf{r}, t), \quad (2.18)$$

the free-particle wave packet of Eq. (2.2) in Chapter 2.

Let us now expand $\psi(\mathbf{r}, t)$ in terms of a set of stationary-state wave functions $\psi_\mathbf{k}(\mathbf{r})$. We will determine their form by the requirement that $\psi(\mathbf{r}, t)$ be expressed as

$$\psi(\mathbf{r}, t) = (2\pi)^{-3} \int d^3k\, A(\mathbf{k})\, \psi_\mathbf{k}(\mathbf{r})\, \exp[-i(\hbar k^2/2m)t], \quad (2.19)$$

with $A(\mathbf{k})$ the Fourier coefficient of the incident wave, as in Eq. (2.18). Inserting this expansion and the closed form of the Green's function, Eq. (2.14), into the second term of Eq. (2.16), we obtain

$$(i\hbar)^{-1} \int_{-T}^{t} dt' \int d^3r'\, G(\mathbf{r}, t; \mathbf{r}', t')\, V(\mathbf{r}')\, \psi(\mathbf{r}', t')$$

$$= (i\hbar)^{-1} (2\pi)^{-3} \int d^3k\, A(\mathbf{k}) \int_{-T}^{t} dt' \int d^3r'$$

$$\times \{-im/2\pi\hbar(t-t')\}^{3/2} \exp[im(\mathbf{r}-\mathbf{r}')^2/2\hbar(t-t')]$$

$$\times V(\mathbf{r}')\, \psi_\mathbf{k}(\mathbf{r}')\, \exp[-i(\hbar k^2/2m)t']. \quad (2.20)$$

To evaluate this we perform the time integration. In the limit as the initial time T goes to infinity, the time integral becomes

$$\int_{-T}^{t} dt'\, \{-im/2\pi\hbar(t-t')\}^{3/2} \exp[i\{m(\mathbf{r}-\mathbf{r}')^2/2\hbar(t-t') - (\hbar k^2/2m)t'\}]$$

$$\xrightarrow[T\to\infty]{} (im/2\pi\hbar|\mathbf{r}-\mathbf{r}'|)\, \exp[i\{k|\mathbf{r}-\mathbf{r}'| - (\hbar k^2/2m)t\}]. \quad (2.21)$$

The scattered wave then becomes

$$(i\hbar)^{-1} \int_{-\infty}^{t} dt' \int d^3r'\, G(\mathbf{r}, t; \mathbf{r}', t')\, V(\mathbf{r}')\, \psi(\mathbf{r}', t')$$

$$= -(2\pi)^{-3}(2m/4\pi\hbar^2) \int d^3k\, A(\mathbf{k}) \int d^3r'\, (\exp(ik|\mathbf{r}-\mathbf{r}'|)/|\mathbf{r}-\mathbf{r}'|) V(\mathbf{r}')$$

$$\times \psi_\mathbf{k}(\mathbf{r'})\, \exp[-i(\hbar k^2/2m)t]. \quad (2.22)$$

3. The Born Approximation and the Fredholm Method

The integral equation for $\psi(\mathbf{r}, t)$, Eq. (2.16), must hold for any initial condition, that is, for any choice of $A(\mathbf{k})$. Then, using the integral expressions, Eqs. (2.18), (2.19), and (2.22), we see that the stationary-state functions $\psi_\mathbf{k}(\mathbf{r})$ must satisfy the integral equation

$$\psi_\mathbf{k}(\mathbf{r}) = e^{i\mathbf{k}\cdot\mathbf{r}} - (2m/4\pi\hbar^2)\int d^3r' \, (\exp(ik|\mathbf{r}-\mathbf{r}'|)/|\mathbf{r}-\mathbf{r}'|) V(\mathbf{r}')\psi_\mathbf{k}(\mathbf{r}'). \quad (2.23)$$

This is just Eq. (1.12), the time-independent integral equation. We see that this equation is a direct result of the Schrödinger equation, together with the treatment of a particle as a localized wave packet and the causality condition, imposed through the retarded Green's function. It is these assumptions that lead to the use of the outgoing-wave, time-independent Green's function in this integral equation.

3. Methods of Solution: The Born Approximation and the Fredholm Method

Only in very special cases can the exact solution of the integral equation for $\psi_\mathbf{k}(\mathbf{r})$ be found. Obviously, the integral equation can be solved for those potentials for which solutions of the differential equation are known. There are also several special potentials for which the integral equation can be more easily solved. In these cases the integral equation generally reduces to a set of linear algebraic equations. In cases other than these, it is necessary to resort to some approximation scheme.

The Separable Potential

One class of potentials for which a solution can be found is the separable potential. This is a special case of a "nonlocal" potential, that is, a potential in which the ordinary potential term in the Schrödinger equation, $V(\mathbf{r})\psi(\mathbf{r})$, is replaced by the integral $\int d^3r' \, V(\mathbf{r}, \mathbf{r}')\psi(\mathbf{r}')$. For a nonlocal potential the Schrödinger equation is an integrodifferential equation of the form

$$\{E + (\hbar^2/2m)\nabla^2\}\psi(\mathbf{r}) = \int d^3r' \, V(\mathbf{r}, \mathbf{r}')\psi(\mathbf{r}'), \quad (3.1)$$

and the integral equation, Eq. (1.12), is

$$\psi_\mathbf{k}^{(+)}(\mathbf{r}) = e^{i\mathbf{k}\cdot\mathbf{r}} + \int d^3r' \, d^3r'' \, G^{(+)}(\mathbf{r}, \mathbf{r}') V(\mathbf{r}', \mathbf{r}'') \psi_\mathbf{k}(\mathbf{r}''). \quad (3.2)$$

When the exact interaction is not known, it is frequently convenient to represent it by a separable potential, thereby allowing the use of a simple exact formula for the scattering amplitude. One type of separable potential is

$$V(\mathbf{r}, \mathbf{r}') = v(\mathbf{r}) \, v(\mathbf{r}'). \tag{3.3}$$

With such a potential the integral equation becomes

$$\psi_\mathbf{k}^{(+)}(\mathbf{r}) = e^{i\mathbf{k}\cdot\mathbf{r}} + \left\{ \int d^3r' \, G^{(+)}(\mathbf{r}, \mathbf{r}') \, v(\mathbf{r}') \right\} \left\{ \int d^3r'' \, v(\mathbf{r}'') \, \psi_\mathbf{k}^{(+)}(\mathbf{r}'') \right\}. \tag{3.4}$$

The second bracket in this equation can be evaluated by multiplying this equation by $v(\mathbf{r})$ and integrating over $\mathbf{r}$. Introducing

$$I_k = \int d^3r \int d^3r' \, v(\mathbf{r}) \, G^{(+)}(\mathbf{r}, \mathbf{r}') \, v(\mathbf{r}'), \tag{3.5}$$

we then find the solution for the wave function

$$\psi_\mathbf{k}^{(+)}(\mathbf{r}) = e^{i\mathbf{k}\cdot\mathbf{r}} + \left\{ \int d^3r' \, G^{(+)}(\mathbf{r}, \mathbf{r}') \, v(\mathbf{r}') \right\}$$
$$\times \left\{ \int d^3r'' \, v(\mathbf{r}'') \exp(i\mathbf{k}\cdot\mathbf{r}'') \right\} \Big/ (1 - I_k). \tag{3.6}$$

The asymptotic behavior of this solution shows that the scattering amplitude is

$$f(\theta) = -(1/4\pi) \left\{ \int d^3r' \exp(-i\mathbf{k}'\cdot\mathbf{r}') \, v(\mathbf{r}') \right\}$$
$$\times \left\{ \int d^3r'' \, v(\mathbf{r}'') \exp(i\mathbf{k}\cdot\mathbf{r}'') \right\} \Big/ (1 - I_k), \tag{3.7}$$

where, as in Section 1, $\mathbf{k}' \cdot \mathbf{k} \equiv k^2 \cos\theta$.

The solution for the nonlocal potential

$$V(\mathbf{r}, \mathbf{r}') = (4\pi rr')^{-1} \sum_{l=0}^{\infty} (2l+1) \, v_l(r) \, v_l(r') \, P_l(\cos\theta_{rr'}), \tag{3.8}$$

a superposition of separable potentials, can be found in a similar fashion by introducing the partial-wave expansion of the wave function. If the wave function is expanded as

$$\psi_\mathbf{k}^{(+)}(\mathbf{r}) = (kr)^{-1} \sum_{l=0}^{\infty} (2l+1) i^l \, u_l(r) \, P_l(\cos\theta), \tag{3.9}$$

the integral equation for $u_l(r)$ becomes

$$u_l(r) = F_l(kr) + \left\{ \int_0^\infty dr'\, g_l(r, r')\, v_l(r') \right\} \left\{ \int_0^\infty dr''\, v_l(r'')\, u_l(r'') \right\}. \quad (3.10)$$

This, of course, is just radial integral equation (2.4) of Chapter 4 for a separable potential. In the same manner as above, the solution of this equation is found to be

$$u_l(r) = F_l(kr) + \left\{ \int_0^\infty dr'\, g_l(r, r')\, v_l(r') \right\}$$
$$\times \left\{ \int_0^\infty dr''\, v_l(r'')\, F_l(kr'') \right\} \Big/ (1 - I_{k,l}), \quad (3.11)$$

with

$$I_{k,l} = \int_0^\infty dr \int_0^\infty dr'\, v_l(r)\, g_l(r, r')\, v_l(r'), \quad (3.12)$$

and the phase shift is given by

$$e^{i\delta_l} \sin \delta_l = -k^{-1} \left\{ \int_0^\infty dr'\, F_l(kr')\, v_l(r') \right\}$$
$$\times \left\{ \int_0^\infty dr''\, v_l(r'')\, F_l(kr'') \right\} \Big/ (1 - I_{k,l}). \quad (3.13)$$

THE BORN EXPANSION

In the more general case, there is no simple way to find exact solutions of the integral equation. A straightforward but approximate approach that can be used for any potential, under the proper conditions, is to proceed by an iteration procedure. One takes as the zeroth approximation to the wave function the plane wave

$$\psi_{\mathbf{k}}^{(0)}(\mathbf{r}) = e^{i\mathbf{k}\cdot\mathbf{r}} \quad (3.14)$$

and then iterates Eq. (1.12) according to the rule

$$\psi_{\mathbf{k}}^{(n+1)}(\mathbf{r}) = e^{i\mathbf{k}\cdot\mathbf{r}} - (1/4\pi) \int d^3r'\, (\exp(ik|\mathbf{r} - \mathbf{r}'|) / |\mathbf{r} - \mathbf{r}'|)\, U(\mathbf{r}')\, \psi_{\mathbf{k}}^{(n)}(\mathbf{r}'). \quad (3.15)$$

(To prevent undue complication, we have suppressed the outgoing-wave superscript on $\psi_{\mathbf{k}}$. The superscript now represents the order of the iteration.)

The first iterate is explicitly

$$\psi_k^{(1)}(\mathbf{r}) = e^{i\mathbf{k}\cdot\mathbf{r}} - (1/4\pi)\int d^3r' \, (\exp(ik|\mathbf{r} - \mathbf{r}'|)/|\mathbf{r} - \mathbf{r}'|) \, U(\mathbf{r}') \, e^{i\mathbf{k}\cdot\mathbf{r}'} \quad (3.16)$$

The second iterate is

$$\psi_k^{(2)}(\mathbf{r}) = e^{i\mathbf{k}\cdot\mathbf{r}} - (1/4\pi)\int d^3r' \, (\exp(ik|\mathbf{r} - \mathbf{r}'|)/|\mathbf{r} - \mathbf{r}'|) \, U(\mathbf{r}') \, e^{i\mathbf{k}\cdot\mathbf{r}'}$$

$$+ (1/4\pi)^2 \int d^3r' \int d^3r'' \, (\exp(ik|\mathbf{r} - \mathbf{r}'|)/|\mathbf{r} - \mathbf{r}'|)$$

$$\times U(\mathbf{r}') \, (\exp(ik|\mathbf{r}' - \mathbf{r}''|)/|\mathbf{r}' - \mathbf{r}''|) \, U(\mathbf{r}'') \, e^{i\mathbf{k}\cdot\mathbf{r}''}. \quad (3.17)$$

This process can obviously be continued, generating what is commonly called the "Born expansion." This expansion bears a close similarity to ordinary stationary-state perturbation theory, so that we may consider the incident wave function as describing the "unperturbed" state, while $\psi_k^{(+)}(\mathbf{r})$ describes the exact "perturbed" scattering state.

Generalizing from Eqs. (3.16) and (3.17), we see that Eq. (3.15) can be rewritten as

$$\psi_k^{(n+1)}(\mathbf{r}_0) = \psi_k^{(n)}(\mathbf{r}_0) + (-1/4\pi)^{n+1}\int d^3r_1 \cdots d^3r_{n+1}$$

$$\times \left\{\prod_{m=0}^{n+1}(\exp(ik|\mathbf{r}_m - \mathbf{r}_{m+1}|)/|\mathbf{r}_m - \mathbf{r}_{m+1}|) \, U(\mathbf{r}_{m+1})\right\} e^{i\mathbf{k}\cdot\mathbf{r}_{n+1}}. \quad (3.18)$$

If we write

$$U(\mathbf{r}) = \{\lambda \, U(\mathbf{r})\}_{\lambda=1}, \quad (3.19)$$

we see that the Born expansion is a power series in λ, evaluated at λ equals unity. The convergence of this series can be discussed in terms of the radius of convergence of the power series in λ. We will see later that the series converges for λ equal to unity only if the potential is sufficiently weak or the energy sufficiently high.

The Born Approximation

A more important practical question is whether the first term, $\psi_k^{(0)}(\mathbf{r})$, is itself a good approximation to the wave function. Rarely if ever are higher-order terms computed, since the complications then become so great that one might as well use a numerical method to obtain the exact solution. The first term is always a good approximation at sufficiently high energies, where

3. The Born Approximation and the Fredholm Method

the series converges to the first term. This approximation, in which the exact wave function $\psi_{\mathbf{k}}(\mathbf{r})$ is replaced by the plane wave $e^{i\mathbf{k}\cdot\mathbf{r}}$, is the "first Born approximation," often abbreviated to the "Born approximation." It was used by Professor Max Born in his original exposition of scattering theory.

This approximation is most useful when calculating the scattering amplitude

$$f(\theta) = -(1/4\pi)\int d^3r\, e^{-i\mathbf{k}'\cdot\mathbf{r}}\, U(\mathbf{r})\, \psi_{\mathbf{k}}^{(+)}(\mathbf{r}). \tag{3.20}$$

If we insert into this the Born expansion for the wave function, we obtain the Born expansion for the scattering amplitude,

$$f(\theta) = -(1/4\pi)\int d^3r\, e^{-i\mathbf{k}'\cdot\mathbf{r}}\, U(\mathbf{r})\, e^{i\mathbf{k}\cdot\mathbf{r}} + (1/4\pi)^2 \int d^3r \int d^3r'$$
$$\times e^{-i\mathbf{k}'\cdot\mathbf{r}}\, U(\mathbf{r})\, \big(\exp(ik|\mathbf{r}-\mathbf{r}'|)\big)\big/|\mathbf{r}-\mathbf{r}'|\, U(\mathbf{r}')\, e^{i\mathbf{k}\cdot\mathbf{r}'} + \cdots, \tag{3.21}$$

or, equivalently,

$$f^{(n)}(\theta) = -(1/4\pi)\int d^3r\, e^{-i\mathbf{k}'\cdot\mathbf{r}}\, U(\mathbf{r})\, \psi_{\mathbf{k}}^{(n)}(\mathbf{r}). \tag{3.22}$$

The first approximation is the Born approximation to the scattering amplitude

$$f^{(1)}(\theta) \equiv f^{(\text{Born})}(\theta) = -(1/4\pi)\int d^3r\, e^{-i\mathbf{k}'\cdot\mathbf{r}}\, U(\mathbf{r})\, e^{i\mathbf{k}\cdot\mathbf{r}}. \tag{3.23}$$

For a central potential this reduces to

$$f^{(\text{Born})}(\theta) = -(1/4\pi)\int d^3r\, e^{i\mathbf{q}\cdot\mathbf{r}}\, U(r) = -(1/q)\int_0^\infty r\, dr\, \sin qr\, U(r), \tag{3.24}$$

with the momentum transfer $\mathbf{q}$ defined as $\mathbf{q} = \mathbf{k} - \mathbf{k}'$. This approximation provides a simple explicit formula for calculating the scattering amplitude. It is especially useful at high energies since the Born expansion converges to this first term for sufficiently high energies, when the potential energy is much less than the kinetic energy and the scattering amplitude is correspondingly small.

A convenient, although nonrigorous, criterion for the validity of the Born approximation can be obtained by requiring that the first-order correction to the wave function be small compared to the incident wave in the region of the potential. Taking the origin as a typical point in this region, this implies the condition

$$\left| (1/4\pi) \int d^3r \, (e^{ikr}/r) \, U(\mathbf{r}) \, e^{i\mathbf{k}\cdot\mathbf{r}} \right| \ll 1. \tag{3.25}$$

For a square well of radius R and depth V_0, this implies

$$\left| (mV_0/\hbar^2 k^2)(e^{ikR} \sin kR - kR) \right| \ll 1, \tag{3.26}$$

or, for low energies,

$$\left| mV_0 R^2/\hbar^2 \right| \ll 1, \tag{3.27}$$

and, for high energies,

$$\left| mV_0 R/\hbar^2 k \right| \ll 1. \tag{3.28}$$

Since a bound state exists when $\left| mV_0 R^2/\hbar^2 \right| \gtrsim 1$, the Born approximation will not be valid at low energies if the potential is so strong that it has a bound state. On the other hand, the criterion can be satisfied for any potential by going to a sufficiently high energy.

The Born approximation to the scattering amplitude does not satisfy the optical theorem $(4\pi/k) \operatorname{Im} f(0) = \sigma_{\text{TOT}}$. In fact, $f^{(\text{Born})}(\theta)$ is *real* so that, although it gives a nonzero total cross section, the optical theorem cannot hold. The reason is that the Born approximation uses an unscattered plane wave function, and the removal of the scattered particles from the incident flux is not included in the wave function. This deficiency is inherent in the iteration method; if the scattering amplitude is computed to nth order in the potential, the resulting total cross section will be of order $2n$, and the optical theorem will not in general be satisfied. Of course the optical theorem does hold consistently to each order in λ. For instance, if $f^{(1)}(\theta)$ is used to compute $\sigma_{\text{TOT}}^{(1)}$, the result will equal $(4\pi/k) \operatorname{Im} f^{(2)}(\theta)$, etc. This can easily be seen by introducing the eigenfunction expansion of the Green's function into the second term of $f^{(2)}(\theta)$, Eq. (3.21).

If the phase shift expansion of the scattering amplitude is adopted, it is possible to take advantage of the calculational simplification afforded by the Born approximation while still obtaining a result that satisfies the optical theorem. This is important when the potential is sufficiently strong that the phase shifts for low angular momenta are large and must be computed precisely. However, the high-l phase shifts will still be small and can be calculated by the Born approximation.

The radial wave function satisfies the integral equation

$$u_l(r) = F_l(kr) + \int_0^\infty dr' \, g_l(r, r') \, U(r') \, u_l(r'). \tag{3.29}$$

3. The Born Approximation and the Fredholm Method

Under proper conditions the iterative solution of this equation converges and is

$$u_l(r) = F_l(kr) + \int_0^\infty dr'\, g_l(r, r')\, U(r')\, F_l(kr') + \cdots. \tag{3.30}$$

The exact phase shift is given by

$$\tan \delta_l = -k^{-1} \int_0^\infty dr\, F_l(kr)\, U(r)\, u_l(r) \tag{3.31}$$

provided the standing-wave solution is used, that is,

$$g_l(r, r') = -k^{-1} F_l(kr_<)\, G_l(kr_>). \tag{3.32}$$

The Born approximation to the phase shift is found by taking for $u_l(r)$ only the incident wave of Eq. (3.30). Then Eq. (3.31) gives

$$\tan \delta_l^{(\text{Born})} = -k^{-1} \int_0^\infty dr\, F_l^2(kr)\, U(r). \tag{3.33}$$

As before, this approximation is expected to be valid when the correction to the wave function near the origin is small. This implies

$$\left| k^{-1} \int_0^\infty dr\, G_l(kr)\, F_l(kr)\, U(r) \right| << 1. \tag{3.34}$$

Taking the square well as an example, this requires

$$(2l + 1)^{-1} \left| mV_0 R^2/\hbar^2 \right| << 1$$

when $l >> kR$ and

$$\left| mV_0/\hbar^2 k^2 \right| << 1$$

when $l << kR$. Thus the Born approximation for the phase shift will be valid for sufficiently large values of l; the centrifugal barrier reduces the scattering so that the iterative procedure becomes rapidly convergent. On the other hand, for small angular momenta this approximation is valid only for weak potentials on high energies. In either case the phase shift will be sufficiently small so that $\tan \delta_l^{(\text{Born})} \simeq \delta_l^{(\text{Born})}$.

The Distorted-Wave Born Approximation

When these criteria are not met, it is often possible to use a generalization of the Born approximation, sometimes called the "distorted-wave Born approximation." One supposes there is a potential $V_0(\mathbf{r})$ that differs by a small amount from $V(\mathbf{r})$ and for which exact solutions can be found. Instead of using the integral equation in the form of a plane wave plus a scattered wave, that is, Eq. (1.12), we use the equivalent equation, Eq. (1.17):

$$\psi_\mathbf{k}^{(+)}(\mathbf{r}) = \chi_\mathbf{k}^{(+)}(\mathbf{r}) + \int d^3r' \, G_0(\mathbf{r}, \mathbf{r}') V_1(\mathbf{r}') \psi_\mathbf{k}^{(+)}(\mathbf{r}'). \tag{3.35}$$

Here $\chi_\mathbf{k}^{(+)}(\mathbf{r})$, the "distorted" incident wave, is the outgoing-wave solution of

$$\{E + (\hbar^2/2m) \nabla^2 - V_0(\mathbf{r})\} \chi_\mathbf{k}^{(+)}(\mathbf{r}) = 0 \tag{3.36}$$

and is assumed to be known, $G_0^{(+)}(\mathbf{r}, \mathbf{r}')$ is the Green's function for the same potential,

$$\{E + (\hbar^2/2m) \nabla^2 - V_0(\mathbf{r})\} G_0^{(+)}(\mathbf{r}, \mathbf{r}') = \delta(\mathbf{r} - \mathbf{r}'), \tag{3.37}$$

and

$$V_1(\mathbf{r}) = V(\mathbf{r}) - V_0(\mathbf{r}). \tag{3.38}$$

As shown in earlier chapters and in Eqs. (1.17)–(1.23), the scattering amplitude consists of two terms,

$$f(\theta) = f_0(\theta) - (1/4\pi) \int d^3r' \, \chi_{\mathbf{k}'}^{(-)*}(\mathbf{r}') U_1(\mathbf{r}') \psi_\mathbf{k}^{(+)}(\mathbf{r}'). \tag{3.39}$$

The first term is the scattering amplitude for the potential $V_0(\mathbf{r})$ and is assumed to be known.

If $V_1(\mathbf{r})$ is sufficiently weak, we may consider iterating Eq. (3.35), thereby generating a series in powers of the strength of $V_1(\mathbf{r})$. Keeping only the first term in this iteration process,

$$\psi_\mathbf{k}^{(+)}(\mathbf{r}) \simeq \chi_\mathbf{k}^{(+)}(\mathbf{r}), \tag{3.40}$$

we obtain the distorted-wave Born approximation

$$f(\theta) \simeq f_0(\theta) - (1/4\pi) \int d^3r' \, \chi_{\mathbf{k}'}^{(-)*}(\mathbf{r}') U_1(\mathbf{r}') \chi_\mathbf{k}^{(+)}(\mathbf{r}'). \tag{3.41}$$

Clearly, Eq. (3.41) will be a good approximation if $V_1(\mathbf{r})$ is sufficiently small, so that the additional scattering that it generates does not significantly modify the wave function. Some examples in which this method is useful include nuclear scattering, in which $V_1(\mathbf{r})$ may be a spin-orbit potential or a perturbation due to many-particle excitations, and atomic scattering, where $V_1(\mathbf{r})$

3. The Born Approximation and the Fredholm Method

may be the deviation from the Coulomb potential or from a Hartree average potential.

One can also perform a partial-wave analysis of Eq. (3.41) to obtain an approximate expression for the phase shift. This result is

$$e^{i\delta_l} \sin \delta_l = e^{i\delta_{0l}} \sin \delta_{0l} - k^{-1} \int_0^\infty dr\, v^2(r)\, U(r). \tag{3.42}$$

THE FREDHOLM METHOD

Quite frequently these methods prove inadequate, since the potentials may be too strong for them to be valid. One must then resort to numerical techniques or to analytical methods which can provide more exact answers. For all potentials that go to zero faster than r^{-1} as $r \to \infty$ and are less singular than r^{-2} at $r = 0$, the Fredholm method provides a way of finding solutions to an arbitrary accuracy, limited only by the time and effort devoted to the calculation. The following is a development of this method.

Let us consider a central potential and find the solution for the radial wave function. Define a discrete set of positions r_i extending from the origin out to radii at which the potential is negligible. Defining the "area" between two radii as $W_i = (r_i - r_{i-1}) V(r_i)$, we approximate the potential by the expression

$$V^{(N)}(r) = \sum_{i=1}^{N} W_i\, \delta(r - r_i). \tag{3.43}$$

If we allow N to approach infinity and $(r_i - r_{i-1}) = \Delta r_i$ to approach zero, this sum becomes

$$V^{(N)}(r) = \sum_{i=1}^{N} (r_i - r_{i-1}) V(r_i) \delta(r - r_i) \xrightarrow[\Delta r_i \to \infty]{N \to \infty} \int_0^\infty dr'\, V(r') \delta(r - r') = V(r); \tag{3.44}$$

in this limit $V^{(N)}(r)$ is just the potential $V(r)$. This limit will be used below in deriving the Fredholm formulas.

Let us first consider the case of finite N. The radial integral equation is

$$u_l(r) = F_l(kr) + \int_0^\infty dr'\, g_l(r, r')\, V(r')\, u_l(r'). \tag{3.45}$$

For the potential of Eq. (3.43), this becomes

$$u_l^{(N)}(r) = F_l(kr) + \sum_{i=1}^{N} g_l(r, r_i)\, W_i\, u_l^{(N)}(r_i). \tag{3.46}$$

Setting $r = r_j$, we obtain the set of linear algebraic equations

$$u_l^{(N)}(r_j) - \sum_{i=1}^{N} g_l(r_j, r_i) W_i u_l^{(N)}(r_i) = F_l(kr_j), \qquad (3.47)$$

which can be solved in the conventional way using Cramer's rule. The solution is

$$u_l^{(N)}(r_j) = \sum_{i=1}^{N} F_l(kr_i) D_{l,ij}^{(N)}/D_l^{(N)}, \qquad (3.48)$$

where $D_l^{(N)}$, called the "Fredholm determinant," is the determinant of the coefficients:

$$D_l^{(N)} = \begin{vmatrix} \{1 - g_l(r_1, r_1) W_1\} & \{-g_l(r_1, r_2) W_2\} & \{-g_l(r_1, r_3) W_3\} \cdots \\ \{-g_l(r_2, r_1) W_1\} & \{1 - g_l(r_2, r_2) W_2\} & \{-g_l(r_2, r_3) W_3\} \cdots \\ \cdot & \cdot & \cdot \\ \cdot & \cdot & \cdot \\ \cdot & \cdot & \cdot \end{vmatrix} \qquad (3.49)$$

and $D_{l,ij}^{(N)}$ is the cofactor of the (ij)th term in $D_l^{(N)}$; that is, the numerator in Eq. (3.48) is the determinant formed by replacing the jth column of the determinant $D_l^{(N)}$ by $F_l(kr_1), F_l(kr_2), \ldots$.

If this solution is inserted into Eq. (3.46), we find for the wave function

$$u_l^{(N)}(r) = F_l(kr) + \sum_{i,j=1}^{N} g_l(r, r_j) W_j D_{l,ij}^{(N)} F_l(kr_i)/D_l^{(N)}. \qquad (3.50)$$

The phase shift is then given by

$$\tan \delta_l^{(N)} = -k^{-1} \sum_{i,j=1}^{N} F_l(kr_j) W_j D_{l,ij}^{(N)} F_l(kr_i)/D_l^{(N)}, \qquad (3.51)$$

if the standing-wave Green's function

$$\begin{aligned} g_l(r, r') &= -k^{-1} F_l(kr_<) G_l(kr_>) \\ &= \frac{2}{\pi} \mathscr{P} \int_0^\infty dk' \, \frac{F_l(k'r) F_l(k'r')}{k^2 - k'^2} \end{aligned} \qquad (3.52)$$

is selected. Equation (3.51) is an exact solution for the potential of Eq. (3.43).

Before taking the limit $N \to \infty$, we may note that this provides another means of obtaining approximate solutions of the scattering equations. If the potential $V(r)$ is smoothly varying, it may be represented by a small number of terms N, and the approximate solution for the wave function is given by Eq. (3.50).

3. The Born Approximation and the Fredholm Method

If the exact solution is desired, we must take the limit $N \to \infty$. To accomplish this, let us arrange the Nth-order determinant of Eq. (3.49) as a series in ascending powers of the strength of the potential, which we again denote by λ. We find for the Fredholm determinant

$$D_l^{(N)} = 1 - \lambda \sum_{i=1}^{N} g_l(r_i, r_i) W_i + \frac{\lambda^2}{2!} \sum_{i,j=1}^{N} \begin{vmatrix} g_l(r_i, r_i) W_i & g_l(r_i, r_j) W_j \\ g_l(r_j, r_i) W_i & g_l(r_j, r_j) W_j \end{vmatrix}$$

$$- \frac{\lambda^3}{3!} \sum_{i,j,k=1}^{N} \begin{vmatrix} g_l(r_i, r_i) W_i & g_l(r_i, r_j) W_j & g_l(r_i, r_k) W_k \\ g_l(r_j, r_i) W_i & g_l(r_j, r_j) W_j & g_l(r_j, r_k) W_k \\ g_l(r_k, r_i) W_i & g_l(r_k, r_j) W_j & g_l(r_k, r_k) W_k \end{vmatrix}$$

$$+ \cdots + (-1)^N \frac{\lambda^N}{N!} \sum_{i,j,\ldots N=1}^{N} \begin{vmatrix} g_l(r_i, r_i) W_i & \cdots & g_l(r_i, r_N) W_N \\ \vdots & & \vdots \\ g_l(r_N, r_i) W_i & \cdots & g_l(r_N, r_N) W_N \end{vmatrix}. \quad (3.53)$$

In the limit as $N \to \infty$ and $\Delta r_i \to 0$, Eq. (3.53) becomes an infinite series and the sums become integrals:

$$D_l = 1 - \lambda \int_0^\infty dr \, g_l(r, r) V(r) + \frac{\lambda^2}{2!} \iint dr \, dr' \begin{vmatrix} g_l(r, r) V(r) & g_l(r, r') V(r') \\ g_l(r', r) V(r) & g_l(r', r') V(r') \end{vmatrix}$$

$$- \frac{\lambda^3}{3!} \iiint dr \, dr' \, dr'' \begin{vmatrix} g_l(r, r) V(r) & g_l(r, r') V(r') & g_l(r, r'') V(r'') \\ g_l(r', r) V(r) & g_l(r', r') V(r') & g_l(r', r'') V(r'') \\ g_l(r'', r) V(r) & g_l(r'', r') V(r') & g_l(r'', r'') V(r'') \end{vmatrix} + \cdots. \quad (3.54)$$

One important feature of this series is that it converges for all values of λ. The radius of convergence of a power series of the form $\sum_{n=0}^{\infty} C_n \lambda^n$ is $[\lim_{n\to\infty} (C_n)^{1/n}]^{-1}$; in the present case the limit is zero, giving an infinite radius of convergence. To see this we use Hadamard's theorem, which states that, if the elements of an nth-order determinant are bounded by $|a|$, the absolute magnitude of the determinant is bounded by $|a|^n n^{n/2}$. If we think of the elements a_{ij} of the determinant as the components of the n vectors a_i, the determinant is equal to the generalized volume of a parallelopiped in n-dimensional space. This volume is a maximum when the vectors a_i are mutually orthogonal. Then, since

$$|\mathbf{a}_i| = (\sum_j |a_{ij}|^2)^{1/2} \leq (\sum_j |a|^2)^{1/2} = |a| n^{1/2}, \quad (3.55)$$

the determinant is less than $(|a| n^{1/2})^n = |a|^n n^{n/2}$. As a result of this theorem, if the potential vanishes for $r > R$ and $g_l(r, r') V(r')$ is less than some limit $|M|$, the coefficient of λ^n is bounded by $1/n! \,|M|^n R^n n^{n/2}$. Since $n^n/n! < e^n$, the nth root of this coefficient is less than $|M| R\, e/n^{1/2}$, which vanishes as $n \to \infty$. Hence the power series for the Fredholm determinant converges for all values of λ.

Now we recognize that the quantity $\sum_{j=1}^{N} g_l(r, r_j) W_j D_{l,ij}^{(N)}$ appearing in the solution for $u_l^{(N)}(r)$, Eq. (3.50), is the determinant formed by replacing the ith row in $D_l^{(N)}$ by $g_l(r, r_1) W_1, g_l(r, r_2) W_2, \ldots$. We will denote this determinant by $D_l^{(N)}(r, r_i)$. Rearranging it as in Eq. (3.53), we find

$$D_l^{(N)}(r, r_i) \equiv \sum_{j=1}^{N} g_l(r, r_j) W_j D_{l,ij}^{(N)}$$

$$= g_l(r, r_i) W_i - \lambda \sum_{j=1}^{N} \begin{vmatrix} g_l(r, r_i) W_i & g_l(r, r_j) W_j \\ g_l(r_j, r_i) W_i & g_l(r_j, r_j) W_j \end{vmatrix}$$

$$+ \frac{\lambda^2}{2!} \sum_{j,k=1}^{N} \begin{vmatrix} g_l(r, r_i) W_i & g_l(r, r_j) W_j & g_l(r, r_k) W_k \\ g_l(r_j, r_i) W_i & g_l(r_j, r_j) W_j & g_l(r_j, r_k) W_k \\ g_l(r_k, r_i) W_i & g_l(r_k, r_j) W_j & g_l(r_k, r_k) W_k \end{vmatrix} + \cdots.$$

(3.56)

Taking the limit as $N \to \infty$ and $\Delta r_i \to 0$, this becomes

$$D_l(r, r') = g_l(r, r') V(r') - \lambda \int_0^\infty dr'' \begin{vmatrix} g_l(r, r') V(r') & g_l(r, r'') V(r'') \\ g_l(r'', r') V(r') & g_l(r'', r'') V(r'') \end{vmatrix}$$

$$+ \frac{\lambda^2}{2!} \int_0^\infty \int_0^\infty dr'' \, dr'''$$

$$\times \begin{vmatrix} g_l(r, r') V(r') & g_l(r, r'') V(r'') & g_l(r, r''') V(r''') \\ g_l(r'', r') V(r') & g_l(r'', r'') V(r'') & g_l(r'', r''') V(r''') \\ g_l(r''', r') V(r') & g_l(r''', r'') V(r'') & g_l(r''', r''') V(r''') \end{vmatrix}$$

$$+ \cdots, \quad (3.57)$$

and Eq. (3.50) for the radial wave function becomes

$$u_l(r) = F_l(kr) + \int_0^\infty dr' \, \{D_l(r, r')/D_l\} F_l(kr'). \tag{3.58}$$

As in the case of the Fredholm determinant D_l, the series for $D_l(r, r')$ converges for all values of λ.

Equation (3.58) is an exact, explicit solution of the integral equation for the wave function. The quantity $\{D_l(r, r')/D_l\}$, commonly called the "resolvent kernel," is the ratio of two convergent power series which, within practical limits, can be calculated to arbitrary accuracy. It may be noted that the

3. The Born Approximation and the Fredholm Method

structure of Eq. (3.58) can be obtained by introducing the partial-wave expansion for the Green's function into the "explicit" solution for $\psi_{\mathbf{k}}(\mathbf{r})$,

$$\psi_{\mathbf{k}}^{(+)}(\mathbf{r}) = \phi_{\mathbf{k}}(\mathbf{r}) + \int d^3 r' \, \bar{G}^{(+)}(\mathbf{r}, \mathbf{r}') \, V(\mathbf{r}') \, \phi_{\mathbf{k}}(\mathbf{r}'). \tag{3.59}$$

This reveals that the resolvent kernel is identical with the product $\bar{g}_l(r, r')$ $V(r')$, where $\bar{g}_l(r, r')$ is the radial Green's function describing propagation in the presence of the potential $V(r)$. We thus have yet another means of finding the Green's function, via the Fredholm determinant.

The resolvent kernel can be found exactly in some special cases. For the potential

$$V(r) = \frac{C}{a^2} \frac{e^{-r/a}}{1 - e^{-r/a}},$$

one can show that the Fredholm determinant for S-wave motion is

$$D_0 = \prod_{n=1}^{\infty} \left\{ 1 - \frac{\lambda C}{n(n - 2ika)} \right\}. \tag{3.60}$$

Likewise, for the separable potential $V(r, r') = v_l(r) \, v_l(r')$, the Fredholm determinant is found to be simply

$$D_l = 1 - \lambda \int\!\!\int_0^\infty dr \, dr' \, v_l(r) \, g_l(r, r') \, v_l(r'), \tag{3.61}$$

while

$$D_l(r, r') = \int_0^\infty dr'' \, g_l(r, r'') \, V_l(r'') \, V_l(r'). \tag{3.62}$$

Inserted into Eq. (3.58), these yield our earlier solution, Eq. (3.11).

Convergence of the Born Expansion

The Fredholm solution may be used to determine (or, more usually, to estimate) the radius of convergence of the Born expansion. This expansion is a power series in λ, while the Fredholm solution is the ratio of two such power series. Thus the Born expansion will be valid only if $(1/D_l)$ can be expanded in a power series in λ. Considering D_l as a function of λ, such an expansion is possible only if the absolute value of λ is smaller than the smallest zero of D_l, say $|\lambda_1|$. Thus, the Fredholm determinant can be written in general as $D_l = \prod_{n=1}^{\infty} (1 - \lambda/\lambda_n)$, where $|\lambda_1| < |\lambda_2| < \cdots$; for

$|\lambda| < |\lambda_1|$, $D_l^{-1} = 1 + (\sum_{n=1}^{\infty} \lambda_n^{-1})\lambda + \ldots$. Since λ is eventually set equal to 1, this expansion will converge only if $|\lambda_1|$ is greater than 1.

For instance, in the case of Eq. (3.60), the Born expansion will converge if

$$|C| < |1 - 2ika| = [1 + 4(ka)^2]^{1/2}. \tag{3.63}$$

Similarly, for the separable potential, the Born expansion will converge if

$$\left|\int_0^\infty \int_0^\infty dr\, dr'\, v_l(r)\, g_l(r, r')\, v_l(r')\right| < 1. \tag{3.64}$$

As these examples indicate, the Born expansion converges when the potential is sufficiently weak, but the convergence is improved as the energy is increased.

The convergence condition can also be related to the behavior of the scattering phase shift. We shall see below that, at a zero of D_l computed using the standing-wave Green's function, the tangent of the phase shift is infinite, corresponding to a scattering resonance. Thus there will be a resonance at an energy for which $|\lambda_1| = 1$. The Born expansion will converge at all energies greater than this, or, conversely, the Born expansion at a given energy will converge provided there are no scattering resonances at that energy or at any higher energy.

The Fredholm Expression for the Phase Shift

We now derive an expression for the phase shift, analogous to the Fredholm solution for the wave function. A useful and interesting result can be obtained if we introduce the integral expression for the standing-wave Green's function, Eq. (3.52). Using Eq. (3.51) or, equivalently, using the asymptotic behavior of Eq. (3.58), we find for the phase shift

$$\tan \delta_l = -(kD_l)^{-1} \int_0^\infty dr' \left\{ F_l(kr')\, V(r') - \lambda \int_0^\infty dr'' \right.$$

$$\left. \times \begin{vmatrix} F_l(kr')\, V(r') & F_l(kr'')\, V(r'') \\ g_l(r'', r')\, V(r') & g_l(r', r'')\, V(r'') \end{vmatrix} + \cdots \right\} F_l(kr')$$

$$= -(kD_l)^{-1} \left\{ V_l(k, k) - \lambda \frac{2}{\pi} \mathscr{P} \int_0^\infty \frac{dk'}{k^2 - k'^2} \right.$$

$$\left. \times \begin{vmatrix} V_l(k, k) & V_l(k, k') \\ V_l(k', k) & V_l(k', k') \end{vmatrix} + \cdots \right\}, \tag{3.65}$$

3. The Born Approximation and the Fredholm Method

where we have introduced the notation

$$V_l(k', k) = \int_0^\infty dr \, F_l(k'r) \, V(r) \, F_l(kr). \tag{3.66}$$

We can also introduce the integral expression for the Green's function into the Fredholm determinant, Eq. (3.54):

$$D_l = 1 - \lambda \frac{2}{\pi} \mathcal{P} \int_0^\infty \frac{dk'}{k^2 - k'^2} V_l(k', k') + \frac{\lambda^2}{2!} \left(\frac{2}{\pi}\right)^2$$

$$\times \mathcal{P} \int_0^\infty \int_0^\infty \frac{dk' \, dk''}{(k^2 - k'^2)(k^2 - k''^2)}$$

$$\times \begin{vmatrix} V_l(k', k') & V_l(k', k'') \\ V_l(k'', k') & V_l(k'', k'') \end{vmatrix} - \cdots. \tag{3.67}$$

This result can be used to simplify the result obtained in Eq. (3.65). The denominator in the λ^2 term of Eq. (3.67) can be decomposed as

$$\frac{1}{(k^2 - k'^2)(k^2 - k''^2)} = \frac{1}{(k^2 - k'^2)(k'^2 - k''^2)} + \frac{1}{(k^2 - k''^2)(k''^2 - k'^2)}. \tag{3.68}$$

Each of these terms gives the same result in Eq. (3.67), since k' and k'' can be interchanged. A similar manipulation can be performed on each term in the power series. As a result, if we denote the expression in brackets in Eq. (3.65) by $\mathcal{D}_l(k)$, that is,

$$\mathcal{D}_l(k) = V_l(k, k) - \lambda \frac{2}{\pi} \mathcal{P} \int_0^\infty \frac{dk'}{k^2 - k'^2} \begin{vmatrix} V_l(k, k) & V_l(k, k') \\ V_l(k', k) & V_l(k', k') \end{vmatrix}$$

$$+ \frac{\lambda^2}{2!} \left(\frac{2}{\pi}\right)^2 \mathcal{P} \int_0^\infty \int_0^\infty \frac{dk' \, dk''}{(k^2 - k'^2)(k^2 - k''^2)}$$

$$\times \begin{vmatrix} V_l(k, k) & V_l(k, k') & V_l(k, k'') \\ V_l(k', k) & V_l(k', k') & V_l(k', k'') \\ V_l(k'', k) & V_l(k'', k') & V_l(k'', k'') \end{vmatrix} - \cdots, \tag{3.69}$$

then the Fredholm determinant is simply

$$D_l = 1 - \lambda \frac{2}{\pi} \mathcal{P} \int_0^\infty \frac{dk'}{k^2 - k'^2} \mathcal{D}_l(k'). \tag{3.70}$$

5. INTEGRAL EQUATIONS OF SCATTERING THEORY

Our final result for the phase shift is

$$\tan \delta_l = -\frac{1}{k} \frac{\mathscr{D}_l(k)}{1 - \frac{2}{\pi} \mathscr{P} \int_0^\infty \frac{dk'}{k^2 - k'^2} \mathscr{D}_l(k')} \tag{3.71}$$

or, in a more suggestive form,

$$-\frac{1}{k} \mathscr{D}_l(k) \cot \delta_l = 1 - \frac{2}{\pi} \mathscr{P} \int_0^\infty \frac{dk'}{k^2 - k'^2} \mathscr{D}_l(k'). \tag{3.72}$$

This result is valuable because it relates the phase shift to a quantity $\mathscr{D}_l(k)$ which has a convergent power series expansion, regardless of the strength of the potential. Equation (3.72) forms the basis for what is sometimes called the "determinantal method." It can be used both in potential theory and in field theory, where a potential may not exist, since it is still possible in that case to define the Fredholm determinant.

Equation (3.72) has the appearance of a generalized effective range formula and, in fact, $(1/k)\mathscr{D}_l(k)$ does behave at low energies as k^{2l+1}, as we expect. The effective range parameters can be computed from $\mathscr{D}_l(k)$. The positions of any scattering resonances or bound states can be determined from the zeros of D_l, that is, from the right-hand side of Eq. (3.72).

THE THREE-DIMENSIONAL CASE

The Fredholm solution for the three-dimensional equation requires special care because of the singularity in the Green's function $G(\mathbf{r}, \mathbf{r}')$ at $\mathbf{r} = \mathbf{r}'$. If the foregoing method were used, then, as in Eq. (3.54), the Fredholm determinant would contain $G(\mathbf{r}, \mathbf{r})$, which is infinite. This difficulty can be avoided if the integral equation is iterated once. If we insert into the integral equation the second term in the Born expansion we obtain

$$\begin{aligned}
\psi_\mathbf{k}(\mathbf{r}) &= e^{i\mathbf{k}\cdot\mathbf{r}} + \int d^3r' \, G(\mathbf{r}, \mathbf{r}') V(\mathbf{r}') \psi_\mathbf{k}(\mathbf{r}') \\
&= e^{i\mathbf{k}\cdot\mathbf{r}} + \int d^3r' \, G(\mathbf{r}, \mathbf{r}') V(\mathbf{r}') e^{i\mathbf{k}\cdot\mathbf{r}'} \\
&\quad + \int d^3r' \, G(\mathbf{r}, \mathbf{r}') V(\mathbf{r}') \{\psi_\mathbf{k}(\mathbf{r}') - e^{i\mathbf{k}\cdot\mathbf{r}'}\} \\
&= \psi_\mathbf{k}^{(1)}(\mathbf{r}) + \int d^3r' \, G_2(\mathbf{r}, \mathbf{r}') V(\mathbf{r}') \psi_\mathbf{k}(\mathbf{r}'),
\end{aligned} \tag{3.73}$$

where $\psi_\mathbf{k}^{(1)}(\mathbf{r})$ is given by Eq. (3.16) and $G_2(\mathbf{r}, \mathbf{r}')$ is defined by

$$G_2(\mathbf{r}, \mathbf{r}') = \int d^3r''\, G''(\mathbf{r}, \mathbf{r}'')\, V(\mathbf{r}'')\, G(\mathbf{r}'', \mathbf{r}'). \tag{3.74}$$

The first iterate of the Green's function is not singular at $\mathbf{r} = \mathbf{r}'$, and the Fredholm solution can now be obtained by following the same steps as led to Eq. (3.58), generalized to three dimensions.

The result is

$$\psi_\mathbf{k}(\mathbf{r}) = \psi_\mathbf{k}^{(1)}(\mathbf{r}) + \int d^3r'\, \frac{D(\mathbf{r}, \mathbf{r}')}{D}\, \psi_\mathbf{k}^{(1)}(\mathbf{r}'), \tag{3.75}$$

with

$$D(\mathbf{r}, \mathbf{r}') = G_2(\mathbf{r}, \mathbf{r}')\, V(\mathbf{r}')$$
$$- \frac{\lambda}{1!} \int d^3r'' \begin{vmatrix} G_2(\mathbf{r}, \mathbf{r}')\, V(\mathbf{r}') & G_2(\mathbf{r}, \mathbf{r}'')\, V(\mathbf{r}'') \\ G_2(\mathbf{r}'', \mathbf{r}')\, V(\mathbf{r}') & G_2(\mathbf{r}'', \mathbf{r}'')\, V(\mathbf{r}'') \end{vmatrix} + \ldots \tag{3.76}$$

and the Fredholm determinant

$$D = 1 - \lambda \int d^3r\, G_2(\mathbf{r}, \mathbf{r})\, V(\mathbf{r})$$
$$+ \frac{\lambda^2}{2!} \int d^3r\, d^3r' \begin{vmatrix} G_2(\mathbf{r}, \mathbf{r})\, V(\mathbf{r}) & G_2(\mathbf{r}, \mathbf{r}')\, V(\mathbf{r}') \\ G_2(\mathbf{r}', \mathbf{r})\, V(\mathbf{r}) & G_2(\mathbf{r}', \mathbf{r}')\, V(\mathbf{r}') \end{vmatrix} - \ldots \tag{3.77}$$

These power series will converge for any potential strength if $V(r)$ goes to zero faster than r^{-2} as $r \to \infty$ and is less singular than r^{-2} as $r \to 0$.

4. The Integral Equation for the Scattering Amplitude

In the previous section we concerned ourselves primarily with the integral equation satisfied by the wave function. We shall now derive another interesting integral equation in which the wave function does not appear explicitly. This integral equation relates the scattering amplitude directly to the interaction that produces the scattering, without requiring the determination of the wave function.

Let us begin with the integral equation for the wave function, Eq. (1.12). Asymptotically, for large values of r, Eq. (1.12) becomes

$$\psi_\mathbf{k}^{(+)}(\mathbf{r}) \xrightarrow[r \to \infty]{} e^{i\mathbf{k} \cdot \mathbf{r}} + f(\theta)(e^{ikr}/r), \tag{4.1}$$

where the scattering amplitude is given by Eq. (1.16). It is convenient to define a quantity proportional to the scattering amplitude

$$T(\mathbf{k}', \mathbf{k}) = -(4\pi\hbar^2/2m) f(\theta) = \int d^3r \, e^{-i\mathbf{k}'\cdot\mathbf{r}} V(\mathbf{r}) \psi_\mathbf{k}^{(+)}(\mathbf{r}). \quad (4.2)$$

We shall call $T(\mathbf{k}', \mathbf{k})$ the T matrix, or transition amplitude. Similarly, we define $V(\mathbf{k}', \mathbf{k})$ to be

$$V(\mathbf{k}', \mathbf{k}) = \int d^3r \, e^{-i\mathbf{k}'\cdot\mathbf{r}} V(\mathbf{r}) e^{i\mathbf{k}\cdot\mathbf{r}}. \quad (4.3)$$

The quantity $V(\mathbf{k}', \mathbf{k})$ may be called the V matrix or the potential matrix. If the plane wave $e^{i\mathbf{k}\cdot\mathbf{r}}$ is substituted for the wave function $\psi_\mathbf{k}^{(+)}(\mathbf{r})$ in the definition of the T matrix, the V matrix is obtained. The V matrix is then just the Born approximation to the T matrix.

If we multiply the integral equation for the wave function, Eq. (1.12), by $\exp(-i\mathbf{k}'\cdot\mathbf{r}) V(\mathbf{r})$ and integrate over all space, we obtain an equation for the T matrix:

$$T(\mathbf{k}', \mathbf{k}) = V(\mathbf{k}', \mathbf{k}) + \int d^3r \, \exp(-i\mathbf{k}'\cdot\mathbf{r}) V(\mathbf{r}) G^{(+)}(\mathbf{r}, \mathbf{r}') V(\mathbf{r}') \psi_\mathbf{k}^{(+)}(\mathbf{r}'). \quad (4.4)$$

Substituting into Eq. (4.4) the integral form of the Green's function,

$$G^{(+)}(\mathbf{r}, \mathbf{r}') = -\int \frac{d^3k''}{(2\pi)^3} \frac{\exp[i\mathbf{k}''\cdot(\mathbf{r} - \mathbf{r}')]}{E(k) - E(k'') + i\varepsilon}, \quad (4.5)$$

with $E(k) = \hbar^2 k^2/2m$, we find immediately the integral equation

$$T(\mathbf{k}', \mathbf{k}) = V(\mathbf{k}', \mathbf{k}) + \int \frac{d^3k''}{(2\pi)^3} \frac{V(\mathbf{k}', \mathbf{k}'') T(\mathbf{k}'', \mathbf{k})}{E(k) - E(k'') + i\varepsilon}. \quad (4.6)$$

This result provides a direct relation between the T matrix and the V matrix. It expresses the scattering amplitude in terms of the interaction, eliminating the intermediate step of finding the wave function. The price we pay for this is, obviously, that Eq. (4.6) involves all energies, so that we must find $T(\mathbf{k}', \mathbf{k})$ not only when energy is conserved, that is, when $k' = k$, but also for all other values of k'. We shall further explore this below.

Under suitable conditions we may solve Eq. (4.6) by iteration. The lowest-order approximation is

$$T^{(1)}(\mathbf{k}', \mathbf{k}) = V(\mathbf{k}, \mathbf{k}). \quad (4.7)$$

4. The Integral Equation for the Scattering Amplitude

This is equivalent to

$$f^{(1)}(\theta) = -\frac{2m}{4\pi\hbar^2}\int d^3r\, e^{-\mathbf{k}'\cdot\mathbf{r}}\, V(\mathbf{r})\, e^{i\mathbf{k}\cdot\mathbf{r}} \tag{4.8}$$

and is thus just the first Born approximation to the T matrix. The recurrence relation for successive iterations is

$$T^{(n+1)}(\mathbf{k}', \mathbf{k}) = V(\mathbf{k}', \mathbf{k}) + \int \frac{d^3k}{(2\pi)^3}\, \frac{V(\mathbf{k}, \mathbf{k}'')\, T^{(n)}(\mathbf{k}'', \mathbf{k})}{E(k) - E(k'') + i\varepsilon}. \tag{4.9}$$

Thus, we have

$$T^{(2)}(\mathbf{k}', \mathbf{k}) = V(\mathbf{k}', \mathbf{k}) + \int \frac{d^3k''}{(2\pi)^3}\, \frac{V(\mathbf{k}', \mathbf{k}'')\, V(\mathbf{k}'', \mathbf{k})}{E(k) - E(k'') + i\varepsilon} \tag{4.10}$$

and

$$\begin{aligned}
T^{(3)}(\mathbf{k}', \mathbf{k}) &= V(\mathbf{k}', \mathbf{k}) + \int \frac{d^3k''}{(2\pi)^3}\, \frac{V(\mathbf{k}', \mathbf{k}'')\, V(\mathbf{k}'', \mathbf{k})}{E(k) - E(k'') + i\varepsilon} \\
&\quad + \int \frac{d^3k''}{(2\pi)^3}\frac{d^3k'''}{(2\pi)^3} \\
&\quad \times \frac{V(\mathbf{k}', \mathbf{k}'')\, V(\mathbf{k}'', \mathbf{k}''')\, V(\mathbf{k}''', \mathbf{k})}{(E(k) - E(k'') + i\varepsilon)(E(k) - E(k''') + i\varepsilon)}.
\end{aligned} \tag{4.11}$$

Within the domain of convergence of the resulting series this is an explicit solution of the integral equation, Eq. (4.6). However, the convergence properties of the Born series are poor, as we have seen, and in many practical cases the series diverges. Furthermore, calculation of the higher Born approximations is extraordinarily laborious. For the case of a strong potential, the most practical method for calculating the T matrix is still to integrate the Schrödinger differential equation numerically to find the wave function, and then to examine its asymptotic form or to construct the integral given in Eq. (4.2).

THE T MATRIX OFF THE ENERGY SHELL

A central problem in particle physics is to infer from the scattering data the interaction giving rise to the observed scattering. It might be thought that this is the inverse of the problem of solving the integral equation we have constructed, that is, that this involves reconstructing the V matrix from a knowledge of the T matrix. It will be recognized at once, however, that such an inversion is not equivalent to reconstructing the interaction from the scattering data. Since energy is conserved in all scattering processes, a complete analysis of the scattering data can at most yield the scattering amplitude "on the energy shell," that is, for values of k' equal to k. But we require a complete knowledge of the T matrix both on and off this "energy shell" in order to be able to obtain the V matrix from Eq. (4.6).

It is instructive to relate this to the previously discussed approach involving solving for the wave function. Let us introduce the integral form of the Green's function into the integral equation for the wave function:

$$\psi_{\mathbf{k}}^{(+)}(\mathbf{r}) = e^{i\mathbf{k}\cdot\mathbf{r}} + \int \frac{d^3k'}{(2\pi)^3} d^3r' \frac{\exp[i\mathbf{k}'\cdot(\mathbf{r}-\mathbf{r}')]}{E(k) - E(k') + i\varepsilon} V(\mathbf{r}') \psi_{\mathbf{k}}^{(+)}(\mathbf{r}). \quad (4.12)$$

We assume, as usual, that the order of the integrations may be freely exchanged. The integration over the spatial coordinates is simply

$$\int d^3r' \exp(i\mathbf{k}'\cdot\mathbf{r}') V(\mathbf{r}') \psi_{\mathbf{k}}^{(+)}(\mathbf{r}') = T(\mathbf{k}', \mathbf{k}). \quad (4.13)$$

Thus we may rewrite Eq. (4.12) as

$$\psi_{\mathbf{k}}^{(+)}(\mathbf{r}) = e^{i\mathbf{k}\cdot\mathbf{r}} + \int \frac{d^3k'}{(2\pi)^3} \frac{\exp(i\mathbf{k}'\cdot\mathbf{r}) T(\mathbf{k}', \mathbf{k})}{E(k) - E(k') + i\varepsilon}. \quad (4.14)$$

We see explicitly that a knowledge of the T matrix $T(\mathbf{k}', \mathbf{k})$ for all values of k' is equivalent to a knowledge of the wave function itself. However, we must know the T matrix for all values of k', not merely those for which $k' = k$ or, equivalently, $E(k') = E$.

Let us examine the relation of the T matrix on the energy shell to the wave function given in Eq. (4.14). On the energy shell, when $k' = k$, we may express the T matrix as $T(k\hat{\mathbf{k}}', \mathbf{k})$ with $\hat{\mathbf{k}}'$ a unit vector in the direction $\mathbf{k}'$. Then we can separate the "on-the-energy-shell" contribution from the "off-the-energy-shell" contribution by writing the wave function as

$$\psi_{\mathbf{k}}^{(+)}(\mathbf{r}) = e^{i\mathbf{k}\cdot\mathbf{r}} + \int \frac{d^3k'}{(2\pi)^3} \frac{\exp(i\mathbf{k}'\cdot\mathbf{r}) T(k\hat{\mathbf{k}}', \mathbf{k})}{E - E(k') + i\varepsilon}$$
$$+ \int \frac{d^3k'}{(2\pi)^3} \frac{\exp(i\mathbf{k}', \mathbf{r}) \{T(\mathbf{k}'\mathbf{k}) - T(k\hat{\mathbf{k}}', \mathbf{k})\}}{E - E(k') + i\varepsilon}. \quad (4.15)$$

We now want to perform the integral in the second term, involving only the on-the-energy-shell T matrix. To accomplish this we introduce the Legendre expansions of the plane waves that enter the integral. Including the plane

4. The Integral Equation for the Scattering Amplitude

wave from the T matrix [see Eq. (4.13)], we obtain

$$\int d\Omega_{\mathbf{k}'} \exp(i\mathbf{k}' \cdot \mathbf{r}) \exp(-i k \hat{\mathbf{k}}' \cdot \mathbf{r}') = \frac{4\pi}{k' k r r'} \sum_{l=0}^{\infty} (2l+1) F_l(k'r)$$
$$\times F_l(kr') P_l(\cos \theta_{rr'}). \quad (4.16)$$

The remaining integral to be performed is

$$I_l = \int_0^\infty k' \, dk' \, \frac{F_l(k'r)}{k^2 - k'^2 + i\varepsilon}. \quad (4.17)$$

As usual, it will be convenient to extend the integration over negative values of k'. Since $k' F_l(k'r)$ is an odd function of k' for odd values of l, we must modify the integrand in some manner. If we insert the function $(k'/|k'|)^l$ into this integral, we have an even integrand that is unchanged on the positive real axis. Then we find

$$I_l = \frac{1}{2} \int_{-\infty}^{\infty} \frac{dk' \, (k')^{l+1} F_l(k'r)}{|k'|^l (k^2 - k'^2 + i\varepsilon)}$$
$$= \frac{1}{4i} \int_{-\infty}^{\infty} \frac{dk' \, (k')^{l+1} \{H_l^{(+)}(k'r) - H_l^{(-)}(k'r)\}}{|k'|^l (k^2 - k'^2 + i\varepsilon)}$$
$$= -\frac{1}{2} \pi H_l^{(+)}(kr), \quad (4.18)$$

where the integral has been performed by closing the contour in the upper half plane in the first integration and in the lower half plane in the second, and we have used $(-1)^l H_l^{(-)}(-x) = H_l^{(+)}(x)$. Combining this with Eq. (4.16), we have

$$\int \frac{d^3 k' \, \exp(i\mathbf{k}' \cdot \mathbf{r}) \exp(-i k \hat{\mathbf{k}}' \cdot \mathbf{r}')}{(2\pi)^3 \, E - E(k') + i\varepsilon} = -\frac{2m}{4\pi \hbar^2} \frac{1}{k r r'} \sum_{l=0}^{\infty} (2l+1)$$
$$\times H_l^{(+)}(kr) F_l(kr') P_l(\cos \theta_{rr'}). \quad (4.19)$$

Comparing this result with Eq. (4.34) of Chapter 4, we see that this is just the Legendre expansion of the outgoing-wave Green's function when r is greater than r'. Outside the range of the potential, where r must be greater than r', the second term in Eq. (4.15) is then simply

$$\int \frac{d^3 k' \, \exp(i\mathbf{k}' \cdot \mathbf{r}) \, T(k\hat{\mathbf{k}}', \mathbf{k})}{(2\pi)^3 \, E - E(k') + i\varepsilon} = \int d^3 r' \, G^{(+)}(\mathbf{r}, \mathbf{r}') V(\mathbf{r}') \psi_\mathbf{k}^{(+)}(\mathbf{r}'), \quad (4.20)$$

which is the full scattered wave. As a consequence, knowledge of the T matrix "on-the-energy-shell" implies a knowledge of the wave function everywhere beyond the range of the interaction. Conversely, knowledge of the wave function within the range of the interaction requires knowledge of the last term in Eq. (4.15), containing the "off-the-energy-shell" contributions. The need to include these physically unmeasurable contributions in the integral equation for $T(\mathbf{k}', \mathbf{k})$ is thus equivalent to the need to evaluate the wave function inside the potential, when using the wave function approach to compute the scattering amplitude.

Chapter

6

The Operator Formalism in Two-Particle Scattering Theory

Techniques similar to those of the previous chapters could be used to describe more complicated problems involving many-body systems, such as inelastic scattering, rearrangement collisions, and particle production processes. However, the abundance of coordinates, momenta, and eigenstate labels would quickly overwhelm us and obscure the physics of these problems. We therefore turn to a discussion of techniques that permit us to discard extraneous coordinates and labels and yet to generalize the possible reaction processes, while still keeping in view the central features of the scattering process.

We shall here briefly review the operator formalism of Dirac and see how it may be applied to the two-body problem. In the following chapter this formalism is extended to a general scattering situation. The reader is referred to Dirac's "The Principles of Quantum Mechanics" for a careful and penetrating treatment of this aspect of quantum mechanics. Our discussion in Section 1 is intended primarily to establish the notation. For this reason we are deliberately brief and, so far as mathematical details are concerned, regrettably superficial.

1. Operator Formalism

In the Dirac operator formalism a geometrical analogy is made between the states of a system and vectors in a many-dimensional space. The analogy proves fruitful because of the correspondence between the expansion techniques of quantum mechanics and the resolution of a vector into its components along a set of coordinate axes.

In quantum mechanics the set of eigenstates associated with a physical observable, or a compatible group of observables, constitutes a complete set of states. That is, any arbitrary state can be considered to be a superposition of the states in this set. This is analogous to the resolution of a vector into its components along a set of coordinate axes. Invoking this analogy, we may represent an arbitrary state by a state vector in a many-dimensional vector space. A complete set of eigenstates is pictured as a set of coordinate axes in this space, and an arbitrary state vector is characterized by its projection or component along the direction defined by each eigenstate. As we shall see, this analogy provides a convenient language to describe physical processes.

If the states can be represented by vectors of finite length, a rigorous mathematical description of the resulting abstract space is readily developed. This is the case when the states are bound states, where the associated wave functions extend over a finite region of coordinate space and can be normalized. For unbound states the eigenfunctions extend to infinite distances and cannot be normalized in the usual way. In this case, which is most important in scattering theory, the corresponding abstract vectors do not have finite lengths, and some special device must be used to develop a mathematically consistent theory.

Several different approaches are possible, including ones which avoid completely the use of state vectors in favor of suitable projection operators, and ones which rely on specially defined normalizable state vectors which are necessarily quite complicated. In practice we prefer to use continuum eigenvectors which are not normalizable in the ordinary way but which have a very simple form. Here then, we adopt an approach which uses normalizable state vectors in the form of wave packets to derive relations between experimental observables and the continuum eigenvectors. These continuum eigenvectors will be "normalized" using the delta-function normalization. By beginning with normalizable "wave packet" vectors, we will show that such continuum eigenvectors can be used consistently and correctly to compute physical quantities. This approach, which is described in detail in Chapter 7, is a generalization to abstract vector space of the wave packet approach we have used for the two body problem in Chapters 2 and 5.

NOTATION

A state, or state vector — these terms may be used interchangeably — will be denoted by a Greek letter. When taking the projection of one state vector on another, we will use the bra and ket symbolism of Dirac. In this notation a state vector is denoted by ψ or, equivalently, by $|\psi\rangle$. In the bra-ket notation we can often suppress the Greek letter for the state when it is identified by other labels; for instance, $\psi_a \equiv |\psi_a\rangle \equiv |a\rangle$.

As in the case of ordinary three-dimensional vectors, in the space of state vectors the sum of two vectors ψ_1 and ψ_2 is a vector ψ_3 and is written $\psi_1 + \psi_2 = \psi_3$. A state vector may also be multiplied by a scalar c so that $c(\psi_1 + \psi_2) = c\psi_1 + c\psi_2$.

In order to discuss the relations between different states, it is necessary to consider operators that transform one state into another. In vector space an operator transforms one state vector into another. The effect of the operator $\mathcal{O}$ acting upon the state ψ is symbolized by

$$\psi' = \mathcal{O}\psi = \mathcal{O}|\psi\rangle = |\mathcal{O}\psi\rangle; \qquad (1.1)$$

the state ψ is transformed into ψ' by the operator $\mathcal{O}$. This transformation can be undone by the inverse operator $\mathcal{O}^{-1}$, if it exists, so that

$$\mathcal{O}^{-1}\psi' = \mathcal{O}^{-1}\mathcal{O}\psi = \psi \qquad (1.2)$$

or, in operator language,

$$\mathcal{O}^{-1}\mathcal{O} = \mathcal{O}\mathcal{O}^{-1} = 1. \qquad (1.3)$$

Here 1 is the unit operator.

Many of the usual manipulations that can be performed on three-dimensional vectors can be performed on vectors in this space. For instance, one can form projections or scalar products if an adjoint or dual vector space is introduced. However, the adjoint vector is not simply related to the original vector because of the presence in quantum mechanics of the imaginary unit i, reflecting the fundamental importance of complex quantities in the basic equations of quantum mechanics.

The vector adjoint to ϕ is denoted by $\phi^\dagger$ or $\langle\phi|$, and the scalar product of ϕ and ψ is written $\langle\phi|\psi\rangle$. The relationship between the original space and the adjoint space is such that $\langle\psi|\phi\rangle$ is the complex conjugate of $\langle\phi|\psi\rangle$, that is,

$$\langle\psi|\phi\rangle = \langle\phi|\psi\rangle^*. \qquad (1.4)$$

The use of the bra vector $\langle\phi|$ and the ket vector $|\psi\rangle$ allows us unambiguously to write the scalar product, a scalar or ordinary number, as $\langle\phi|\psi\rangle$; the quantity $|\psi\rangle\langle\phi|$ obtained by inverting the order of these two vectors is not a number at all, but is in fact an operator.

We can define a set of eigenstates ϕ_a for the operator A, corresponding to a given observable, by the eigenvalue equation

$$A\phi_a = a\phi_a, \tag{1.5}$$

where, of course, a is the eigenvalue. When several observables are required to specify a state, the corresponding operators will commute. The states ϕ_a can then be selected to be simultaneous eigenstates of each of these operators, and the symbol a will refer to the full set of eigenvalues (for instance, momentum and spin quantum numbers).

The set of eigenstates associated with a maximal set of commuting observables forms a complete set of states, so that any normalizable state can be expanded in terms of them. They also form an orthogonal set; the corresponding eigenvectors are orthogonal in vector space. Thus if a given observable, or set of observables, has distinct eigenvalues denoted by a, the eigenstates satisfy the orthogonality relation

$$\langle \phi_a | \phi_{a'} \rangle = 0, \quad a \neq a'. \tag{1.6}$$

It is convenient to normalize the state vectors, if they have finite length, so that their length is unity:

$$\langle \phi_a | \phi_a \rangle = 1. \tag{1.7}$$

If the eigenvalues a are discrete, these can be combined into the statement

$$\langle \phi_a | \phi_{a'} \rangle = \delta_{aa'}, \tag{1.8}$$

where the Kronecker delta $\delta_{aa'}$ is defined in the usual way as

$$\delta_{aa'} = \begin{cases} 1, & a = a', \\ 0, & a \neq a'. \end{cases} \tag{1.9}$$

If the variable a is continuous, $\delta_{aa'}$ is replaced by a multiple of the Dirac delta function $\delta(a - a')$. In general, we will use the Kronecker delta and summations over the variables, with the understanding that these are replaced by delta functions and integrations when the variables are continuous. Even when dealing with continuous variables, it is convenient to begin with discrete variables and subsequently to make the transition to the continuous case. If the discrete variable is identified by an integer n, then the transition is made by the replacements

$$\sum_a \rightarrow \int (da/dn)^{-1} \, da \tag{1.10}$$

1. Operator Formalism

and

$$\delta_{aa'} \to (da/dn)\,\delta(a - a'). \tag{1.11}$$

For instance, if a is the momentum k_n, then for periodic boundary conditions in one dimension $k_n = 2\pi n/L$ and $dk/dn = 2\pi/L$. For three dimensions, then,

$$\sum_{\mathbf{k}} \to (2\pi)^{-3} \int d^3k \tag{1.12}$$

and

$$\delta_{\mathbf{k},\mathbf{k}'} \to (2\pi)^3\,\delta(\mathbf{k} - \mathbf{k}'), \tag{1.13}$$

assuming a unit normalization volume.

Eigenvector Expansions

The utility of the Dirac notation is most clearly illustrated in the concise way in which different expansions of ψ are obtained. The set of numerical coefficients in any expansion of ψ serves as a "representative" of ψ, and different "representations" prove useful in different contexts. The operator formalism is independent of representation and thus readily permits the adoption of the most useful representation. Suppose we expand the state ψ in terms of the eigenstates ϕ_a:

$$\psi = \sum_a C_a \phi_a. \tag{1.14}$$

The set of coefficients C_a is the A-representative of ψ. In the probabilistic interpretation of quantum mechanics, C_a is the probability amplitude for finding a system simultaneously in the state ψ and in the eigenstate ϕ_a; that is, $|C_a|^2$ is the probability that a measurement of the variable A applied to a system in the state ψ will yield the value a.

Using the orthonormality of the set ϕ_a, we find that the expansion coefficients are given by

$$C_a = \langle \phi_a | \psi \rangle. \tag{1.15}$$

Thus, the projection of the vector ψ on the eigenvector ϕ_a is just the scalar product, as expected. If we insert this result into the expansion, we have

$$|\psi\rangle = \sum_a |\phi_a\rangle \langle \phi_a | \psi \rangle. \tag{1.16}$$

Thus a mathematical statement of the completeness of the set ϕ_a, that is, of its ability to provide an expansion of any state, is the relation

$$1 = \sum_a | \phi_a \rangle \langle \phi_a |. \qquad (1.17)$$

This complicated way of writing the unit operator is useful in a great many problems. It provides a straightforward means of expressing ψ in any desired representation.

Two particularly useful representations are the spatial or coordinate representation and the momentum representation. In the coordinate representation eigenstates of position, denoted by $| \phi_r \rangle$ or $| \mathbf{r} \rangle$, are used. Here $\mathbf{r}$ may be the position of a single particle, or it may be the ensemble of the position vectors of many particles. Since $\mathbf{r}$ is a continuous variable, these states are normalized according to

$$\langle \mathbf{r} | \mathbf{r}' \rangle = \delta(\mathbf{r} - \mathbf{r}'), \qquad (1.18)$$

and the completeness statement is

$$\int | \mathbf{r} \rangle d^3r \langle \mathbf{r} | = 1. \qquad (1.19)$$

In this representation an arbitrary state vector ψ is written

$$| \psi \rangle = \int | \mathbf{r} \rangle d^3r \langle \mathbf{r} | \psi \rangle. \qquad (1.20)$$

This has a simple interpretation in terms of our previous work. The ket-vector $| \mathbf{r} \rangle$ represents a system localized at $\mathbf{r}$. The scalar product $\langle \mathbf{r} | \psi \rangle$ is then the amplitude for finding the system at $\mathbf{r}$, which is the Schrödinger wave function. Thus we have

$$\langle \mathbf{r} | \psi \rangle \equiv \psi(\mathbf{r}), \qquad (1.21)$$

in our previous notation.

Momentum eigenstates are denoted by $| \mathbf{k} \rangle$. Conventionally, they are normalized so that

$$\langle \mathbf{k} | \mathbf{k}' \rangle = (2\pi)^3 \delta(\mathbf{k} - \mathbf{k}'). \qquad (1.22)$$

The completeness relation is then

$$(2\pi)^{-3} \int | \mathbf{k} \rangle d^3k \langle \mathbf{k} | = 1, \qquad (1.23)$$

1. Operator Formalism

and the momentum expansion of the state vector is

$$|\psi\rangle = (2\pi)^{-3} \int |\mathbf{k}\rangle \, d^3k \, \langle \mathbf{k} | \psi \rangle. \tag{1.24}$$

The relation between these two representations is easy to see. Let us suppose ψ to be a momentum eigenstate $|\mathbf{k}\rangle$; then $\langle \mathbf{r} | \mathbf{k} \rangle$ is the wave function for a system having a wave number $\mathbf{k}$, or explicitly

$$\langle \mathbf{r} | \mathbf{k} \rangle = A_\mathbf{k} \, e^{i\mathbf{k} \cdot \mathbf{r}}. \tag{1.25}$$

The normalization constant $A_\mathbf{k}$ is obtainable from the normalizations of the vectors $|\mathbf{r}\rangle$ and $|\mathbf{k}\rangle$. If we use the completeness relation for position eigenstates and the normalization condition for the momentum eigenstates, we have

$$(2\pi)^3 \, \delta(\mathbf{k} - \mathbf{k}') = \langle \mathbf{k} | \mathbf{k}' \rangle = \int \langle \mathbf{k} | \mathbf{r} \rangle \, d^3r \, \langle \mathbf{r} | \mathbf{k}' \rangle$$

$$= A_\mathbf{k}^* A_{\mathbf{k}'} \int d^3r \, \exp[i(\mathbf{k}' - \mathbf{k}) \cdot \mathbf{r}]. \tag{1.26}$$

The last integral in Eq. (1.26) is $(2\pi)^3 \, \delta(\mathbf{k} - \mathbf{k}')$, so that $A_\mathbf{k}$ can be chosen to have a phase such that $A_\mathbf{k} = 1$, from which it follows that

$$\langle \mathbf{r} | \mathbf{k} \rangle = e^{i\mathbf{k} \cdot \mathbf{r}}. \tag{1.27}$$

If we insert the completeness relation for momentum eigenstates into the coordinate representation of a general state, we find

$$\langle \mathbf{r} | \psi \rangle = (2\pi)^{-3} \int \langle \mathbf{r} | \mathbf{k} \rangle \, d^3k \, \langle \mathbf{k} | \psi \rangle$$

$$= (2\pi)^{-3} \int d^3k \, e^{i\mathbf{k} \cdot \mathbf{r}} \, \langle \mathbf{k} | \psi \rangle. \tag{1.28}$$

Hence the function $\langle \mathbf{r} | \psi \rangle$, the coordinate representative of ψ, is just the Fourier transform of the function $\langle \mathbf{k} | \psi \rangle$, the momentum representative.

OPERATORS

The vector analogy gives us a practical method for describing an operator. We simply let the operator act upon each of the unit vectors associated with some complete set of eigenstates and find the effect of the operator upon them. Thus we can take the complete set $\{|\phi_a\rangle\} \equiv \{|a\rangle\}$ containing, say, N vectors and obtain N new vectors $|\mathcal{O} a\rangle$ by allowing $\mathcal{O}$ to operate on each of them. If we evaluate the projections of these new vectors on each of the original basis vectors, we obtain the set of N^2 numbers $\{\langle a' | \mathcal{O} | a \rangle\}$. In this way we can obtain a set of numbers that completely characterize an abstract operator, in the sense that its effect on an arbitrary state can be determined from these numbers. This set of numbers, based on a particular complete set of states, gives a representative of the operator $\mathcal{O}$.

As with state vectors, the ability to use different representations in different contexts proves very useful in practical applications. To transform to a different representation one simply inserts a new complete set of states on each side of the operator, using Eq. (1.17) and the fact that $\mathcal{O}1 = 1\mathcal{O} = \mathcal{O}$. Thus we have the identity

$$\langle a' | \mathcal{O} | a \rangle = \sum_{b,b'} \langle a' | b' \rangle \langle b' | \mathcal{O} | b \rangle \langle b | a \rangle, \qquad (1.29)$$

which relates the A representative of $\mathcal{O}$ to the B representative.

It is illuminating to arrange the numbers $\langle a' | \mathcal{O} | a \rangle$ in a square array, ordered according to the labels of the states $| a \rangle$:

$$\begin{vmatrix} \langle a_1 | \mathcal{O} | a_1 \rangle & \langle a_1 | \mathcal{O} | a_2 \rangle & \langle a_1 | \mathcal{O} | a_3 \rangle & \cdots \\ \langle a_2 | \mathcal{O} | a_1 \rangle & \langle a_2 | \mathcal{O} | a_2 \rangle & & \cdots \\ \langle a_3 | \mathcal{O} | a_1 \rangle & & & \cdots \\ \vdots & & & \end{vmatrix}. \qquad (1.30)$$

It is easy to show that this array is a matrix. The quantities $\langle a_i | \mathcal{O} | a_j \rangle$ are the matrix elements. An operator is then represented by a matrix and is completely known if all its matrix elements are known. When describing the operators that occur in scattering theory, we shall often speak of the $\mathcal{O}$ matrix rather than the $\mathcal{O}$ operator. While the concept of an abstract operator is quite general, the related matrices provide the tools for practical manipulations and computations. One can, of course, recover the abstract operator from its matrix elements using the completeness relation,

$$\mathcal{O} = \sum_{a,a'} | a' \rangle \langle a' | \mathcal{O} | a \rangle \langle a |, \qquad (1.31)$$

so that there is no loss of information or lack of precision in the use of this language.

The elements $\langle a_i | \mathcal{O} | a_i \rangle$ are the diagonal elements of the matrix $\langle a_i | \mathcal{O} | a_j \rangle$; the matrix itself is said to be diagonal if all other matrix elements vanish, that is, if $\langle a_i | \mathcal{O} | a_j \rangle = \delta_{ij} \mathcal{O}_j$. In this case the numbers $\mathcal{O}_j$

are the eigenvalues of the operator, and $|a_j\rangle$ are its eigenstates. A central problem of quantum mechanics, although not of scattering theory, is to find those states that will yield a diagonal representation for a given operator. These states are the eigenstates of $\mathcal{O}$. Diagonalization of the matrix of $\mathcal{O}$ yields the eigenvectors and eigenvalues of the operator $\mathcal{O}$.

The adjoint operator $\mathcal{O}^\dagger$ is defined by the requirement that the adjoint of the vector $\mathcal{O}|a\rangle$ be $\langle a|\mathcal{O}^\dagger$. From this it follows that a general matrix element of the adjoint operator satisfies

$$\langle a'|\mathcal{O}^\dagger|a\rangle = \langle a|\mathcal{O}|a'\rangle^*. \tag{1.32}$$

The adjoint matrix is the transposed complex-conjugate matrix. It follows readily that the adjoint of a product satisfies $(\mathcal{O}_1\mathcal{O}_2)^\dagger = \mathcal{O}_2^\dagger \mathcal{O}_1^\dagger$. If $\mathcal{O}^\dagger = \mathcal{O}$, the operator is self-adjoint or Hermitian and its matrix elements satisfy

$$\langle a'|\mathcal{O}|a\rangle = \langle a|\mathcal{O}|a'\rangle^*. \tag{1.33}$$

Such a matrix, whose transpose is equal to its complex conjugate, is a Hermitian matrix. A Hermitian operator will also satisfy

$$\langle \psi_1|(\mathcal{O}^\dagger - \mathcal{O})|\psi_2\rangle = \langle \mathcal{O}\psi_1|\psi_2\rangle - \langle \psi_1|\mathcal{O}\psi_2\rangle = 0. \tag{1.34}$$

An operator whose adjoint and inverse are identical, that is, which satisfies $\mathcal{O}^\dagger = \mathcal{O}^{-1}$, is a unitary operator. Such an operator is important in physics because the states produced when it acts upon a set of orthonormal states are themselves orthonormal. The scalar products between states formed by a unitary operator $\mathcal{O}$ acting upon the set ϕ_a are

$$\langle \mathcal{O}a'|\mathcal{O}a\rangle = \langle a'|\mathcal{O}^\dagger\mathcal{O}|a\rangle = \langle a'|\mathcal{O}^{-1}\mathcal{O}|a\rangle = \langle a'|a\rangle$$
$$= \delta_{a'a}. \tag{1.35}$$

Let us now apply the operator notation to the results of the previous chapters and then to more general problems.

2. Operator Form of the Scattering Equations

The central results of the earlier chapters may be summarized in the following equations:

(a) Differential equation for the wave function:

$$\{E + (\hbar^2/2m)\nabla^2\}\psi(\mathbf{r}) = V(\mathbf{r})\psi(\mathbf{r}); \tag{2.1}$$

(b) Integral equation for the wave function:

$$\psi_{\mathbf{k}}^{(+)}(\mathbf{r}) = \phi_{\mathbf{k}}(\mathbf{r}) + \int d^3 r' \, G^{(+)}(\mathbf{r}, \mathbf{r}') \, V(\mathbf{r}') \, \psi_{\mathbf{k}}^{(+)}(\mathbf{r}'), \qquad (2.2)$$

with

$$G^{(+)}(\mathbf{r}, \mathbf{r}') = -\left(\frac{2m}{4\pi\hbar^2}\right) \frac{\exp(ik|\mathbf{r} - \mathbf{r}'|)}{|\mathbf{r} - \mathbf{r}'|} \qquad (2.3)$$

and

$$\phi_{\mathbf{k}}(\mathbf{r}) = e^{i\mathbf{k} \cdot \mathbf{r}}; \qquad (2.4)$$

(c) Integral expression for the scattering amplitude:

$$f(\theta) = -\left(\frac{2m}{4\pi\hbar^2}\right) \int d^3 r \, \phi_{\mathbf{k}}^*(\mathbf{r}) \, V(\mathbf{r}) \, \psi_{\mathbf{k}}^{(+)}(\mathbf{r}); \qquad (2.5)$$

(d) Integral equation for the transition amplitude:

$$T(\mathbf{k}', \mathbf{k}) = V(\mathbf{k}', \mathbf{k}) + (2m/\hbar^2)(2\pi)^{-3} \int d^3 k'' \, \frac{V(\mathbf{k}', \mathbf{k}'') \, T(\mathbf{k}'', \mathbf{k})}{k^2 - k''^2 + i\varepsilon}, \qquad (2.6)$$

with

$$V(\mathbf{k}', \mathbf{k}) = \int d^3 r \, e^{-i\mathbf{k}' \cdot \mathbf{r}} \, V(\mathbf{r}) \, e^{i\mathbf{k} \cdot \mathbf{r}}. \qquad (2.7)$$

We now express these equations in operator language.

THE SCHRÖDINGER EQUATION

The wave function $\psi(\mathbf{r})$ is the spatial representative of a state vector ψ, and we write

$$\psi(\mathbf{r}) = \langle \mathbf{r} | \psi \rangle. \qquad (2.8)$$

The potential $V(\mathbf{r})$ is related to the spatial representative of a potential operator V. The state vector $V\psi$ has the spatial representative

$$\langle \mathbf{r} | V | \psi \rangle = \int \langle \mathbf{r} | V | \mathbf{r}' \rangle \, d^3 r' \, \langle \mathbf{r}' | \psi \rangle$$
$$= \int d^3 r' \, \langle \mathbf{r} | V | \mathbf{r}' \rangle \, \psi(\mathbf{r}'). \qquad (2.9)$$

Since we want the operator V to reproduce the potential energy term in the

2. Operator Form of the Scattering Equations

Schrödinger equation, Eq. (2.1), we must require

$$\int d^3r' \langle \mathbf{r} | V | \mathbf{r}' \rangle \psi(\mathbf{r}') = V(\mathbf{r}) \psi(\mathbf{r}). \tag{2.10}$$

Thus the spatial representative of the potential operator V is

$$\langle \mathbf{r} | V | \mathbf{r}' \rangle = \delta(\mathbf{r} - \mathbf{r}') V(\mathbf{r}). \tag{2.11}$$

This can be derived by inspection or by multiplying Eq. (2.10) by $\psi^*(\mathbf{r}'')$ and using the completeness of the set of eigenfunctions of the Hamiltonian.

The matrix element $\langle \mathbf{r} | V | \mathbf{r}' \rangle$ may be thought of as a general nonlocal potential. As such it is usually denoted by $V(\mathbf{r}, \mathbf{r}')$. The potential $V(\mathbf{r})$ that we have used until now is the special case of a local potential. We can see from the operator formalism that a nonlocal potential is the most general potential that can be written; it is the spatial representative of an arbitrary operator. Exchange forces, velocity-dependent forces, and separable potentials are some familiar examples of nonlocal potentials.

We might note here one important property of nonlocal potentials. The operator V must be Hermitian in order for the energy eigenvalues to be real so that

$$\langle \mathbf{r} | V | \mathbf{r}' \rangle^* = \langle \mathbf{r}' | V^\dagger | \mathbf{r} \rangle = \langle \mathbf{r}' | V | \mathbf{r} \rangle, \tag{2.12}$$

or

$$V(\mathbf{r}, \mathbf{r}')^* = V(\mathbf{r}', \mathbf{r}). \tag{2.13}$$

The kinetic energy operator H_0 can be written in a form analogous to that of the potential operator:

$$\langle \mathbf{r} | H_0 | \mathbf{r}' \rangle = \delta(\mathbf{r} - \mathbf{r}')(-\hbar^2/2m)\nabla^2. \tag{2.14}$$

If this is allowed to act upon the wave function, it reproduces the kinetic energy term of the Schrödinger equation.

We can now write the Schrödinger equation as

$$E\langle \mathbf{r} | \psi \rangle - \langle \mathbf{r} | H_0 | \psi \rangle = \langle \mathbf{r} | V | \psi \rangle. \tag{2.15}$$

This is just the spatial representative of the operator equation

$$(E - H_0)\psi = V\psi \tag{2.16}$$

or

$$(E - H)\psi = 0. \tag{2.17}$$

This result looks deceptively simple, since in this particular case a similar form could have been obtained merely by suppressing the dependence on **r** in Eq. (2.1). The fact that this result could have been obtained so easily simply reflects the fact that the operator formalism leads to little additional insight or simplification for scattering by a single fixed potential. However, we have given a precise and quite general meaning to the symbols in this equation and will be able to use them later to discuss more general and complex problems.

INTEGRAL EQUATION FOR THE WAVE FUNCTION

The integral equation for $\psi_{\mathbf{k}}^{(+)}(\mathbf{r})$ contains two new elements. One is very simple and is just the coordinate representative of a state vector $\phi_{\mathbf{k}} = |\mathbf{k}\rangle$ for the initial unperturbed state:

$$\phi_{\mathbf{k}}(\mathbf{r}) = \langle \mathbf{r} | \phi_{\mathbf{k}} \rangle = \langle \mathbf{r} | \mathbf{k} \rangle = e^{i\mathbf{k} \cdot \mathbf{r}}. \tag{2.18}$$

The second element, the Green's function for the unperturbed Hamiltonian H_0, is more interesting.

We recall that the Green's function $G(\mathbf{r}, \mathbf{r}')$ describes the propagation of a particle from $\mathbf{r}'$ to $\mathbf{r}$. It is thus natural to represent it by an operator G which has the effect of transforming a state localized at $\mathbf{r}'$ into one localized at $\mathbf{r}$. This is a nonlocal operator, and its spatial representative is written

$$\langle \mathbf{r} | G | \mathbf{r}' \rangle = G(\mathbf{r}, \mathbf{r}'). \tag{2.19}$$

We now want to discuss a way of expressing the operator G in terms of a Hamiltonian H_0. The differential equation for the Green's function is

$$\{E + (\hbar^2/2m) \nabla^2\} G(\mathbf{r}, \mathbf{r}') = \delta(\mathbf{r} - \mathbf{r}'). \tag{2.20}$$

If we note that the delta function is the coordinate representative of the unit operator and use the definition of H_0 given by Eq. (2.14), we see that this is equivalent to the operator equation

$$(E - H_0)G = 1. \tag{2.21}$$

Thus formally the operator G is

$$G = (E - H_0)^{-1} = 1/(E - H_0), \tag{2.22}$$

provided that the inverse operator exists.

To examine whether this operator exists, let us consider G in the momentum representation. The states $|\mathbf{k}'\rangle$ are eigenstates of H_0 having energy

2. Operator Form of the Scattering Equations

$E' = \hbar^2 k'^2/2m$, so that
$$(E' - H_0)|\mathbf{k}'\rangle = 0. \tag{2.23}$$

In this representation G is

$$\langle \mathbf{k}''|G|\mathbf{k}'\rangle = \langle \mathbf{k}''|\frac{1}{E - H_0}|\mathbf{k}'\rangle = (2\pi)^3 \, \delta(\mathbf{k}'' - \mathbf{k}) \frac{1}{E - E'}. \tag{2.24}$$

Since the eigenvalues E' span all real positive numbers, the energy denominator can vanish when E is real. The right-handside of Eq. (2.24) is therefore singular and is undefined when $E = E'$.

To more closely define the inverse operator, we must specify the behavior of the inverse matrix in the neighborhood of the singularity. This can be done if we use the Fourier expansion of the Green's function discussed in Chapter 4. There we found that the outgoing-wave Green's function can be written

$$G^{(+)}(\mathbf{r}, \mathbf{r}') = (2m/\hbar^2)(2\pi)^{-3} \int d^3k' \, \frac{\exp[i\mathbf{k}' \cdot (\mathbf{r} - \mathbf{r}')]}{k^2 - k'^2 + i\varepsilon}. \tag{2.25}$$

In this and all subsequent formulas it is understood that the limit $\varepsilon \to 0$ is to be taken *after* the integral is performed. Written in the Dirac notation, this integral becomes

$$G^{(+)}(\mathbf{r}, \mathbf{r}') = (2m/\hbar^2)(2\pi)^{-3} \int d^3k' \, \frac{\langle \mathbf{r}|\mathbf{k}'\rangle \langle \mathbf{k}'|\mathbf{r}'\rangle}{k^2 - k'^2 + i\varepsilon} \tag{2.26}$$

or, equivalently,

$$G^{(+)}(\mathbf{r}, \mathbf{r}') = (2m/\hbar^2)(2\pi)^{-3} \int d^3k'$$
$$\times \frac{\langle \mathbf{r}|\mathbf{k}'\rangle \, \delta(\mathbf{k}' - \mathbf{k}'') \, d^3k'' \, \langle \mathbf{k}''|\mathbf{r}'\rangle}{k^2 - k''^2 + i\varepsilon}. \tag{2.27}$$

We now recognize that, as in Eq. (2.24), we can express the denominator as the matrix element of an operator:

$$\left(\frac{2m}{\hbar^2}\right) \frac{(2\pi)^3 \, \delta(\mathbf{k}' - \mathbf{k}'')}{k^2 - k''^2 + i\varepsilon} = \langle \mathbf{k}'|\frac{1}{E - H_0 + i\varepsilon}|\mathbf{k}''\rangle. \tag{2.28}$$

Then the Green's function is

$$G^{(+)}(\mathbf{r}, \mathbf{r}') = (2\pi)^{-6} \int \langle \mathbf{r}|\mathbf{k}'\rangle \, d^3k' \, \langle \mathbf{k}'|\frac{1}{E - H_0 + i\varepsilon}|\mathbf{k}''\rangle \, d^3k'' \, \langle \mathbf{k}''|\mathbf{r}'\rangle$$
$$= \langle \mathbf{r}|\frac{1}{E - H_0 + i\varepsilon}|\mathbf{r}'\rangle. \tag{2.29}$$

Our final well-defined result for the operator form of the outgoing-wave Green's function is therefore

$$G^{(+)} = \frac{1}{E - H_0 + i\varepsilon}. \qquad (2.30)$$

While this result was derived for the special case of the free-particle Green's function, it is easily applied to the more general case of a Green's function satisfying

$$\{E + (\hbar^2/2m)\nabla^2 - V(\mathbf{r})\} G(\mathbf{r}, \mathbf{r}') = \delta(\mathbf{r} - \mathbf{r}'). \qquad (2.31)$$

If this Green's function is expanded in terms of a set of properly normalized regular solutions of the equation

$$\{E + (\hbar^2/2m)\nabla^2 - V(\mathbf{r})\} \psi_\mathbf{k}(\mathbf{r}) = 0, \qquad (2.32)$$

we find the expansion

$$G(\mathbf{r}, \mathbf{r}') = (2\pi)^{-3} \int_C d^3k' \frac{\psi_{\mathbf{k}'}(\mathbf{r}) \psi_{\mathbf{k}'}^*(\mathbf{r}')}{E - E'}. \qquad (2.33)$$

This is identical with our earlier result, Eq. (4.50) of Chapter 4. The contour C depends upon the boundary conditions satisfied by the Green's function. For outgoing-wave boundary conditions the proper choice is, as in the free-particle case,

$$G^{(+)}(\mathbf{r}, \mathbf{r}') = (2\pi)^{-3} \int d^3k' \frac{\psi_{\mathbf{k}'}(\mathbf{r}) \psi_{\mathbf{k}'}^*(\mathbf{r}')}{E - E' + i\varepsilon}. \qquad (2.34)$$

The wave function $\psi_\mathbf{k}(\mathbf{r})$ is the spatial representative of the state vector $\psi_\mathbf{k}$, that is, $\psi_\mathbf{k}(\mathbf{r}) = \langle \mathbf{r} | \psi_\mathbf{k} \rangle$. Then Eq. (2.34) may be expressed as

$$G^{(+)}(\mathbf{r}, \mathbf{r}') = (2\pi)^{-3} \int d^3k' \frac{\langle \mathbf{r} | \psi_{\mathbf{k}'} \rangle \langle \psi_{\mathbf{k}'} | \mathbf{r}' \rangle}{E - E' + i\varepsilon}$$

$$= \langle \mathbf{r} | \frac{1}{E - H + i\varepsilon} | \mathbf{r}' \rangle, \qquad (2.35)$$

where $(E - H)\psi_\mathbf{k} = 0$ and the completeness relation

$$(2\pi)^{-3} \int d^3k | \psi_\mathbf{k} \rangle \langle \psi_\mathbf{k} | = 1 \qquad (2.36)$$

has been used in obtaining Eq. (2.35). Hence, in general, the outgoing-wave

2. Operator Form of the Scattering Equations

Green's function is represented by the operator

$$G^{(+)} = \frac{1}{E - H + i\varepsilon}. \tag{2.37}$$

This form makes explicit the energy, the Hamiltonian, and the boundary condition. This will prove especially useful in our later work where more complicated situations arise and this explicit labeling is most helpful.

The inverse operator appearing here or in Eq. (2.30) can be written formally as

$$\frac{1}{E - H_0 + i\varepsilon} = \mathscr{P} \frac{1}{E - H_0} - i\pi \delta(E - H_0), \tag{2.38}$$

where $\mathscr{P}$ denotes the Cauchy principal value. The momentum-space matrix element of this is identical with the result obtained in Chapter 4, Eq. (3.106), and illustrated in Fig. 4.5.

Whittaker & Watson P.102

We now have an operator for the Green's function and can write the state vector $\psi_\mathbf{k}^{(+)}$ in abstract operator language as

$$\psi_\mathbf{k}^{(+)} = \phi_\mathbf{k} + \frac{1}{E - H_0 + i\varepsilon} V \psi_\mathbf{k}^{(+)}. \tag{2.39}$$

This integral equation is, of course, equivalent to the Schrödinger equation but, as we saw in Chapter 5, it also incorporates the boundary condition on the wave function and provides a convenient starting point for approximate calculations.

Uses of the Integral Equation for the Scattering State Vector

The chief advantage of the operator technique is that manipulations on the wave function in this form are now algebraic rather than analytic, resulting in simplifications for complicated scattering problems. The state vector can be expanded in powers by V by simply iterating the algebraic equation, Eq. (2.39):

$$\psi_\mathbf{k}^{(+)} = \phi_\mathbf{k} + \frac{1}{E - H_0 + i\varepsilon} V \phi_\mathbf{k} + \frac{1}{E - H_0 + i\varepsilon} V \frac{1}{E - H_0 + i\varepsilon} V \phi_\mathbf{k} + \cdots. \tag{2.40}$$

This is just the Born expansion which was discussed in Chapter 5. Here we have obtained it by algebraic means. To test the convergence of the series it is still necessary to use ordinary analytic methods in a particular representation, but the derivation and manipulation of the series is very much easier in this abstract formalism.

178   6. OPERATOR FORMALISM IN TWO-PARTICLE SCATTERING THEORY

A formal solution to Eq. (2.39) can also be obtained by algebraic means. We use the algebraic identity

$$\frac{1}{E - H + i\varepsilon} - \frac{1}{E - H_0 + i\varepsilon} = \frac{1}{E - H + i\varepsilon} V \frac{1}{E - H_0 + i\varepsilon}, \quad (2.41)$$

which can be proved by algebraic manipulation, if one keeps in mind that V, H_0, and H do not commute and that the order of operators is important. If this identity is substituted into Eq. (2.39), we find that the right-hand side no longer contains $\psi_{\mathbf{k}}^{(+)}$ and is an explicit form for the scattering state vector:

$$\psi_{\mathbf{k}}^{(+)} = \phi_{\mathbf{k}} + \frac{1}{E - H + i\varepsilon} V \psi_{\mathbf{k}}^{(+)} - \frac{1}{E - H + i\varepsilon} V \frac{1}{E - H_0 + i\varepsilon} V \psi_{\mathbf{k}}^{(+)}$$

$$= \phi_{\mathbf{k}} + \frac{1}{E - H + i\varepsilon} V \psi_{\mathbf{k}}^{(+)} - \frac{1}{E - H + i\varepsilon} V (\psi_{\mathbf{k}}^{(+)} - \phi_{\mathbf{k}})$$

$$= \phi_{\mathbf{k}} + \frac{1}{E - H + i\varepsilon} V \phi_{\mathbf{k}}. \quad (2.42)$$

This is just the abstract form of Eq. (1.35) of Chapter 5, if $\chi_{\mathbf{k}}^{(+)}(\mathbf{r})$ in that equation is taken to be the free-particle wave function $\phi_{\mathbf{k}}(\mathbf{r})$. As we observed there, this formal "solution" brings us no closer to a solution of the scattering problem than we were before. The operator $(E - H + i\varepsilon)^{-1}$ is the Green's function for the total Hamiltonian and is as difficult to obtain as a solution of the original integral equation. Many algebraic manipulations are simplified by this form, however, and it frequently proves useful in theoretical derivations. We shall see one example shortly.

The Green's function $(E - H + i\varepsilon)^{-1}$ can be expanded in powers of V by using the binomial theorem or, equivalently, the identity given by Eq. (2.41). Thus

$$\frac{1}{E - H + i\varepsilon} = \frac{1}{E - H_0 + i\varepsilon} + \frac{1}{E - H_0 + i\varepsilon} V \frac{1}{E - H_0 + i\varepsilon}$$

$$+ \frac{1}{E - H_0 + i\varepsilon} V \frac{1}{E - H_0 + i\varepsilon} V \frac{1}{E - H_0 + i\varepsilon} + \cdots. \quad (2.43)$$

If this is substituted into Eq. (2.42), the Born expansion is again reproduced.

ORTHOGONALITY OF THE STATE VECTORS

These algebraic solutions permit us to obtain a simple proof of the orthogonality of the eigenstates $\psi_{\mathbf{k}}^{(+)}$. Let us for convenience denote the initial

2. Operator Form of the Scattering Equations

or unperturbed states by ϕ_i, where the label i symbolizes the momentum of the state (i.e., $i \equiv \mathbf{k}$). With this notation

$$\psi_i^{(+)} = \phi_i + (E_i - H_0 + i\varepsilon)^{-1} V \psi_i^{(+)} \tag{2.44}$$

or, equivalently,

$$\psi_i^{(+)} = \phi_i + (E_i - H + i\varepsilon)^{-1} V \phi_i. \tag{2.45}$$

We want to evaluate the scalar product $\langle \psi_i^{(+)} | \psi_j^{(+)} \rangle$.
For $\psi_i^{(+)}$ we use the explicit form, Eq. (2.45), so that

$$\langle \psi_i^{(+)} | \psi_j^{(+)} \rangle = \langle \phi_i | \psi_j^{(+)} \rangle + \langle \phi_i | V \frac{1}{E_i - H - i\varepsilon} | \psi_j^{(+)} \rangle. \tag{2.46}$$

In obtaining Eq. (2.46) we have used the fact that V and H are Hermitian, but that $(i\varepsilon)^\dagger = -i\varepsilon$. Since $H\psi_j^{(+)} = E_j \psi_j^{(+)}$, the second term is simply

$$\langle \phi_i | V \frac{1}{E_i - H - i\varepsilon} | \psi_j^{(+)} \rangle = \frac{1}{E_i - E_j - i\varepsilon} \langle \phi_i | V | \psi_j^{(+)} \rangle. \tag{2.47}$$

Using the implicit form for $\psi_j^{(+)}$, Eq. (2.44), the first term is

$$\langle \phi_i | \psi_j^{(+)} \rangle = \langle \phi_i | \phi_j \rangle + \langle \phi_i | \frac{1}{E_j - H_0 + i\varepsilon} V | \psi_j^{(+)} \rangle$$

$$= \langle \phi_i | \phi_j \rangle + \frac{1}{E_j - E_i + i\varepsilon} \langle \phi_i | V | \psi_j^{(+)} \rangle, \tag{2.48}$$

with $H_0 \phi_i = E_i \phi_i$. Substitution of these results into Eq. (2.46) yields, in the limit as $\varepsilon \to 0$, the orthogonality relation

$$\langle \psi_i^{(+)} | \psi_j^{(+)} \rangle = \langle \phi_i | \phi_j \rangle = \delta_{ij}. \tag{2.49}$$

It may appear that, when $E_i = E_j$, the cancellation between Eq. (2.47) and the second term of Eq. (2.48) occurs only if the infinitesimal ε has the same value in each term. Since the limit $\varepsilon \to 0$ is always understood in every independent term, such a condition would be meaningless. However, the time-dependent definition of ψ developed in Chapter 8 can be used to show that this result is correct for all values of E_i and E_j. For instance, one need merely write the scattered wave in terms of exponential operators as in Chapter 8:

$$\frac{1}{E_j - H_0 + i\varepsilon} V | \psi_j^{(+)} \rangle = (i/\hbar) \int_{-\infty}^{0} dt \, \exp[(\varepsilon/\hbar)t] \exp[-(i/\hbar)H_0 t]$$

$$\times V \exp[(i/\hbar)Ht] | \psi_j^{(+)} \rangle \tag{2.50}$$

and similarly

$$\frac{1}{E_i - H + i\varepsilon} V |\phi_i\rangle = (i/\hbar) \int_{-\infty}^{0} dt \, \exp[(\varepsilon/\hbar)t] \exp[-(i/\hbar)Ht]$$
$$\times V \exp[(i/\hbar)H_0 t] |\phi_i\rangle. \quad (2.51)$$

Taking the adjoint of Eq. (2.51), we see that the cancellation is unambiguous.

THE WAVE MATRIX

In many applications the algebraic manipulations are simplified and made more transparent by the introduction of an operator to represent the effects of the scattering process upon the state vector. A scattering experiment begins with the initial state ϕ_i, which then develops under the influence of the scattering potential V. We would thus like an operator that acts upon the state ϕ_i to create the scattering state $\psi_i^{(+)}$.

Let us then define a wave operator or wave matrix $\Omega^{(+)}$ so that

$$\psi_i^{(+)} = \Omega^{(+)} \phi_i. \quad (2.52)$$

The wave operator $\Omega^{(+)}$ transforms a free-particle state ϕ_i into a scattering state $\psi_i^{(+)}$. In the spatial representation it is a nonlocal function $\langle \mathbf{r} | \Omega^{(+)} | \mathbf{r}' \rangle$ which produces the wave function $\psi_i^{(+)}(\mathbf{r})$ by the integral operation

$$\psi_i^{(+)}(\mathbf{r}) = \int d^3 r' \, \langle \mathbf{r} | \Omega^{(+)} | \mathbf{r}' \rangle \phi_i(\mathbf{r}'). \quad (2.53)$$

It describes the change in the relative two-particle wave function due to the mutual interaction $V(\mathbf{r})$.

The wave matrix $\Omega^{(+)}$ satisfies the operator equation

$$\Omega^{(+)} = 1 + \frac{1}{E - H_0 + i\varepsilon} V \Omega^{(+)}, \quad (2.54)$$

as we see immediately from Eq. (2.39). From Eq. (2.42) an explicit formula for $\Omega^{(+)}$ is

$$\Omega^{(+)} = 1 + \frac{1}{E - H + i\varepsilon} V. \quad (2.55)$$

At this point the reader may be wondering whether this is a legitimate way of describing the scattering process, since the energy E in these formulas refers to a particular state ϕ_i, while an operator should apply to the entire

2. Operator Form of the Scattering Equations

vector space. Indeed, the quantity defined by Eq. (2.54) or (2.55) gives a proper state vector only when acting on one particular set of eigenstates of H_0, namely, those having an energy E. In practice this turns out to be satisfactory since $\Omega^{(+)}$ always acts on an unperturbed state ϕ_i. If it were to act on an arbitrary state that is not an eigenstate of H_0, this definition would of course not be appropriate.

A more general definition can be obtained by using exponential operators. The wave matrix may be defined by

$$\Omega^{(+)} = 1 - (i/\hbar) \int_{-\infty}^{0} dt \, \exp[(\varepsilon/\hbar)t] \exp[(i/\hbar)Ht] \, V \exp[-(i/\hbar)H_0 t]. \quad (2.56)$$

We shall see in Chapter 8 that this definition of $\Omega^{(+)}$ is obtained directly in the time-dependent formalism. If we let this operator act on a state ϕ_i and perform the integration, we obtain our previous result, Eq. (2.55). However, this operator does not refer to any particular state and can act legitimately on any vector in the space. In most cases it is more convenient to use the first definition.

Since the set of vectors $\psi_i^{(+)}$ satisfies the same orthonormality relation as the unperturbed set ϕ_i, we might expect $\Omega^{(+)}$ to be a unitary operator. However, this is not so in general. To see this, let us recall that an operator can be characterized by its effect upon each vector in some complete set. Thus we can write the wave matrix as

$$\Omega^{(+)} = \sum_j |\psi_j^{(+)}\rangle \langle \phi_j|. \quad (2.57)$$

This expression, together with the orthonormality condition $\langle \phi_i | \phi_j \rangle = \delta_{ij}$, implies that

$$\Omega^{(+)} \phi_i = \sum_j |\psi_j^{(+)}\rangle \langle \phi_j | \phi_i \rangle = \psi_i^{(+)}. \quad (2.58)$$

so that Eq. (2.57) can be used in place of our earlier expressions for $\Omega^{(+)}$.

Now the adjoint of the wave matrix is

$$\Omega^{(+)\dagger} = \sum_j |\phi_j\rangle \langle \psi_j^{(+)}|, \quad (2.59)$$

and the product is

$$\Omega^{(+)\dagger} \Omega^{(+)} = \sum_{j,k} |\phi_j\rangle \langle \psi_j^{(+)} | \psi_k^{(+)} \rangle \langle \phi_k| = \sum_j |\phi_j\rangle \langle \phi_j| = 1, \quad (2.60)$$

using Eq. (2.49) and the completeness of the set of free-particle states. On the other hand, the product in the reverse order is

$$\Omega^{(+)}\Omega^{(+)\dagger} = \sum_{j,k} |\psi_j^{(+)}\rangle \langle \phi_j | \phi_k \rangle \langle \psi_k^{(+)}| = \sum_j |\psi_j^{(+)}\rangle \langle \psi_j^{(+)}|, \qquad (2.61)$$

which is not necessarily equal to 1. The vectors $\psi_j^{(+)}$ represent only the scattering states of the system. If there are bound states, these are not included in the sum of Eq. (2.61) since only unbound states can develop from initial free-particle states. Then if the bound states are denoted by ψ_b,

$$\Omega^{(+)}\Omega^{(+)\dagger} = 1 - \sum_b |\psi_b\rangle \langle \psi_b|, \qquad (2.62)$$

and $\Omega^{(+)}$ is unitary only if H has no bound states.

THE SCATTERING AMPLITUDE AND THE T MATRIX

The previous results enable us immediately to write the scattering amplitude in operator language. If the initial state is ϕ_i and the final state is ϕ_j, we have

$$f(\theta) = -(2m/4\pi\hbar^2) \langle \phi_j | V | \psi_i^{(+)} \rangle. \qquad (2.63)$$

The transition amplitude is just

$$T(\mathbf{k}', \mathbf{k}) = -(4\pi\hbar^2/2m) f(\theta) = \langle \phi_j | V | \psi_i^{(+)} \rangle. \qquad (2.64)$$

We can use the wave matrix to write this as

$$T(\mathbf{k}', \mathbf{k}) = \langle \phi_j | V\Omega^{(+)} | \phi_i^{(+)} \rangle, \qquad (2.65)$$

so that the transition amplitude is simply the matrix element of the operator $V\Omega^{(+)}$ between free-particle states. We then denote the transition operator by

$$T = V\Omega^{(+)}, \qquad (2.66)$$

so that

$$T\phi_i = V\psi_i^{(+)}. \qquad (2.67)$$

The transition amplitude is just the matrix element of this operator:

$$T(\mathbf{k}', \mathbf{k}) = \langle \phi_j | T | \phi_i \rangle. \qquad (2.68)$$

One often discusses scattering processes in terms of the "matrix element for the process," meaning the matrix element of the transition operator between unperturbed states ϕ_i. The complete matrix involving transitions between all possible initial and final states is the T matrix. (Often the operator T itself is called the T matrix.) We emphasize that the T matrix gives directly the scattering amplitude whose determination is the goal of scattering theory.

The transition operator T is, of course, closely related to the potential operator V. If Eqs. (2.54) and (2.55) for the wave matrix are introduced into the definition of T, we find that the transition operator satisfies the operator equations

$$T = V + V \frac{1}{E - H_0 + i\varepsilon} T \qquad (2.69)$$

or

$$T = V + V \frac{1}{E - H + i\varepsilon} V. \qquad (2.70)$$

Eq. (2.6) is just (2.69) written in the momentum representation, so we have completed our task of writing Eqs. (2.1)–(2.7) in operator notation.

In the first Born approximation the wave operator $\Omega^{(+)}$ is just the unit operator, so that $T^{(\text{Born})} = V$. The exact solution for $\Omega^{(+)}$ introduces the modifications that transform ϕ_i into the exact solution $\psi_i^{(+)}$. Hence T contains both the potential and the alterations in the wave function introduced by the potential. In coordinate space it is a nonlocal complex function $\langle \mathbf{r} | T | \mathbf{r}' \rangle$, usually called a pseudopotential.

Superficially, the matrix element $\langle \phi_j | T | \phi_i \rangle$ has a form identical with that obtained by treating an ordinary potential in Born approximation, but it must be emphasized that the pseudopotential gives the *exact* scattering amplitude. Because the T matrix includes within itself the wave function modifications, no further changes need be made in the wave function to obtain exact results. This then provides a convenient formulation of the scattering problem, since all operators can act solely on free-particle states to yield the correct wave functions through operator transformations. It also is a convenient way of handling many-particle problems, especially when singular potentials such as hard cores are present. In these cases it is often simpler to make the appropriate modifications in the two-body potentials (changing V into T) than to attempt to introduce them into the many-particle wave function.

3. The Optical Theorem

Before leaving the two-body problem, we will show how the optical theorem is contained in these algebraic equations. This theorem relates the imaginary part of the scattering or transition amplitude to its absolute square. The imaginary part of $f(\theta)$ is one half the difference between $f(\theta)$ and $f^*(\theta)$;

6. OPERATOR FORMALISM IN TWO-PARTICLE SCATTERING THEORY

the operator analog of this is half the difference between T and $T^\dagger$. Using Eq. (2.69) we have

$$T^\dagger = V + T^\dagger \frac{1}{E - H_0 - i\varepsilon} V. \tag{3.1}$$

If we use Eq. (2.69) to substitute for V in the second term, we have

$$T^\dagger = V + T^\dagger \frac{1}{E - H_0 - i\varepsilon}\left(T - V \frac{1}{E - H_0 + i\varepsilon} T\right). \tag{3.2}$$

Using Eq. (3.1) similarly to substitute for V in Eq. (2.69), we find the result

$$T - T^\dagger = T^\dagger \left(\frac{1}{E - H_0 + i\varepsilon} - \frac{1}{E - H_0 - i\varepsilon}\right) T. \tag{3.3}$$

From Eq. (2.38) we obtain

$$\left(\frac{1}{E - H_0 + i\varepsilon} - \frac{1}{E - H_0 - i\varepsilon}\right) = -2\pi i\, \delta(E - H_0), \tag{3.4}$$

so that this is

$$T - T^\dagger = -2\pi i\, T^\dagger\, \delta(E - H_0)\, T. \tag{3.5}$$

Alternatively, from Eq. (2.70) we derive

$$T - T^\dagger = -2\pi i\, V\, \delta(E - H)\, V. \tag{3.6}$$

Either of these is the operator form of the optical theorem, which is then a simple algebraic consequence of the Schrödinger equation in integral form.

To see that Eq. (3.5), for instance, does indeed yield the optical theorem, we must use the representation of the operator $\delta(E - H_0)$ in the eigenstates of H_0. Using $\langle \mathbf{k}' | \delta(E - H_0) | \mathbf{k}'' \rangle = \langle \mathbf{k}' | \mathbf{k}'' \rangle \delta(E - E')$, we have

$$\delta(E - H_0) = (2\pi)^{-3} \int |\mathbf{k}'\rangle d^3k'\, \delta(E - E') \langle \mathbf{k}' |. \tag{3.7}$$

If we take the diagonal matrix element of Eq. (3.5) and use

$$\langle \mathbf{k}' | T^\dagger | \mathbf{k} \rangle \equiv \langle \mathbf{k} | T | \mathbf{k}' \rangle^*, \tag{3.8}$$

we find

$$\begin{aligned}\mathrm{Im}\,\langle \mathbf{k} | T | \mathbf{k} \rangle &= -\pi \langle \mathbf{k} | T^\dagger\, \delta(E - H_0)\, T | \mathbf{k} \rangle \\ &= -\pi(2\pi)^{-3} \int d^3k'\, \langle \mathbf{k} | T^\dagger | \mathbf{k}' \rangle \delta(E - E') \langle \mathbf{k}' | T | \mathbf{k} \rangle.\end{aligned} \tag{3.9}$$

3. The Optical Theorem

The energy-momentum relation implies that

$$d^3k' = k'^2 \, dk' \, d\Omega' = (m/\hbar^2) k' \, dE' \, d\Omega'. \tag{3.10}$$

Inserting this into Eq. (3.9) and using the fact that

$$\langle \mathbf{k} | T^\dagger | \mathbf{k}' \rangle \equiv \langle \mathbf{k}' | T | \mathbf{k} \rangle^* = -(4\pi\hbar^2/2m) \, f^*(\theta) \tag{3.11}$$

when $E = E'$, we obtain the optical theorem

$$(4\pi/k) \operatorname{Im} f(0) = \int d\Omega \, |f(\theta)|^2. \tag{3.12}$$

A generalization of the optical theorem, involving nonforward directions, can be derived if time reversal and space reflection invariance are assumed. As shown in Chapter 10, these imply that

$$\langle \mathbf{k}' | T^\dagger | \mathbf{k} \rangle \equiv \langle \mathbf{k} | T | \mathbf{k}' \rangle^* = \langle \mathbf{k}' | T | \mathbf{k} \rangle^*. \tag{3.13}$$

We then have from Eq. (3.5)

$$\operatorname{Im} \langle \mathbf{k}' | T | \mathbf{k} \rangle$$
$$= -\pi(2\pi)^{-3} \int d^3k'' \, \langle \mathbf{k}' | T | \mathbf{k}'' \rangle^* \delta(E - E'') \langle \mathbf{k}'' | T | \mathbf{k} \rangle, \tag{3.14}$$

from which the generalized theorem follows immediately:

$$(4\pi/k) \operatorname{Im} f(\theta_{\mathbf{k}'\mathbf{k}}) = \int d\Omega_{\mathbf{k}''} \, f^*(\theta_{\mathbf{k}'\mathbf{k}''}) f(\theta_{\mathbf{k}''\mathbf{k}}). \tag{3.15}$$

We can see that the origin of the optical theorem, as in Chapters 1 and 2, is probability conservation. Suppose we construct a scattering operator S, commonly called the S matrix and defined by

$$S = 1 - 2\pi i \, \delta(E - H_0) \, T. \tag{3.16}$$

Then the operator product $S^\dagger S$ is

$$S^\dagger S = (1 + 2\pi i \, \delta(E - H_0) \, T^\dagger)(1 - 2\pi i \, \delta(E - H_0) \, T)$$
$$= 1 - 2\pi i \, \delta(E - H_0)(T - T^\dagger + 2\pi i \, T^\dagger \, \delta(E - H_0) \, T) = 1 \tag{3.17}$$

by virtue of the optical theorem, Eq. (3.5). Similarly, $SS^\dagger = 1$, and the operator S is a unitary operator. We saw earlier that unitarity implies the conservation of normalization or of probability. Therefore, we have an operator S which creates a state $S\phi_i$ having, as a result of the optical theorem, the same normalization as ϕ_i. It is of interest to ask what is the nature of the state $S\phi_i$.

Since this state clearly is related to the scattering process, one might think that it is $\psi_i^{(+)}$, the time-independent scattering state. But we have already seen that $\psi_i^{(+)} = \Omega^{(+)}\phi_i$, and $\Omega^{(+)}$ is a quite different operator from S. This difference can give us a clue, though. If we compare $S\phi_i$ with $\Omega^{(+)}\phi_i$, we see one striking difference: $\Omega^{(+)}\phi_i$ in general has a nonzero overlap with all states ϕ_j, while $S\phi_i$ has a projection only onto unperturbed states having an energy E_i. Thus S is an operator that couples only those eigenstates of H_0 that have equal energies.

If we view scattering as a process in which different momentum states, that is, different eigenstates of H_0 are mixed, we see that during a scattering process the wave function acquires, in general, all momentum components. However, we know that long after the scattering process has ended, the outgoing system must have an energy equal to the initial energy of the system. Over short time spans the uncertainty principle permits energy fluctuations due to the interaction, but for long time spans energy must be conserved. This suggests that the energy-conserving state $S\phi_i$ is related to the behavior of the system long after the scattering has occurred. We shall show in Chapter 7 that $S\phi_i$ is in fact the limit as $t \to \infty$ of the time-dependent state vector describing the scattering process. This limit is of course the one which is of interest in a scattering experiment, so that S contains all the information obtainable from such an experiment. We shall learn more about this important operator in succeeding chapters.

The formalism developed in this chapter will be used extensively in later chapters, especially in deriving general formulas for scattering amplitudes and cross sections. However, we shall feel free to use whatever notation is most convenient in each particular situation. While the Dirac formalism is convenient for general derivations, the physical content in some particular applications is most clearly shown in the usual coordinate representation. We will adopt the most convenient formulation in each case, in order to illustrate the flexibility of the scattering formalism and to emphasize the freedom that is available to select different descriptions in different cases.

Chapter

7

Cross Sections for General Collision Processes

We have seen how the elastic scattering of a single particle by a center of force, or of two particles by their mutual interaction, can be described in terms of differential and integral equations, and we have written these equations in operator language. In this chapter we shall apply the operator formalism to more general collision processes.

All collision processes initially involve a projectile and a target. The target may be a single elementary particle, an atom, a complex molecule, or an atomic crystal. In all cases the general principles describing the collision process are the same. Suppose the target is composed of two particles denoted by 2 and 3, while the projectile is a single particle denoted by 1. Some of the possible collision processes are:

$1 + (2, 3) \rightarrow 1 + (2, 3)$ Elastic scattering   ⎫ Ordinary
$ \rightarrow 1 + (2, 3)^*$ Inelastic scattering   ⎬ scattering
                               (target remains in an   ⎭ processes.
                                 excited state)

$1 + (2, 3) \rightarrow 2 + (1, 3)$ Exchange or pickup reaction   ⎫ Rearrangement
$ \rightarrow 3 + (1, 2)$ Exchange or pickup reaction   ⎬ processes.
$ \rightarrow 1 + 2 + 3$ Breakup reaction   ⎭

It will prove convenient to refer to those scattering processes in which the incident particles are the same as the scattered particles as ordinary scattering, while those processes in which the incident and scattered particles differ will be referred to as rearrangement processes.

There is a basic difference between ordinary scattering and the rearrangement processes, a difference that suggests a separate theoretical development for each. In elastic and inelastic scattering the final state and the initial state are eigenstates of the same free-particle Hamiltonian H_0, whereas in rearrangement collisions the final state is a solution for a different unperturbed Hamiltonian. For instance, in the cases listed above the initial unperturbed Hamiltonian H_0 is the sum of the three kinetic energy operators plus the interaction potential V_{23}, while the final-state Hamiltonian for exchange scattering is the sum of the three kinetic operators plus V_{13}. In the first section of this chapter we shall treat elastic and inelastic scattering. A discussion of rearrangement collisions will appear in the later sections of this chapter.

1. General Scattering Formalism for Ordinary Scattering Processes

Our first task is to generalize the formalism of the previous chapter to permit it to handle complex particles. It will quickly become apparent that little generalization is required, since the operator formalism used earlier can be directly applied to most situations. We then review the central results of Chapter 6, keeping in mind the more general meaning to be given to the quantities introduced there.

DEFINITIONS

Any scattering experiment begins with two separated systems which we shall, for convenience, call *particles*. Of course these particles may be complex systems; for example, they may be electrons, nucleons, nuclei, atoms, molecules, entire crystal lattices, or amorphous solids. When the two systems are widely separated, they are not under the influence of their mutual finite-ranged interaction. Hence the two particles may be considered to be in an eigenstate of an unperturbed or free-particle Hamiltonian Hg. The unperturbed Hamiltonian H_0 is of course, just the total Hamiltonian minus the mutual interaction between the two systems, represented as before by the operator V. It may be a central two-body force as we have discussed earlier, or it may be nonlocal or spin dependent. It may be the sum of interactions between the constituents of each particle (as in electron-atom scattering, when it is the sum of Coulomb forces between the incident electron and the constituents of the target atom), or it may contain complicated many-body forces between these constituents. It may create or annihilate particles as, for instance, photons or mesons.

The initial separated particles are described by a free-particle state vector ϕ_i, where the index i describes not only the relative momentum of the two

1. General Scattering Formalism

particles but also their spin orientation, energy eigenvalues (whether they are in their ground states or in some excited state), and any other quantum numbers needed to specify the initial states of the two particles. The free-particle state is an eigenstate of the Hamiltonian H_0 satisfying

$$(E_i - H_0)\phi_i = 0, \tag{1.1}$$

where the eigenvalues E_i form a continuous spectrum.

The total Hamiltonian for the interacting system is

$$H = H_0 + V, \tag{1.2}$$

and the exact time-independent solution of the scattering problem satisfies the equation

$$(E_i - H)\psi_i = 0. \tag{1.3}$$

In this case, however, E_i may include discrete energies for which the two particles can form bound states, in addition to the continuum values. As in the previous chapter the formal solution to this equation satisfying the boundary condition that it asymptotically contain the free-particle state ϕ_i plus only outgoing waves is

$$\psi_i^{(+)} = \phi_i + \frac{1}{E_i - H_0 + i\varepsilon} V\psi_i^{(+)}. \tag{1.4}$$

This is just the integral form of the differential equation, subject to the outgoing-wave scattering boundary condition. It will be recalled from Chapter 6 that $+i\varepsilon$ defines the singularity of the Green's function and goes to zero after any integral has been performed.

ASYMPTOTIC FORM OF THE STATE VECTOR

The use of the outgoing-wave boundary condition was justified for the case of two structureless particles in Chapter 2, where we saw that this guaranteed the correct form of the wave function before and after the scattering takes place. In Chapter 6 we showed that the use of $+i\varepsilon$ yields outgoing waves. Here we shall use the operator formalism to justify the use of this prescription in the general case of the scattering of two arbitrarily complex particles.

To accomplish this we return briefly to a description of the scattering process in terms of wave packets. This permits us to localize the particles at a given time and to ask whether the outgoing-wave condition guarantees in general that the wave packet becomes, for very early times, a free-particle state having no scattered wave.

In constructing the wave packet we will, as in Chapter 2, use a linear superposition of states having different initial momenta. These eigenstates will be identified by the initial momentum $\mathbf{k}$ and by a complete set of internal variables denoted by α; thus, $i \equiv (\mathbf{k}, \alpha)$. The state vector describing the scattering process is postulated to be of the form

$$\psi(t) = (2\pi)^{-3} \int d^3k \, A(\mathbf{k}) \, \psi_{\mathbf{k},\alpha}^{(+)} \exp[-(i/\hbar)E_k t]. \tag{1.5}$$

Naturally, this must satisfy the time-dependent Schrödinger equation

$$i\hbar \frac{\partial \psi(t)}{\partial t} = H\psi(t). \tag{1.6}$$

The wave packet of Chapter 2, Eq. (2.6), is the spatial representation of this state for the special case of two structureless particles. For the general case of the scattering of two complex particles, we want to show that, with $\psi_{\mathbf{k},\alpha}^{(+)}$ given by Eq. (1.4), the scattered wave vanishes as $t \to -\infty$. We will then show how the scattering cross section may be obtained for this general case.

The state vector describing the incident beam is

$$\phi(t) = (2\pi)^{-3} \int d^3k \, A(\mathbf{k}) \, \phi_{\mathbf{k},\alpha} \exp[-(i/\hbar)E_k t]. \tag{1.7}$$

Using the assumed form of $\psi_{\mathbf{k},\alpha}^{(+)}$, the scattered wave is then

$$\psi_{\text{sc}}(t) = (2\pi)^{-3} \int d^3k \, A(\mathbf{k}) \frac{1}{E_k - H_0 + i\varepsilon} V \psi_{\mathbf{k},\alpha}^{(+)} \exp[-(i/\hbar)E_k t]. \tag{1.8}$$

As always, the limit $\varepsilon \to 0$ is taken after the integral is performed. In performing this integral the operator H_0 may be treated as a real number. This can be justified by introducing a complete set of eigenstates, $\phi_{\mathbf{k}',\alpha'}$, so that

$$\frac{1}{E_k - H_0 + i\varepsilon} = (2\pi)^{-3} \sum_{\alpha'} \int d^3k' \frac{1}{E_k - H_0 + i\varepsilon} | \phi_{\mathbf{k}',\alpha'} \rangle \langle \phi_{\mathbf{k}',\alpha'} |$$

$$= (2\pi)^{-3} \sum_{\alpha'} \int d^3k' \, | \phi_{\mathbf{k}',\alpha'} \rangle \frac{1}{E_k - E_{k'} + i\varepsilon} \langle \phi_{\mathbf{k}',\alpha'} |, \tag{1.9}$$

and interchanging the orders of integration in Eq. (1.8). To emphasize the energy dependence of the integrand, we write the scattered wave in the form

$$\psi_{\text{sc}}(t) = \int_0^\infty dE_k \frac{1}{E_k - H_0 + i\varepsilon} \gamma(E_k) \exp[-(i/\hbar)E_k t], \tag{1.10}$$

where

$$\gamma(E_k) = (2\pi)^{-3} \int d\Omega_k \, k^2 \, \frac{dk}{dE_k} \, A(\mathbf{k}) \, V\psi^{(+)}_{\mathbf{k},\alpha}. \qquad (1.11)$$

Because of the presence of the wave packet amplitude $A(\mathbf{k})$, the energy-dependent state vector $\gamma(E_k)$ vanishes for large values of E_k.

Let us assume for the moment that $\gamma(E_k)$ is a smoothly varying function of E_k. The energy integral can then be performed by closing the contour along an arc in the upper half of the complex energy plane and down the imaginary axis (Fig. 7.1). In the limit $t \to -\infty$ there will be no contribution

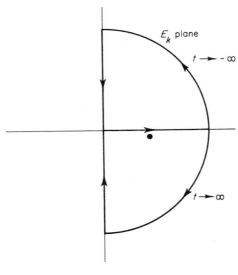

FIG. 7.1

from this path since the exponential $\exp[-(i/\hbar)E_k t]$ is vanishingly small everywhere on it. Since the only singularity is in the lower half plane, the integral vanishes. Thus we find

$$\lim_{t \to -\infty} \psi_{sc}(t) = 0. \qquad (1.12)$$

We see that the use of $+i\varepsilon$ to define the outgoing-wave Green's function is a completely general device for ensuring that the scattered wave vanishes for very early times.

In deriving this result we have assumed that $\gamma(E_k)$ varies smoothly on the real axis and is not too large along the contour in the upper half plane. It might have singularities in the upper half plane but, because of the exponential factor $\exp[-(i/\hbar)E_k t]$, their contributions would vanish as $t \to -\infty$. The amplitude $A(\mathbf{k})$ can be chosen to be nonsingular for real energies and to give only a finite contribution along the contour in the upper half plane. The state vector $\psi_{\mathbf{k},\alpha}^{(+)}$, whose spatial representation is just the scattering wave function, varies smoothly on the real axis for "normal" scattering potentials. Furthermore, an examination of $\gamma(E_k)$ using the spatial representation shows that the vector $V\psi_{\mathbf{k},\alpha}^{(+)}$ will not dominate the time-dependent exponential on this portion of the path, provided that the potential has a finite range. For instance, the wave function contains a term of the form $e^{i\mathbf{k}\cdot\mathbf{r}}$ which will not dominate the exponential $\exp[-(i/\hbar)E_k t]$ so long as the distance $|\mathbf{r}|$ remains finite.

At very late times ($t \to \infty$) the contour can be closed in the lower half plane (Fig. 7.1). In this case the integrand does have a singularity within the contour—otherwise there would be no scattered wave at all—and the total contribution comes from the residue at this singularity. As $\varepsilon \to 0$, this singularity moves onto the real axis; the contributions of any singularities in the lower half plane would vanish as $t \to \infty$. Performing this integral, we find

$$\psi_{sc}(t) \xrightarrow[t \to \infty]{} -2\pi i (2\pi)^{-3} \int d^3k\, A(\mathbf{k})\, \delta(E_k - H_0)\, V\psi_{\mathbf{k},\alpha}^{(+)} \exp[-(i/\hbar)E_k t] \quad (1.13)$$

or

$$\psi(t) \xrightarrow[t \to \infty]{} (2\pi)^{-3} \int d^3k\, A(\mathbf{k})\, \{\phi_{\mathbf{k},\alpha} - 2\pi i\, \delta(E_k - H_0)\, V\psi_{\mathbf{k},\alpha}^{(+)}\} \exp[-(i/\hbar)E_k t]. \quad (1.14)$$

Thus at large times only terms having the same energy as the initial state will contribute. This is simply an illustration of the uncertainty principle $\Delta E \Delta t \sim \hbar$.

The state vector in brackets may be written

$$\phi_{\mathbf{k},\alpha} - 2\pi i\, \delta(E_k - H_0)\, V\psi_{\mathbf{k},\alpha}^{(+)} = S\phi_{\mathbf{k},\alpha}, \quad (1.15)$$

where the scattering operator or S matrix is a generalization of that discussed in the previous chapter,

$$S = 1 - 2\pi i\, \delta(E - H_0)\, T. \quad (1.16)$$

As in Chapter 6, E is the energy of the free-particle state to its right. The

1. General Scattering Formalism

transition operator or T matrix can be defined by the same equation as in the previous chapter:

$$T = V + V \frac{1}{E - H_0 + i\varepsilon} T$$

$$= V + V \frac{1}{E - H + i\varepsilon} V, \qquad (1.17)$$

but here V can be a general interaction operator. As before, the T matrix is related to the state vector through the wave matrix by

$$T = V\Omega^{(+)}, \qquad (1.18)$$

where

$$\psi_i^{(+)} = \Omega^{(+)} \phi_i \qquad (1.19)$$

and

$$\Omega^{(+)} = 1 + \frac{1}{E - H_0 + i\varepsilon} V\Omega^{(+)}$$

$$= 1 + \frac{1}{E - H + i\varepsilon} V. \qquad (1.20)$$

Thus all the operators defined in the previous chapter carry over with identical definitions to the more general case.

CALCULATION OF THE CROSS SECTION

We now use Eq. (1.14) to compute the cross section for a general scattering problem. In Chapter 2 we used the spatial representation of the scattering state to compute the outgoing flux of particles. We cannot compute the spatial representation without introducing a specific form for H_0, and hence for the Green's function, so this technique is not available to us here. Instead of using the spatial representation, we can ask for the scalar products of $\psi(t)$ with various momentum eigenstates $\phi_{\mathbf{k}',\alpha'}$. This will give the same information because we know that a particle moving away from the scatterer in the direction $\mathbf{r}$ has a momentum in the same direction; that part of the outgoing wave that passes through a detector located at $\mathbf{r} = (r, \theta, \phi)$ must have a wave number $\mathbf{k}' = (k', \theta, \phi)$.

If we are going to specify the outgoing wave by its momentum, we must be careful to describe the experiment in a way that is independent of the form of the initial packet. If we ask for the probability that the outgoing wave has a precise momentum, the result will depend on $A(\mathbf{k})$, the packet amplitude. To avoid this we use a detector that measures the probability that the final momentum lies within a range $\Delta k'$ that is much larger than the momentum spread within the packet itself. This assures that the detector will not distinguish between different states within the incident packet.

We first evaluate the overlap of a free-particle eigenstate of momentum $\mathbf{k}'$ with the scattering state $\psi(t)$ in the limit as $t \to \infty$. We also look for the cross section in a direction other than that of the initial beam. If $\mathbf{k}'$ lies outside the momentum range in the initial packet (that is, if $A(\mathbf{k}') = 0$), there will be no overlap with the unscattered component of the final state. Taking the overlap with the stationary state $\phi_{\mathbf{k}',\alpha'} \exp[-(i/\hbar)E_{k'} t]$, we then obtain

$$\lim_{t \to \infty} \exp[(i/\hbar)E_{k'} t] \langle \phi_{\mathbf{k}',\alpha'} | \psi(t) \rangle = (2\pi)^{-3} \int d^3k\, A(\mathbf{k})\, (-2\pi i)\, \delta(E_k - E_{k'})$$
$$\times \langle \phi_{\mathbf{k}',\alpha'} | V | \psi_{\mathbf{k},\alpha}^{(+)} \rangle \quad (1.21)$$

using $H_0 \phi_{\mathbf{k}',\alpha'} = E_{k'\alpha} \phi_{\mathbf{k}',\alpha'}$. (Note that scattering into a state having $\alpha' \neq \alpha$ corresponds to inelastic scattering or spin-flip scattering.) As in Chapter 2 we will assume that the transition amplitude $\langle \phi_{\mathbf{k}',\alpha'} | V | \psi_{\mathbf{k},\alpha}^{(+)} \rangle$ does not vary appreciably over the range of momenta in the packet. Hence, if $\mathbf{k}_0$ is a momentum contained within the packet,

$$\lim_{t \to \infty} \exp[(i/\hbar)E_{k'} t] \langle \phi_{\mathbf{k}',\alpha'} | \psi(t) \rangle \simeq \langle \phi_{\mathbf{k}',\alpha'} | V | \psi_{\mathbf{k}_0,\alpha}^{(+)} \rangle (-2\pi i)$$
$$\times (2\pi)^{-3} \int d^3k\, A(\mathbf{k})\, \delta(E_k - E_{k'}). \quad (1.22)$$

The probability that the scattered particle has a momentum in the range $\Delta k'$ is obtained from the absolute square of Eq. (1.22) integrated over the momentum range $\Delta k'$:

$$P_{\Delta k'} = \lim_{t \to \infty} (2\pi)^{-3} \int_{\Delta k'} d^3 k' \, | \exp[(i/\hbar)E_{k'} t] \langle \phi_{\mathbf{k}',\alpha'} | \psi(t) \rangle |^2$$
$$= (2\pi)^{-3} \int_{\Delta k'} d^3 k' \, | \langle \phi_{\mathbf{k}',\alpha'} | V | \psi_{\mathbf{k}_0,\alpha}^{(+)} \rangle |^2 (2\pi)^2 \int \frac{d^3 k_1}{(2\pi)^3} \frac{d^3 k_2}{(2\pi)^3}$$
$$\times A(\mathbf{k}_1) A^*(\mathbf{k}_2)\, \delta(E_{k_1} - E_{k'})\, \delta(E_{k_2} - E_{k'}). \quad (1.23)$$

The product of delta functions in Eq. (1.23) can be rewritten as

$$\delta(E_{k_1} - E_{k'})\, \delta(E_{k_2} - E_{k'}) = \delta(E_{k'} - E_{k_1})\, \delta(E_{k_1} - E_{k_2}). \quad (1.24)$$

The first delta function requires that the packet energy E_{k_1} equal an energy $E_{k'}$ to which the detector is sensitive. Since the detector accepts a large range

1. General Scattering Formalism

of energies and does not discriminate among values in the packet, it is sufficient to replace $\delta(E_{k'} - E_{k_1})$ with $\delta(E_{k'} - E_{k_0})$. If we define the density of states by

$$\rho(E_{k_0}) \, d\Omega = \int_{\Delta k'} \frac{d^3 k'}{(2\pi)^3} \delta(E_{k'} - E_{k_0}) = \frac{k'^2}{(2\pi)^3} \left. \frac{dk'}{dE'} \right|_{E_{k'} = E_{k_0}} d\Omega, \quad (1.25)$$

Equation (1.23) may be written as

$$\frac{dP_{\Delta k'}}{d\Omega} = |\langle \phi_{k',\alpha'} | V | \psi^{(+)}_{k_0,\alpha} \rangle |^2_{E_{k'} = E_{k_0}} \, \rho(E_{k_0}) \, (2\pi)^2 \int \frac{d^3 k_1}{(2\pi)^3} \frac{d^3 k_2}{(2\pi)^3}$$
$$\times A(\mathbf{k}_1) \, A^*(\mathbf{k}_2) \, \delta(E_{k_1} - E_{k_2}). \quad (1.26)$$

As a result of the assumption that $\Delta k'$, the detector resolution, is larger than the momentum spread in the incident packet, this no longer depends on the resolution.

The differential cross section is the ratio of this probability of detection to the probability that the incident packet crosses a unit area located along the z axis (parallel to $\mathbf{k}_0$). Since we have assumed that the shape of the incident packet does not change, this can be obtained either by integrating along the z axis at a fixed time or by integrating over all time at a fixed position. With $dz = v_0 \, dt$ (v_0 is the speed of the center of the packet), this latter procedure yields for the probability per unit area

$$v_0 \int_{-\infty}^{\infty} dt \, | \langle \mathbf{r} | \phi(t) \rangle |^2_{\mathbf{r}=0} = 2\pi \hbar v_0 \int \frac{d^3 k_1}{(2\pi)^3} \frac{d^3 k_2}{(2\pi)^3} A(\mathbf{k}_1) \, A^*(\mathbf{k}_2) \, \delta(E_{k_1} - E_{k_2}), \quad (1.27)$$

where we use Eq. (1.7) and assume that the initial eigenstates are normalized so that they have unit probability density on the z axis. For instance, the wave function for the initial state may usually be expressed as

$$\phi_{\mathbf{k},\alpha}(\mathbf{r}; \mathbf{r}_1, \mathbf{r}_2, \ldots) = e^{i\mathbf{k}\cdot\mathbf{r}} \, \eta_\alpha(\mathbf{r}_1, \mathbf{r}_2, \ldots), \quad (1.28)$$

where $\mathbf{r}$ is the relative separation of the projectile and the target and $\mathbf{r}_1, \mathbf{r}_2 \ldots$, are the internal coordinates of these two systems. This state is properly normalized if the internal wave function is normalized to unity:

$$\int d^3 r_1 \, d^3 r_2 \ldots | \eta_\alpha(\mathbf{r}_1, \mathbf{r}_2, \ldots) |^2 = 1. \quad (1.29)$$

In Eq. (1.27) we have also used

$$\int_{-\infty}^{\infty} dt \, \exp\left[-\frac{i}{\hbar}(E_{k_1} - E_{k_2})t\right] = 2\pi \hbar \, \delta(E_{k_1} - E_{k_2}). \quad (1.30)$$

Taking the ratio of Eq. (1.26) to Eq. (1.27), we obtain

$$\frac{d\sigma_{fi}}{d\Omega} = \frac{2\pi}{\hbar v_i} |\langle \phi_f | V | \psi_i^{(+)} \rangle|^2_{E_f = E_i} \, \rho_f(E_i). \tag{1.31}$$

This is the central result of our development. We observe that, because of the definition of the cross section and the assumed properties of the detector, the result no longer depends on the shape of the wave packet.

The form of this result is very general, applying to a large class of collision processes. In this derivation we have assumed that H_0 describes both the initial and final free-particle states; in the next section we shall remove this restriction. The form of the result is unchanged, however.

If we use the definition of the transition operator given in Eqs. (1.18) and (1.19), this result may be expressed as

$$\frac{d\sigma_{fi}}{d\Omega} = \frac{2\pi}{\hbar v_i} |\langle \phi_f | T | \phi_i \rangle|^2_{E_f = E_i} \, \rho_f(E_i). \tag{1.32}$$

This result reduces to the form described in previous chapters for the case of scattering of two structureless particles. In that case

$$\langle \phi_f | T | \phi_i \rangle = \langle \phi_f | V | \psi_i^{(+)} \rangle$$
$$= \int d^3r \, \phi_f^*(\mathbf{r}) V(\mathbf{r}) \psi_i^{(+)}(\mathbf{r}) = -\frac{4\pi\hbar^2}{2m} f(\theta), \tag{1.33}$$

and the variable i includes only the momentum since there are no internal variables. Thus we have in this special case

$$\rho_f(E_i) = \frac{1}{(2\pi)^3} \frac{d^3k_f}{dE_f}\bigg|_{E_f = E_i} = \frac{1}{(2\pi)^3} \frac{mk_i}{\hbar^2} \tag{1.34}$$

and

$$d\sigma/d\Omega = |f(\theta)|^2, \tag{1.35}$$

which is our earlier result.

It is often helpful in checking manipulations to know the dimensions of the quantities involved. One can easily check using the completeness relations that, for two-particle states,

$$|\phi_i\rangle \sim L^{3/2}$$
$$|\psi_i^{(+)}\rangle \sim L^{3/2}$$
$$|\mathbf{r}\rangle \sim L^{-3/2}$$

2. Rearrangement Collisions

and naturally

$$V \sim E$$
$$\hbar \sim E \times t$$
$$v \sim L \times t^{-1}$$
$$\rho(E) \sim L^{-3} \times E^{-1}$$
$$\langle \phi_f | T | \phi \rangle \sim L^3 \times E.$$

In relativistic problems it is convenient to use units in which $\hbar = c = 1$. In these units length and time appear to have the same units since v is considered to be dimensionless, and energy has the units of inverse length. Then

$$V \sim L^{-1}$$
$$\hbar = 1$$
$$v \sim 1$$
$$\rho(E) \sim L^{-2}$$
$$\langle \phi_f | T | \phi_i \rangle \sim L^2.$$

2. Rearrangement Collisions

In this section we shall discuss processes in which the outgoing systems are not identical with those in the initial state. The unperturbed Hamiltonian for the final state, describing the final particles when they are well separated, differs from that for the initial state, and the results of the previous section must be re-examined.

"FINAL-STATE" FORM OF THE STATE VECTOR

In Section 1 we computed the cross section using the scalar product of the full scattering wave function with the wave function for the final state. This procedure is correct in general, but in the present case it is exceedingly inconvenient unless some modification is made in the previous approach. The time-dependent state vector, Eq. (1.5), was expressed as a superposition of stationary states

$$\psi_i^{(+)} = \phi_i + \frac{1}{E - H_0 + i\varepsilon} V \psi_i^{(+)}, \qquad (2.1)$$

with ϕ_i the initial free-particle state. It is difficult to evaluate the overlap of $\psi_i^{(+)}$ with a rearranged final state using Eq. (2.1); instead we will rewrite it in a form more suited to the calculation of rearrangement cross sections.

For this purpose we denote the final-state unperturbed Hamiltonian and potential by H_0' and V', respectively. These are related to the initial-state

operators since their sum is the total Hamiltonian, so that
$$H = H_0 + V = H_0' + V'. \tag{2.2}$$

The Green's function in Eq. (2.1) is easily expanded in the complete set ϕ_i, the eigenstates of H_0
$$\frac{1}{E - H_0 + i\varepsilon} = \sum_j \frac{|\phi_j\rangle \langle \phi_j|}{E - E_j + i\varepsilon}. \tag{2.3}$$

Thus the original form of $\psi_i^{(+)}$ may readily be expressed as an expansion in the set of initial eigenstates. To determine the scalar product of this function with an eigenstate of H_0' we must express $\psi_i^{(+)}$ instead as a sum over the eigenstates of H_0'.

The initial-state Green's function may be rewritten in terms of the final-state Green's function by means of the relation

$$\frac{1}{E - H_0 + i\varepsilon} = \frac{1}{E - H_0' + i\varepsilon} + \frac{1}{E - H_0 + i\varepsilon}(V' - V)\frac{1}{E - H_0' + i\varepsilon}$$

$$= \frac{1}{E - H_0' + i\varepsilon} + \frac{1}{E - H_0' + i\varepsilon}(V' - V)\frac{1}{E - H_0 + i\varepsilon}. \tag{2.4}$$

This equation is completely equivalent to the trivial algebraic relation
$$A^{-1} - B^{-1} = A^{-1}(B - A)B^{-1}. \tag{2.5}$$

If we insert the second of the equalities in Eq. (2.4) into Eq. (2.1), we obtain

$$\psi_i^{(+)} = \phi_i + \frac{1}{E - H_0' + i\varepsilon} V \psi_i^{(+)} + \frac{1}{E - H_0' + i\varepsilon}(V' - V)\frac{1}{E - H_0 + i\varepsilon} V \psi_i^{(+)}$$

$$= \phi_i + \frac{1}{E - H_0' + i\varepsilon} V \psi_i^{(+)} + \frac{1}{E - H_0' + i\varepsilon}(V' - V)(\psi_i^{(+)} - \phi_i)$$

$$= \phi_i - \frac{1}{E - H_0' + i\varepsilon}(V' - V)\phi_i + \frac{1}{E - H_0' + i\varepsilon} V' \psi_i^{(+)}. \tag{2.6}$$

We note that, if $V' = V$ and $H_0' = H_0$, Eq. (2.6) reduces to Eq. (2.1).

If the final states associated with H_0' represent a true rearrangement of the initial states, in the last equality of Eq. (2.6) the first two terms will cancel each other. To see this let us denote their sum by Δ' and write it as

$$\Delta' = \phi_i - \frac{1}{E - H_0' + i\varepsilon}(V' - V)\phi_i$$

$$= \left\{1 - \frac{1}{E - H_0' + i\varepsilon}(H_0 - H_0')\right\}\phi_i. \tag{2.7}$$

If $H_0' \equiv H_0$, then the operator in braces is unity and Δ' becomes ϕ_i. If, however, ϕ_i is an eigenfunction of H_0 but not of H_0', then Δ' is zero.

This result can be demonstrated in the coordinate representation, in a manner similar to that employed in Chapter 4. We write Δ' as

$$\Delta' = \left\{(E - H_0')\frac{1}{E - H_0' + i\varepsilon} - \frac{1}{E - H_0' + i\varepsilon}(E - H_0')\right\}\phi_i$$
$$= \{(E - H_0')G'^{(+)} - G'^{(+)}(E - H_0')\}\phi_i, \tag{2.8}$$

which in the coordinate representation may be written in explicit form as

$$\Delta'(\mathbf{r}_1, \ldots, \mathbf{r}_n) = \int d^3r_1', \ldots, d^3r_n' \{(E - H_0') G'^{(+)}(\mathbf{r}_1, \ldots, \mathbf{r}_n; \mathbf{r}_1', \ldots, \mathbf{r}_n')$$
$$- G'^{(+)}(\mathbf{r}_1, \ldots, \mathbf{r}_n; \mathbf{r}_1', \ldots, \mathbf{r}_n') (E - H_0')\} \phi_i(\mathbf{r}_1', \ldots, \mathbf{r}_n'). \tag{2.9}$$

For a spherical surface with an infinite radius, Green's theorem then yields the result that

$$\Delta'(\mathbf{r}_1, \ldots, \mathbf{r}_n) = \sum_{j=1}^{n} (\hbar^2/2m_j) \lim_{r_j' \to \infty} \int dS_j'$$
$$\times \{\nabla_j' G'^{(+)}(\mathbf{r}_1, \ldots, \mathbf{r}_n; \mathbf{r}_1' \ldots, \mathbf{r}_n') \phi_i(\mathbf{r}_1', \ldots, \mathbf{r}_n')$$
$$- G'^{(+)}(\mathbf{r}_1, \ldots, \mathbf{r}_n; \mathbf{r}_1', \ldots, \mathbf{r}_n')$$
$$\times \nabla_j' \phi_i(\mathbf{r}_1', \ldots, \mathbf{r}_n')\}, \tag{2.10}$$

where $dS_j' = d^3r_1' \ldots d^3r_{j-1}' d^3r_{j+1}' \ldots d^3r_n' r_j'^2 dr_j'$ and $G'^{(+)}$ is the Green's function for the final state configuration. An outgoing-wave Green's function satisfies the homogeneous boundary condition

$$\nabla_j' G'^{(+)}(\mathbf{r}_1 \ldots \mathbf{r}_n; \mathbf{r}_1' \ldots \mathbf{r}_n') \xrightarrow[r_j' \to \infty]{} ik_j \frac{\mathbf{r}_j'}{r_j'} G'^{(+)}(\mathbf{r}_1 \ldots \mathbf{r}_n; \mathbf{r}_1' \ldots \mathbf{r}_n'), \tag{2.11}$$

when all other position vectors are fixed and finite. Thus the surface integral in Eq. (2.10) becomes

$$\Delta'(\mathbf{r}_1 \ldots \mathbf{r}_n) = \sum_{j=1}^{n} (\hbar^2/2m_j) \lim_{r_j' \to \infty} G'^{(+)}(\mathbf{r}_1 \ldots \mathbf{r}_n; \mathbf{r}_1' \ldots \mathbf{r}_n')$$
$$\times \int dS_j' \cdot \left\{ik_j \frac{\mathbf{r}_j'}{r_j'} - \nabla_j'\right\} \phi_i(\mathbf{r}_1' \ldots \mathbf{r}_n'). \tag{2.12}$$

As examples, let us examine this result for the three-body rearrangement processes mentioned at the beginning of this chapter. The initial wave function is

$$\phi_i(\mathbf{r}_1, \mathbf{r}_2, \mathbf{r}_3) = \exp\left[i\mathbf{k}\cdot\left(\mathbf{r}_1 - \frac{m_2\mathbf{r}_2 + m_3\mathbf{r}_3}{m_2 + m_3}\right)\right]\eta_\alpha(\mathbf{r}_2 - \mathbf{r}_3), \quad (2.13)$$

where $\eta_\alpha(\mathbf{r}_2 - \mathbf{r}_3)$ is the bound-state wave function and $\hbar\mathbf{k}$ is the relative momentum of the projectile and the target. For $j = 2$ and 3 the initial wave function vanishes on the surface S_j', so that there is no contribution to $\Delta'(\mathbf{r}, \ldots \mathbf{r}_n)$. For $j = 1$, $\phi_i(\mathbf{r}_1, \mathbf{r}_2, \mathbf{r}_3)$ is nonzero on the surface, but we can see that the Green's function vanishes too rapidly to give a nonzero contribution. Those terms in the eigenfunction expansion of $G'^{(+)}(\mathbf{r}_1, \mathbf{r}_2, \mathbf{r}_3; \mathbf{r}_1', \mathbf{r}_2', \mathbf{r}'^3)$, in which, for instance, particle 1 is bound to particle 3, corresponding to exchange or pickup processes, vanish exponentially for large r_1'. At sufficiently high energies there will also be energy-conserving terms in which all particles are unbound, corresponding to breakup processes; these go to zero for large r_1' as $(1/r_1'^4)$ (see Section 5 of Chapter 4). Similar considerations hold in the general case of many particles. Then

$$\Delta'(\mathbf{r}_1 \ldots \mathbf{r}_n) = 0 \quad (2.14)$$

or, in the operator language,

$$\Delta' = \phi_i - \frac{1}{E - H_0' + i\varepsilon}(E - H_0')\phi_i = 0. \quad (2.15)$$

Suppose H_0' were replaced by H_0, so that the Green's function were not a rearrangement Green's function. There would then be a finite contribution on S_1' from the term in the Green's function having particles 2 and 3 bound together and particle 1 outgoing. On S_1', this Green's function behaves as

$$G^{(+)}(\mathbf{r}_1, \mathbf{r}_2, \mathbf{r}_3; \mathbf{r}_1', \mathbf{r}_2', \mathbf{r}_3') \xrightarrow[r_1' \to \infty]{} -\frac{2m_1}{4\pi\hbar^2}\frac{\exp(ikr' - i\mathbf{k}\cdot\mathbf{r}'/r')}{r'}$$
$$\times \eta_\alpha(\mathbf{r}_2 - \mathbf{r}_3)\eta_\alpha(\mathbf{r}_2' - \mathbf{r}_3'), \quad (2.16)$$

with $\mathbf{r}$ the relative coordinate of the projectile and the target:

$$\mathbf{r} = \mathbf{r}_1 - \frac{m_2\mathbf{r}_2 + m_3\mathbf{r}_3}{m_2 + m_3}.$$

Using this and Eq. (2.13), Eq. (2.12) gives us

$$\Delta'(\mathbf{r}_1, \mathbf{r}_2, \mathbf{r}_3) = \phi_i(\mathbf{r}_1, \mathbf{r}_2, \mathbf{r}_3). \quad (2.17)$$

2. Rearrangement Collisions

Clearly one must exercise great care in dealing with singular operators or, in coordinate space, unbounded regions.

One can also see from these results that a Hamiltonian is not self-adjoint when it appears between its outgoing-wave Green's function and one of its own eigenfunctions. The reason, of course, is that there is then a net current across the surface at infinity.

An illustration of this lack of hermiticity is the very existence of a nonzero scattering amplitude. If $(E - H_0)\phi_f = 0$, we have the identity

$$\langle \phi_f | (H_0^\dagger - H_0) | \psi_i^{(+)} \rangle = \langle \phi_f | (E - H_0) | \psi_i^{(+)} \rangle \\ = \langle \phi_f | V | \psi_i^{(+)} \rangle, \qquad (2.18)$$

which is proportional to the scattering amplitude.

In conclusion, we find that for rearrangement collisions the scattering state vector, which satisfies Eq. (2.1), can be rewritten as

$$\psi_i^{(+)} = \frac{1}{E - H_0' + i\varepsilon} V' \psi_i^{(+)}. \qquad (2.19)$$

Several related features of this result are of special interest:

(a) There is no free-particle state in this equation. Such a state has to be an eigenfunction of H_0' if $\psi_i^{(+)}$ is to satisfy the Schrödinger equation $(E - H)\psi_i^{(+)} = 0$. But the eigenstates of the final-state Hamiltonian appear physically only *after* the scattering process has occurred. The incident wave is not an eigenstate of H_0'.

(b) The original boundary condition on $\psi_i^{(+)}$ seems to have been lost. There is no longer in Eq. (2.19) the information that $\psi_i^{(+)}$ contains a free-particle state ϕ_i plus outgoing waves. This information must be incorporated by imposing the required boundary condition on the solution.

(c) In the case of rearrangement collisions the homogeneous integral equation has a solution for all energies. Then the solutions to Eq. (2.1) are not unique, and care must be taken in obtaining the physically correct solution.

(d) It might appear that Eq. (2.19) contains only outgoing waves and that a time-dependent treatment, such as that in the preceding section, could be used to demonstrate that the scattering state vector $\psi_i^{(+)}$ vanishes as $t \to -\infty$. Since this vector must in fact approach $\phi_i^{(+)}$ in that limit, we must be able to show that it still contains the incident wave. In fact, if we were to retrace the preceding steps by inserting the original form for $\psi_i^{(+)}$, Eq. (2.1), into Eq. (2.19), we would recover the initial state ϕ_i. Clearly, then, one must use caution in performing the integration over the wave packet distribution to be certain that the apparent outgoing wave actually exists in every term.

THE CROSS SECTION

The "final-state" form of $\psi_i^{(+)}$, Eq. (2.19), may now be inserted into the wave packet to obtain the cross section for a rearrangement collision. Taking the limit $t \to \infty$, we find

$$\psi(t) \xrightarrow[t \to +\infty]{} -2\pi i \int \frac{d^3k}{(2\pi)^3} A(\mathbf{k}) \delta(E_k - H_0') V' \psi_{\mathbf{k},\alpha}^{(+)} \exp[-(i/\hbar)E_k t]. \qquad (2.20)$$

The form of this result is identical with that of the scattered wave of Eq. (1.13), and the cross section can be computed by a procedure identical with that leading to the nonrearrangement cross section. Since $\psi(t)$ now has a form in which it is a simple matter to find the overlap with a rearranged final state, the result can be obtained by inspection. If we denote the final state eigenfunctions by the free-particle symbol $\phi_{f'}$, we have

$$\frac{d\sigma_{f'i}}{d\Omega} = \frac{2\pi}{\hbar v_i} |\langle \phi_{f'} | V' | \psi_i^{(+)} \rangle|^2_{E_f = E_i} \rho_f(E_i), \qquad (2.21)$$

where, in general, the density of states for an N-particle final state is

$$\rho_{f'}(E) = \prod_{i=1}^{N} \int \frac{d^3 p_i}{(2\pi\hbar)^{3N-3}} \delta(\sum_{j=1}^{N} \mathbf{p}_j) \delta(E - \sum_{j=1}^{N} E_j). \qquad (2.22)$$

The appropriate T-matrix element in the case of rearrangement collisions is then $\langle \phi_{f'} | V' | \psi_i^{(+)} \rangle$ with, in general, V' being the difference between H_0' which defines $\phi_{f'}$ and H.

To summarize, we have shown that $\psi_i^{(+)}$ can be written in a form that permits easy computation of the overlap of the packet state $\psi(t)$ with any final state $\phi_{f'}$. The resulting cross section formula has a structure that is common to both rearrangement and nonrearrangement processes. Equation (2.21) is the exact result for *all* collision processes.

SYMMETRY OF THE T MATRIX

The general result for the differential cross section, Eq. (2.21), does not seem to be symmetric between initial and final states. On the other hand, we expect that there is such a symmetry inherent in the matrix element

2. Rearrangement Collisions

$\langle \phi_{f'} | V' | \psi_i^{(+)} \rangle$, and we can in fact write it in a form that exhibits the symmetry between initial and final states.

The explicit solution for the scattering state vector is

$$\psi_i^{(+)} = \Omega^{(+)} \phi_i = \left(1 + \frac{1}{E - H + i\varepsilon} V\right) \phi_i. \qquad (2.23)$$

Inserting this into the transition matrix element, we have

$$\langle \phi_{f'} | V' | \psi_i^{(+)} \rangle = \langle \phi_{f'} | V' | \phi_i \rangle + \langle \phi_{f'} | V' \frac{1}{E - H + i\varepsilon} V | \phi_i \rangle. \qquad (2.24)$$

The second term has the kind of symmetry we expect, but the first term seems to involve only the final-state potential. However, V' can be replaced by $\frac{1}{2}(V' + V) + \frac{1}{2}(E - H_0') - \frac{1}{2}(E - H_0)$, and the matrix elements vanish

$$\langle \phi_{f'} | E - H_0 | \phi_i \rangle = \langle \phi_{f'} | E - H_0' | \phi_i \rangle = 0. \qquad (2.25)$$

The free-particle states ϕ_i and $\phi_{f'}$ are eigenstates of H_0 and H_0', respectively, with the same eigenvalue, and the hermiticity of H_0' in this context can be easily demonstrated in any particular case. Thus we find, when $E_i = E_f = E$, that

$$\langle \phi_{f'} | V' | \phi_i \rangle = \langle \phi_{f'} | V | \phi_i \rangle = \tfrac{1}{2} \langle \phi_{f'} | V + V' | \phi_i \rangle, \qquad (2.26)$$

so that also in the first term of Eq. (2.24) the initial and final states enter on an equivalent footing, and Eq. (2.24) can be written in the manifestly symmetric form

$$\langle \phi_{f'} | T | \phi_i \rangle = \langle \phi_{f'} | V' | \psi_i^{(+)} \rangle = \tfrac{1}{2} \langle \phi_{f'} | V + V' | \phi_i \rangle$$
$$+ \langle \phi_{f'} | V' \frac{1}{E - H + i\varepsilon} V | \phi_i \rangle. \qquad (2.27)$$

In the Born approximation $\psi_i^{(+)} \approx \phi_i$, so that the Born approximation to the T matrix is given by the identity of Eq. (2.26). When these two forms, V and V', have been used in calculations, they have frequently given different results, leading to what has been called the "post-prior paradox." The differences actually arise from approximations made in calculating the initial and final state wave functions; the formal identity of Eq. (2.26) remains generally correct if true eigenfunctions are used.

We can exploit the symmetry of these results to recast the T-matrix element $\langle \phi_{f'} | V' | \psi_i^{(+)} \rangle$ into a form that emphasizes the initial-state interaction V instead of the final-state potential V'. In deriving Eq. (2.24) we wrote $\psi_i^{(+)}$ in the form of an operator $\Omega^{(+)}$ acting to the right on ϕ_i. Now that we have an equivalent symmetric form, Eq, (2.27), we may suspect that this could have been obtained just as easily from an operator acting to the left on $\phi_{f'}$. To identify this operator, we use Eq. (2.26) and rewrite Eq. (2.27) as

$$\langle \phi_{f'} | T | \phi_i \rangle = \langle \phi_{f'} | V' | \psi_i^{(+)} \rangle$$

$$= \langle \phi_{f'} | \left(1 + V' \frac{1}{E - H + i\varepsilon}\right) V | \phi_i \rangle$$

$$= \langle \phi_{f'} | W^\dagger V | \phi_i \rangle, \qquad (2.28)$$

where the operator W is obviously given by

$$W = 1 + \frac{1}{E - H - i\varepsilon} V', \qquad (2.29)$$

provided H and V' are Hermitian.

The operator W operating on $\phi_{f'}$, the eigenstate of H_0', produces a state $\psi_{f'}^{(-)}$ obeying the integral equation

$$\psi_{f'}^{(-)} = \left(1 + \frac{1}{E - H - i\varepsilon} V'\right) \phi_{f'} = \phi_{f'} + \frac{1}{E - H_0' - i\varepsilon} V' \psi_{f'}^{(-)}. \qquad (2.30)$$

The state $\psi_{f'}^{(-)}$ is an eigenfunction of the total Hamiltonian corresponding asymptotically to a plane wave plus an *incoming* spherical wave in the two-particle arrangement referred to by the prime—that is to say, in the arrangement for which H_0' is the correct asymptotic Hamiltonian. The operator W is therefore the wave matrix, conventionally symbolized by $\Omega'^{(-)}$, viz.,

$$W = \Omega'^{(-)} = 1 + \frac{1}{E - H - i\varepsilon} V' = 1 + \frac{1}{E - H_0' - i\varepsilon} V' \Omega'^{(-)}. \qquad (2.31)$$

In terms of this incoming state we now have the reciprocal relation for the T matrix

$$\langle \phi_{f'} | T | \phi_i \rangle = \langle \phi_{f'} | V' | \psi_i^{(+)} \rangle = \langle \psi_{f'}^{(-)} | V | \phi_i \rangle, \qquad (2.32)$$

with $E_f = E_i = E$.

This result is useful in calculating rearrangement cross sections in cases where it is more convenient to use one potential than the other, or where it is easier to solve for one scattering function than the other. The same type of relation also holds when $V' = V$, that is when there is no rearrangement. In that situation, however, there is no particular advantage to one form over the other.

The T Matrix for Rearrangement Processes

We have seen earlier that the T matrix for ordinary scattering obeys the integral equation

$$T = V + V \frac{1}{E - H_0 + i\varepsilon} T = V + T \frac{1}{E - H_0 + i\varepsilon} V. \quad (2.33)$$

However, the T matrix so defined is not the T matrix for rearrangement collisions. Explicit solution of Eq. (2.33) yields

$$T = V + V \frac{1}{E - H + i\varepsilon} V = V\Omega^{(+)} = \Omega^{(-)\dagger}V. \quad (2.34)$$

However, from Eq. (2.32) we see that for rearrangement collisions the T matrix must be defined as

$$T = V'\Omega^{(+)} = \Omega'^{(-)\dagger}V. \quad (2.35)$$

The T matrix so defined obeys the integral equation

$$T = V' + V' \frac{1}{E - H_0 + i\varepsilon} T = V' + V' \frac{1}{E - H + i\varepsilon} V, \quad (2.36)$$

or, equivalently,

$$T = V + T \frac{1}{E - H_0' + i\varepsilon} V = V + V \frac{1}{E - H + i\varepsilon} V. \quad (2.37)$$

Since we have already seen that $\langle \phi_{f'} | V | \phi_i \rangle = \langle \phi_{f'} | V' | \phi_i \rangle$, we see immediately that the T matrices given by Eqs. (2.36) and (2.37) are equivalent, by which we mean that the matrix elements $\langle \phi_{f'} | T | \phi_i \rangle$ calculated from the solution of Eq. (2.36) or Eq. (2.37) are identical. The T matrix defined by Eq. (2.36) or Eq. (2.37) reduces to that of Eq. (2.33) for the case of ordinary scattering in which $H_0' = H_0$ and $V = V'$. Thus the T matrix given by Eq. (2.35) is quite general and holds for both ordinary and rearrangement scattering.

3. Collisions Involving Identical Particles

The methods that have been discussed heretofore are not entirely adequate to deal with collisions in which the projectile is identical with the target or some constituents within it. For instance, we may wish to study electrons scattering on atoms, neutrons scattering on nuclei, or protons scattering on protons. In each case the particles emerging in the final state may or may not be the very same particles that impinged on the target. When particles are identical, there is in principle no way to discern which of them is emerging. More importantly, there are characteristic effects of this identity that are apparent in the angular distributions of the resulting collision processes. We shall now show that we do not need to develop new methods to deal with this situation, since the methods of the previous sections are easily adapted to take account of the identity of the particles. Our discussion will focus on a particular set of collision processes, but the same approach can be used on any process of arbitrary complexity.

ANTISYMMETRIZATION OF THE STATE VECTOR

For specificity, we will consider the elastic and inelastic scattering of a neutron incident on a nucleus containing N neutrons. The final state will then contain a single outgoing neutron that may be either the incident neutron or one of the neutrons initially contained in the target. While the two cases are experimentally indistinguishable, they are different in that for the latter case the kinetic energy of the incident neutron must be transferred to a target neutron. This process may lead to a different angular distribution than would result from passage of the incident neutron through the target.

For this case, in which we have identical spin-$\frac{1}{2}$ particles, the Pauli principle states that the wave function must be antisymmetric; that is, it must change sign when the coordinates of any two of these particles are exchanged. Since the wave function is simply the spatial representation of the state vector for the system, that vector must be antisymmetric under interchange of any two identical particles.

Let us introduce the permutation operator $P_{\mu\nu}$ which interchanges the identical particles μ and ν. If particle μ is in a state α_1, denoted by $\eta_{\alpha_1}(\mu)$, while ν is in a state $\eta_{\alpha_2}(\nu)$, the effect of $P_{\mu\nu}$ on the resulting two-particle state is

$$P_{\mu\nu}\, \eta_{\alpha_1}(\mu)\, \eta_{\alpha_2}(\nu) = \eta_{\alpha_1}(\nu)\, \eta_{\alpha_2}(\mu). \tag{3.1}$$

In a particular representation all the coordinates of the particles including their positions and spins are exchanged by $P_{\mu\nu}$.

Since we are primarily concerned with the scattering problem, we will assume that the wave function for the target is known and is properly antisymmetrized. If the state vector for the target is denoted by $\eta_\alpha(1, 2, ..., N)$,

3. Collisions Involving Identical Particles

this vector will satisfy

$$P_{\mu\nu}\,\eta_\alpha(1, 2,\ldots, N) = -\eta_\alpha(1, 2,\ldots, N) \tag{3.2}$$

for all μ and ν between 1 and N. One way of constructing such a state vector is to use one or more determinants of single-particle state vectors. If the target can be described by an "independent-particle" or "shell" model Hamiltonian, that is,

$$H = \tilde{H}(1) + \tilde{H}(2) + \cdots \tilde{H}(N), \tag{3.3}$$

then a possible state vector would be

$$\eta_\alpha(1, 2,\ldots, N) = \frac{1}{(N!)^{1/2}} \begin{vmatrix} \eta_{\alpha_1}(1) & \eta_{\alpha_1}(2) & \eta_{\alpha_1}(3) & \cdots & \eta_{\alpha_1}(N) \\ \eta_{\alpha_2}(1) & \eta_{\alpha_2}(2) & & \cdots & \cdot \\ \eta_{\alpha_3}(1) & & & \cdots & \cdot \\ \eta_{\alpha_N}(1) & \eta_{\alpha_N}(2) & & \cdots & \eta_{\alpha_N}(N) \end{vmatrix}. \tag{3.4}$$

In this form the state vector obviously satisfies Eq. (3.2). When this is written in the spatial representation, it becomes the familiar Slater determinant and each of the state vectors becomes $\eta_\alpha(\mathbf{r}_\nu, m_\nu)$, with $\mathbf{r}_\nu$ and m_ν the position and the z component of the spin for the νth particle.

If the incident projectile, denoted by 0, were distinguishable from the bound particles, the initial state would be described by the vector

$$\phi_i = \phi_{\mathbf{k}}(0)\,\eta_\alpha(1, 2, 3,\ldots, N) \equiv \phi_i(0). \tag{3.5}$$

However, when 0 is identical with the N bound neutrons, this description fails to satisfy the Pauli principle. We can, however, easily construct a properly antisymmetrized state vector by introducing the operator

$$\mathscr{A} = (N + 1)^{-1/2}\left(1 - \sum_{\mu=1}^{N} P_{\mu 0}\right). \tag{3.6}$$

The state vector

$$\mathscr{A}\phi_i = \mathscr{A}\phi_{\mathbf{k}}(0)\,\eta_\alpha(1, 2,\ldots, N) = (N+1)^{-1/2}\{\phi_{\mathbf{k}}(0)\,\eta_\alpha(1, 2,\ldots N) \\ - \phi_{\mathbf{k}}(1)\,\eta_\alpha(0, 2,\ldots, N) - \phi_{\mathbf{k}}(2)\,\eta_\alpha(1, 0, 3,\ldots, N) - \cdots\} \tag{3.7}$$

is antisymmetric in all $(N + 1)$ neutrons, as one can easily see by inspection.

To treat scattering processes involving identical particles, we will use the operator $\mathscr{A}$ to construct a solution of the Schrödinger equation that is antisymmetric among all these particles and is properly normalized. We will then compute the differential cross section using this proper scattering state.

The wave packet of Section 1 treated the projectile as distinguishable from the bound neutrons. Thus the packet $\psi(t)$, given by Eq. (1.5), is a superposition of eigenstates

$$\psi_i(0) = \phi_i(0) + \frac{1}{E - H_0(0) + i\varepsilon} V(0) \psi_i^{(+)}(0), \quad (3.8)$$

where $H_0(0)$ is the Hamiltonian describing a free neutron 0 and bound neutrons 1, 2,..., N and $V(0)$ is the interaction between these two systems.

A suitable fully antisymmetric wave packet is then provided by $\mathscr{A}\psi(t)$ with $\psi(t)$ given by Eq. (3.8). The full Hamiltonian H is symmetric among all $(N + 1)$ identical neutrons, so that H commutes with $\mathscr{A}$, which simply interchanges the neutrons. Then this state vector satisfies the Schrödinger equation

$$i\hbar \frac{\partial}{\partial t} (\mathscr{A}\psi(t)) = i\hbar \mathscr{A} \frac{\partial \psi(t)}{\partial t} = \mathscr{A}H\psi(t) = H(\mathscr{A}\psi(t)). \quad (3.9)$$

This packet is also normalized to unity. Since it satisfies the Schrödinger equation, the normalization $\langle \mathscr{A}\psi(t) | \mathscr{A}\psi(t) \rangle$ is time independent. For very early times $\psi(t)$ approaches $\phi(t)$, the free-particle wave packet, which then describes a widely separated projectile and target. In determining the normalization and using the definition of $\mathscr{A}$, terms such as $P_{\mu 0}\phi(t)$ will be orthogonal to $\phi(t)$ since for very early times the former corresponds to neutron μ widely separated from a target while the latter has this particle bound to the target. The normalization is then

$$\langle \mathscr{A}\psi(t) | \mathscr{A}\psi(t) \rangle = \langle \mathscr{A}\phi(t) | \mathscr{A}\phi(t) \rangle |_{t \to -\infty}$$
$$= \frac{1}{(N+1)} \langle \phi(t) | (1 + \sum_{\mu=1}^{N} P_{\mu 0}^2) | \phi(t) \rangle = 1. \quad (3.10)$$

THE CROSS SECTION

To evaluate the cross section, we follow Section 1 but use antisymmetrized states throughout. We then must evaluate the scalar product of $\mathscr{A}\psi(t)$ with the set of antisymmetrized final states $\mathscr{A}\phi_f(0)$, where the state $\phi_f(0)$ has neutron 0 unbound and separated from the target. This computation is simplified if we use

$$P_{\mu 0}\mathscr{A}\phi_f(0) = -\mathscr{A}\phi_f(0) \quad (3.11)$$

to obtain

$$\mathscr{A}^2\phi_f(0) = (N+1)^{1/2}\mathscr{A}\phi_f(0). \quad (3.12)$$

Then

$$\langle \mathscr{A}\phi_f(0) | \mathscr{A}\psi(t)\rangle = (N+1)^{1/2} \langle \mathscr{A}\phi_f(0) | \psi(t)\rangle$$
$$= \langle (1 - \sum_{\mu=1}^{N} P_{\mu 0}) \phi_f(0) | \psi(t)\rangle. \quad (3.13)$$

In this matrix element all N terms in the sum are identical, since both $\phi_f(0)$ and $\psi(t)$ are assumed to be antisymmetric in the set of bound neutrons. Then Eq. (3.13) becomes

$$\langle \mathscr{A}\phi_f(0) | \mathscr{A}\psi(t)\rangle = \langle \phi_f(0) | \psi(t)\rangle - N\langle \phi_f(1) | \psi(t)\rangle, \quad (3.14)$$

where, as the notation indicates, neutron 1 is unbound in $\phi_f(1)$ while neutron 0 remains bound to the target.

The first term is just the scalar product evaluated in Section 1 and shown there to be proportional to $\langle \phi_f(0) | V(0) | \psi_i^{(+)}(0)\rangle$. The second term is the "exchange" term; one of the constituents of the target has been exchanged for the incident particle. As we demonstrated in Section 2, such a scalar product, involving a rearranged final state, is best evaluated by using a rearranged form of $\psi_i^{(+)}(0)$, as in Eq. (2.19). As there, the result is proportional to $\langle \phi_f(1) | V(1) | \psi_i^{(+)}(0)\rangle$, where $V(1)$ is the interaction between neutron 1 and the residual nucleus.

Following the derivation of these earlier sections, we obtain for the differential cross section the result

$$\frac{d\sigma_{fi}}{d\Omega} = \frac{2\pi}{hv_i} | \langle \phi_f(0) | V(0) | \psi_i^{(+)}(0)\rangle$$
$$- N \langle \phi_f(1) | V(1) | \psi_i^{(+)}(0)\rangle |^2_{E_f = E_i} \rho_f(E_i). \quad (3.15)$$

This is our general result. It expresses the cross section for scattering involving identical particles in terms of quantities we have encountered before, in which the incident and outgoing particles are treated as distinguishable from the remaining particles. This permits us to carry over all of our earlier results when dealing with identical particles. It also avoids the necessity to include the Pauli principle in calculating the scattering state vector, aside from the use of an antisymmetrized wave function for the target.

The physical significance of Eq. (3.15) is clear. The first term is the same as we have obtained before, corresponding to emission of the same particle that was incident on the target. In the second or exchange term, however, one of the particles originally bound to the target is emitted, and the projectile is captured. The exchange term represents an additional contribution to the elastic or inelastic scattering of the neutrons. The factor of N in the amplitude follows from the fact that any one of N particles could have been emitted; their contributions are coherent with each other and with the direct scattering term because there is, in principle, no experimental way to distinguish these particles from each other.

Explicit calculations using Eq. (3.15) are no more difficult than in the case of distinguishable particles. However, they require the introduction of spin functions to correctly treat spin-$\frac{1}{2}$ particles. This will be discussed in a later chapter.

This result may also be written in a form which displays more explicitly the symmetry among the $N + 1$ identical particles. We recognize that the following identity holds,

$$V(0)\,\phi_f(0) = \{H - H_0(0)\}\,\phi_f(0) = (H - E_f)\,\phi_f(0), \qquad (3.16)$$

and that the Hamiltonian H is symmetric among the identical particles. Then we can use the properties of the antisymmetrization operator $\mathscr{A}$, as in Eq. (3.14), to transform Eq. (3.15) into the symmetric form

$$\frac{d\sigma_{fi}}{d\Omega} = \frac{2\pi}{\hbar v_i}\,|\langle (H - E_f)\,\mathscr{A}\phi_f(0)\,|\,\mathscr{A}\psi_i^{(+)}(0)\rangle\,|^2_{E_f = E_i}\,\rho_f(Ei). \qquad (3.17)$$

In this result both state vectors are antisymmetrized, and the identical particles are all treated equivalently. This result is also more general than Eq. (3.15), since one can readily see that it applies to rearrangement collisions as well as to ordinary scattering processes. However, it must usually be converted into a form such as Eq. (3.15), in which the projectile is distinguished from the bound particles, in order to carry out practical calculations.

Chapter

8

The Time-Dependent Approach to Scattering Theory

In this chapter we shall briefly outline some features of the time-dependent approach to scattering theory. We have previously shown using wave packets that time-independent methods will give correct results for the scattering cross section. We saw that, in practice, the incident wave could be represented by an energy eigenfunction, that is, by a time-independent solution of the Schrödinger equation. To obtain a localization in space and time, such as might be present in an actual scattering experiment, we constructed a wave packet as a superposition of such eigenfunctions. Nevertheless, we found that the scattering cross section could be expressed in terms of the scattering amplitude at a given energy, so that we were able to focus our attention on the stationary-state solution at that energy.

In this chapter we illustrate methods which are usually termed "time-dependent," since the time enters as an explicit parameter. However, we will see here that one can deal with initial and final states that are eigenfunctions of energy and momentum, so that the complexities of the wave packet need not be present. These results may be physically justified, since they can be obtained as limits from a treatment using wave packets. We will also show that the time-dependent formulation leads to equations identical to those we obtained in the time-independent formulation, so that either method may be used, the choice depending only upon which is the more convenient.

An important reason for considering the "time-dependent" method is its usefulness in relativistic theories. The time-dependent method provides the bridge between nonrelativistic and relativistic scattering theories. This chapter may thus be helpful to the reader as an introduction to the relativistic theory of scattering.

1. The Schrödinger and Interaction Pictures

In Chapter 7 we demonstrated the validity of the time-independent method by considering the state vector $\psi(t)$ to be a wave packet

$$\psi(t) = \int \frac{d^3k}{(2\pi)^3} A(\mathbf{k}) \, \psi_{\mathbf{k},\alpha}^{(+)} \exp[-(i/\hbar)E_k t]. \tag{1.1}$$

The state vector $\psi(t)$ satisfies the Schrödinger equation

$$i\hbar \frac{\partial \psi(t)}{\partial t} = H \psi(t). \tag{1.2}$$

The use of a state vector having this dependence on time is one way of introducing the time into the nonrelativistic theory, but as we shall see, it is not the only possible way. Before examining alternate approaches, however, let us explore this one somewhat further.

The dependence on time of Eqs. (1.1) and (1.2) can be exhibited in a very simple way if we introduce an exponential operator, defined by its associated power series expansion to be

$$\exp[-(i/\hbar)Ht] \equiv \sum_{n=0}^{\infty} \frac{1}{n!} \left(-\frac{i}{\hbar} Ht\right)^n. \tag{1.3}$$

Expressed in terms of such an operator, the first-order homogeneous differential equation in Eq. (1.2) has the solution

$$\psi(t) = \exp[-(i/\hbar) H(t - t_0)] \psi(t_0), \tag{1.4}$$

where $\psi(t_0)$ would be assumed to be known in any application of this formula. Since $\psi_{\mathbf{k}\alpha}^{(+)}$ is an eigenfunction of H, the wave packet of Eq. (1.1) can also be written in this way.

Equation (1.4) exhibits the time dependence as a transformation of the state vector $\psi(t_0)$ into the vector $\psi(t)$,

$$\psi(t) = U(t, t_0) \psi(t_0), \tag{1.5}$$

with the time-translation operator $U(t, t_0)$ given by

$$U(t, t_0) = \exp[-(i/\hbar) H(t - t_0)]. \tag{1.6}$$

This operator is closely related to the time-dependent Green's function. A comparison of Eq. (1.6) with Eq. (2.11) of Chapter 5 shows that the operator

1. The Schrödinger and Interaction Pictures

form of this Green's function is

$$G(t, t_0) = \begin{cases} U(t, t_0), & t > t_0, \\ 0, & t \leq t_0. \end{cases} \quad (1.7)$$

If H is Hermitian, the time-translation operator is unitary, and the length of the state vector remains fixed. Then $U(t, t_0)$ generates a pure rotation in abstract vector space. If we picture $\psi(t)$ as a vector having its base at the origin in vector space, then this vector rotates with respect to any set of fixed eigenvectors as time progress. Only if the system is in an eigenstate of H will the state vector maintain a fixed direction; in general, it will rotate in accordance with the transformation of Eq. (1.6). From Eq. (1.6) we also have the relation

$$U^\dagger(t, t_0) = U^{-1}(t, t_0) = U(t_0, t), \quad (1.8)$$

so that the inverse operator generates the time development in the reverse direction of time.

Direct integration of Eq. (1.2) leads to an integral equation for $\psi(t)$ or $U(t, t_0)$, viz.,

$$\psi(t) = \psi(t_0) - (i/\hbar) \int_{t_0}^{t} dt' \, H \, \psi(t'), \quad (1.9)$$

or

$$U(t, t_0) = 1 - (i/\hbar) \int_{t_0}^{t} dt' \, H \, U(t', t_0). \quad (1.10)$$

If this equation is iterated in powers of H, we obtain the expansion

$$U(t, t_0) = 1 - \frac{i}{\hbar} \int_{t_0}^{t} dt' \, H + \left(-\frac{i}{\hbar}\right)^2 \int_{t_0}^{t} dt' \, H \int_{t_0}^{t'} dt'' \, H + \cdots$$

$$= 1 - \frac{i}{\hbar} H(t - t_0) + \frac{1}{2}\left(-\frac{i}{\hbar}\right)^2 H^2 (t - t_0)^2 + \cdots$$

$$\equiv \exp[-(i/\hbar) H(t - t_0)]. \quad (1.11)$$

Thus the integral equation is completely equivalent to the exponential expression, provided the power series is meaningful.

A description in which the time variation of the state vector is controlled by the full Hamiltonian, as in Eq. (1.2), is called the "Schrödinger picture." This picture is useful when exact solutions can be found for the full Schrödinger equation. However, for many problems perturbation methods must be used, and in these cases the Schrödinger picture is not especially convenient. We would prefer for such problems a description in which expansions in powers of a perturbation $V \equiv H - H_0$ can be more readily obtained.

In scattering problems such a description is appropriate even when perturbation methods are not used. As we saw in our discussion of wave packets, the projectile is described by a free-particle wave function when it is far from the target, even though eigenstates of the total Hamiltonian must be used to describe the complete scattering experiment. Thus, in a scattering process, "perturbed" and "unperturbed" wave functions enter because of the nature of the physical process, regardless of the strength of the interaction.

To make use of this distinction between free-particle and exact eigenstates, one introduces the "interaction picture." In this picture, free-particle states are described by time-independent state vectors, while the time dependence of exact states arises from the effects of the interaction alone. Matrix elements and scalar products of state vectors are independent of the particular picture that is used, since different pictures are related to each other by unitary transformations. As a result, one can simply use the most convenient picture for any particular problem.

To move to the "interaction picture," let us observe that the free-particle state vector $\phi(t)$ is also a solution of a time-dependent Schrödinger equation and may be written in the Schrödinger picture as

$$\phi(t) = \exp[-(i/\hbar) H_0(t - t_0)] \phi(t_0). \tag{1.12}$$

Thus the vector

$$\phi_I(t) \equiv \exp[(i/\hbar) H_0 t] \phi(t) = \exp[(i/\hbar) H_0 t_0] \phi(t_0) \tag{1.13}$$

which we define as the free-particle state vector in the "interaction picture," is independent of the time. In general, a state vector in the interaction picture is related to one in the Schrödinger picture by the relation

$$\psi_I(t) = \exp[(i/\hbar) H_0 t] \psi(t). \tag{1.14}$$

In the transformation to the interaction picture, all vectors in Hilbert space rotate according to the unitary transformation $\exp[(i/\hbar) H_0 t]$. Thus free-particle state vectors remain fixed as time passes, but eigenvectors of any Hamiltonian other than H_0 vary with time. The extent of the rotation depends upon the interaction $V = H - H_0$.

If we introduce the definition of the state vector in the interaction picture

1. The Schrödinger and Interaction Pictures

into Eq. (1.2), we find

$$i\hbar \frac{\partial}{\partial t} (\exp[-(i/\hbar)H_0 t] \psi_I(t)) = H_0 \exp[-(i/\hbar)H_0 t] \psi_I(t)$$

$$+ i\hbar \exp[-(i/\hbar)H_0 t] \frac{\partial \psi_I(t)}{\partial t}$$

$$= (H_0 + V) \exp[-(i/\hbar)H_0 t] \psi_I(t). \quad (1.15)$$

The explicit dependence on H_0 cancels out, and, if we introduce the definition

$$V(t) = \exp[(i/\hbar)H_0 t] \, V \exp[-(i/\hbar)H_0 t], \quad (1.16)$$

Eq. (1.15) becomes

$$i\hbar \frac{\partial \psi_I(t)}{\partial t} = V(t) \psi_I(t). \quad (1.17)$$

As expected, the time dependence of $\psi_I(t)$ arises entirely from the interaction. Any free-particle state vector is a constant vector; in particular, any initial state in a scattering experiment is described by a constant vector in this picture.

This removal of the free-particle time dependence has been achieved at the expense of introducing a time-dependent interaction operator $V(t)$. This is the interaction operator in the rotating coordinate system that defines the interaction picture. The effect of the progression of time cannot be eliminated by unitary transformation; only its description can be changed. If the state vectors are modified, the operators that act upon the vectors must be correspondingly modified. If we differentiate $V(t)$, defined by Eq. (1.16), we obtain

$$\frac{\partial V(t)}{\partial t} = \frac{i}{\hbar} [H_0, V(t)], \quad (1.18)$$

so that in the interaction picture the time dependence of $V(t)$ is controlled by its commutator with the free-particle Hamiltonian H_0.

We may note that one can carry this process to completion and introduce the "Heisenberg picture," in which the full state vector

$$\psi_H = \exp[(i/\hbar)Ht] \psi(t) \quad (1.19)$$

is completely independent of time. In this picture all operators satisfy the Heisenberg equation of motion

$$\frac{\partial \mathcal{O}_H(t)}{\partial t} = (i/\hbar)[H, \mathcal{O}_H(t)] + \frac{\partial \mathcal{O}_H(t)}{\partial t}\bigg|_{\text{intrinsic}}, \quad (1.20)$$

where the last term arises from any intrinsic time dependence that the operator $\mathcal{O}$ may have as a result, for instance, of the variation of some external constraints or potentials.

As in the case of the Schrödinger picture, we can define a unitary operator that generates the time development of the vector $\psi_I(t)$ in the interaction picture. Equations (1.4) and (1.14) imply that this vector obeys the transformation

$$\psi_I(t) = U_I(t, t_0)\,\psi_I(t_0), \qquad (1.21)$$

where the time-translation operator in the interaction picture is

$$U_I(t, t_0) = \exp[(i/\hbar)H_0 t]\,\exp[-(i/\hbar)\,H(t - t_0)]\,\exp[-(i/\hbar)H_0 t_0]. \qquad (1.22)$$

We can integrate Eq. (1.17) to obtain

$$\psi_I(t) = \psi_I(t_0) - \frac{i}{\hbar}\int_{t_0}^{t} dt'\; V(t')\,\psi_I(t'), \qquad (1.23)$$

or

$$U_I(t, t_0) = 1 - \frac{i}{\hbar}\int_{t_0}^{t} dt'\; V(t')\,U_I(t', t_0). \qquad (1.24)$$

These are the fundamental equations of the time-dependent treatment of scattering processes. One way to calculate $U_I(t, t_0)$ is to expand it in powers of the interaction V. Equation (1.24) may be iterated to yield such an expansion:

$$U_I(t, t_0) = 1 - \frac{i}{\hbar}\int_{t_0}^{t} dt_1\; V(t_1)$$
$$+ \left(-\frac{i}{\hbar}\right)^2 \int_{t_0}^{t} dt_1 \int_{t_0}^{t_1} dt_2\; V(t_1)\,V(t_2) + \cdots. \qquad (1.25)$$

This expansion gives the familiar expressions of time-dependent perturbation theory. Thus, suppose the system is in a state ϕ_i at $t = t_0$, so that $\psi_I(t_0) = \phi_i$. To the first order in V, Eq. (1.23) gives

$$\psi_I^{(1)}(t) = \phi_i - \frac{i}{\hbar}\int_{t_0}^{t} dt_1\; V(t_1)\,\phi_i. \qquad (1.26)$$

If we denote the overlap of $\psi_I(t)$ with ϕ_f by $C_{fi}(t)$, we have for this coefficient

$$C_{fi}^{(1)}(t) = \langle\phi_f\,|\,\psi_I^{(1)}(t)\rangle = \delta_{fi} - \frac{i}{\hbar}\int_{t_0}^{t} dt_1\; \langle\phi_f\,|\,V(t_1)\,|\,\phi_i\rangle. \qquad (1.27)$$

Using the definition of $V(t)$ and the Schrödinger equation for ϕ_i, the matrix element entering here is

$$\langle\phi_f| V(t_1)|\phi_i\rangle = \langle\phi_f| \exp[(i/\hbar)H_0t_1] V \exp[-(i/\hbar)H_0t_1]|\phi_i\rangle$$
$$= \exp[(i/\hbar)(E_f - E_i)t] V_{fi}, \quad (1.28)$$

with $V_{fi} = \langle\phi_f| V|\phi_i\rangle$. We then obtain

$$C_{fi}^{(1)}(t) = \delta_{fi} - \frac{i}{\hbar}\int_{t_0}^{t} dt_1 \exp[(i/\hbar)(E_f - E_i)t_1] V_{fi}, \quad (1.29)$$

which is the standard expression given by first-order perturbation theory. Thus Eq. (1.25) is a formal means of writing the perturbation series for the wave function.

The selection of one of these pictures for use in any particular application is solely a matter of convenience. That is, it follows from the definitions of the state vectors and operators in each of the pictures that matrix elements are the same in any picture. (In other words, the transformation between pictures is a unitary transformation.) As a result, we have the identity

$$\langle\psi_1(t)|\mathcal{O}(t)|\psi_2(t)\rangle = \langle\psi_{1I}(t)|\mathcal{O}_I(t)|\psi_{2I}(t)\rangle = \langle\psi_{1H}|\mathcal{O}_H(t)|\psi_{2H}\rangle, \quad (1,30)$$

where the state vectors in each case have been defined above and the operators are given by

$$\mathcal{O}_I(t) \equiv \exp[(i/\hbar)H_0t]\,\mathcal{O}(t)\exp[-(i/\hbar)H_0t] \quad (1.31)$$

and

$$\mathcal{O}_H(t) \equiv \exp[(i/\hbar)Ht]\,\mathcal{O}(t)\exp[-(i/\hbar)Ht]. \quad (1.32)$$

2. Infinite Limits

All the expressions we have dealt with suffer from one practical difficulty: they have no limits for very early times ($t \to -\infty$) or for very late times ($t \to +\infty$). For instance, the operator of $U_I(t, t_0)$ given by Eq. (1.22) oscillates without limit for large $|t|$. In scattering problems we are interested in very large times since, in general, the total elapsed time from emission of the projectile by the source until its entry into the detector is large compared to the transit time across the target. This formalism will be useful in describing scattering experiments only if expressions exist that do not depend upon when the projectile is emitted or detected. Stated another way, the distance of the source and the detector from the target should not affect the calculated cross section. In practice, this means that one should be able to pass to the limits $t \to \pm\infty$.

To achieve this, the foregoing formalism must be modified in some way. Some authors have attempted to do this by introducing a technique known as "adiabatic switching," in which the interaction is turned off at very early times and very late times. If this switching is done adiabatically, that is, over times that are long compared to characteristic times involved in the interaction with the target, there should be no observable consequences ("transients") of the switching. Although this procedure does give a precise meaning to all infinite integrals, no one has succeeded in reproducing with this method the results obtained in Chapter 7 without the use of a perturbation expansion. Since the use of such an expansion hinders the analysis of problems in which the interaction is so strong as to allow the projectile to be bound to the target, this method is not adequate for our purposes.

We shall instead use the wave packet description. We will obtain a definition of the limit $U_I(t, \pm\infty)$ which retains all the formal properties of $U_I(t, t_0)$ and which permits us to rederive the results obtained in the previous chapter.

Let us begin with the U matrix for finite times which, from Eq. (1.22), is

$$U_I(t, t_0) = \exp[(i/\hbar)H_0 t] \exp[-(i/\hbar) H(t - t_0)] \exp[-(i/\hbar)H_0 t_0]. \quad (2.1)$$

If this is differentiated with respect to t, we obtain the operator form of the Schrödinger equation

$$\frac{\partial U_I(t, t_0)}{\partial t} = -\frac{i}{\hbar} V(t) \, U_I(t, t_0). \quad (2.2)$$

Using the boundary condition $U_I(t_0, t_0) = 1$, this can be integrated to give

$$U_I(t, t_0) = 1 - \frac{i}{\hbar} \int_{t_0}^{t} dt' \, V(t_1) \, U_I(t_1, t_0), \quad (2.3)$$

which is just Eq. (1.24). The initial time t_0 enters into $U_I(t, t_0)$ in very much the same way as the later time t. Then if we differentiate Eq. (2.1) with respect to t_0, we find

$$\frac{\partial U_I(t, t_0)}{\partial t_0} = \frac{i}{\hbar} U_I(t, t_0) \, V(t_0). \quad (2.4)$$

Integration of Eq. (2.4) gives the result

$$U_I(t, t_0) = 1 - \frac{i}{\hbar} \int_{t_0}^{t} dt' \, U_I(t, t') \, V(t'). \quad (2.5)$$

Equations (2.3) and (2.5) provide alternative but equivalent equations for $U_I(t, t_0)$. However, neither of these has a limit as $|t|$ or $|t_0|$ increases with-

2. Infinite Limits

out limit, nor do these limits exist when state vectors are formed by acting with these operators on free-particle eigenstates ϕ_i. The difficulty is simply that the free-particle eigenfunctions extend over all space with uniform density, so that the projectile never ceases to interact with the target.

We will avoid this difficulty by using a wave packet to describe the initial state, in much the same way as we did in Chapter 7. Suppose the system is initially in the state ϕ_I. Since this is a state vector for a free particle, it is independent of the time in the interaction picture, and we have the boundary condition on the scattering state vector

$$\lim_{t \to -\infty} \psi_I(t) = \phi_I. \tag{2.6}$$

If ϕ_I is a localized wave packet, we now are able to show that the vector $U_I(t, t_0) \psi_I(t_0)$ will have a well-defined limit as $t_0 \to -\infty$.

From Eq. (2.5) we have

$$\psi_I(t) = U_I(t, t_0) \psi_I(t_0)$$
$$= \psi_I(t_0) - (i/\hbar) \int_{t_0}^{t} dt' \, U_I(t, t') V(t') \psi_I(t_0). \tag{2.7}$$

Allowing t_0 to approach $-\infty$, this becomes

$$\psi_I(t) = \phi_I - (i/\hbar) \int_{-\infty}^{t} dt' \, U_I(t, t') V(t') \phi_I, \tag{2.8}$$

provided that the infinite integral converges at its lower limit. For a finite-range potential the product $V(t') \phi_I$ will vanish as t' approaches $-\infty$ since the wave packet ϕ_I no longer overlaps the potential $V(t')$. We can see this in the more familiar Schrödinger picture, in which we have

$$V(t') \phi_I = \exp[(i/\hbar)H_0 t'] \, V \exp[-(i/\hbar)H_0 t'] \, \phi_I$$
$$= \exp[(i/\hbar)H_0 t'] \, V \phi(t'), \tag{2.9}$$

where $\phi(t')$ is the free-particle wave packet in the Schrödinger picture. In the coordinate representation the product $V \phi(t')$ is $V(\mathbf{r}) \phi(\mathbf{r}, t')$, which vanishes for all positions $\mathbf{r}$ as $t' \to -\infty$. Then the limit in Eq. (2.8) does exist, and this result provides an explicit formula for the scattering state vector in the interaction picture.

While this result is perfectly well defined, it cannot yet be written as $U_I(t, -\infty) \phi_I$, since this time-translation operator is not defined. However, we can give this limit a precise meaning, and can at the same time avoid the use of wave packets in performing practical calculations, if we introduce here a particular form for the wave packet. Since we will ultimately pass to a limit in which this form no longer matters, this will not restrict the validity of our result.

The packet we will use to describe the relative motion is given, in the Schrödinger picture, by

$$\phi(\mathbf{r}, t) = \exp(i\mathbf{k} \cdot \mathbf{r}) \exp[-(i/\hbar)E_k t] \exp[-(|x|/D)] \exp[-(|y|/D)] \\ \times \exp[-(|z - vt|/D)]; \qquad (2.10)$$

in practice, this must be supplemented by the internal wave functions and spin functions of the projectile and target. This packet has the general form found in Chapter 2 for a free-particle packet moving with a velocity v along the z axis. If the packet is large compared to the range of the force ($D \gg R$), the product $V(\mathbf{r}) \phi(\mathbf{r}, t)$ is nonzero only for $|\mathbf{r}| \ll D$. In that region we have the approximate result

$$V(\mathbf{r}) \phi(\mathbf{r}, t) \simeq V(\mathbf{r}) \exp(i\mathbf{k} \cdot \mathbf{r}) \exp[-(i/\hbar)E_k t] \exp[-(v/D)|t|] \\ = V(\mathbf{r}) \phi_i(\mathbf{r}, t) \exp[-(v/D)|t|], \qquad (2.11)$$

with $\phi_i(\mathbf{r}, t)$ as usual being a plane-wave eigenfunction. As a result, we can replace the wave packet $\phi(t)$ by the damped plane wave

$$\phi(t) \simeq \phi_i(t) \exp[-(v/D)|t|] = \exp[-(v/D)|t|] \exp[-(i/\hbar)H_0 t] \phi_i$$

(cf. Eq. (1.12)). Returning to the interaction picture, we then have

$$V(t') \phi_I = \exp[(i/\hbar)H_0 t'] V \phi(t') \simeq \exp[-(v/D)|t'|] V(t') \phi_i. \qquad (2.12)$$

Making this replacement in Eq. (2.8) and passing to the limit $D \to \infty$, we have

$$\psi_I(t) = \phi_i - (i/\hbar) \lim_{\varepsilon \to 0} \int_{-\infty}^{t} dt' \exp[-(\varepsilon/\hbar)|t'|] U_I(t, t') V(t') \phi_i, \qquad (2.13)$$

where we define $\varepsilon = (\hbar v/D)$. (This may be viewed as the characteristic spread of energies in the packet.) This presents the state vector in the form of an operator acting on the initial free-particle eigenstate. If we define the limit of the time-translation operator for very early or very late times by

$$U_I(t, \pm\infty) \equiv 1 - (i/\hbar) \lim_{\varepsilon \to 0} \int_{\pm\infty}^{t} dt' \exp[-(\varepsilon/\hbar)|t'|] U_I(t, t') V(t'), \qquad (2.14)$$

we then have the desired result

$$\psi_I(t) = U_I(t, -\infty) \phi_i. \qquad (2.15)$$

2. Infinite Limits

The definition of Eq. (2.14) may appear rather arbitrary, since it followed from the use of the particular wave packet given in Eq. (2.10). However, it is in fact perfectly general, in much the same way as the definition of the outgoing-wave Green's function, $\lim_{\varepsilon \to 0} 1/(E - H_0 + i\varepsilon)$, is perfectly general. We saw in Chapter 4 that there are many equivalent ways to allow the spherical boundary surface to go to infinity, but that they all lead to the same result. Similarly, there are other ways to define $U_I(t, \pm\infty)$ (for instance, any other convergence function could replace $\exp[-(\varepsilon/\hbar)|t|]$), but the resulting operator will give the same results, independent of which technique is used.

One may follow a similar limiting procedure for the final state to arrive at the definition

$$U_I(\pm\infty, t_0) \equiv 1 - (i/\hbar) \lim_{\varepsilon \to 0} \int_{t_0}^{\pm\infty} dt' \exp[-(\varepsilon/\hbar)|t'|] V(t') U_I(t', t_0). \quad (2.16)$$

Both this and Eq. (2.14) involve $U_I(t, t_0)$ for only finite times, and they both have the properties exhibited by time-translation operators with finite arguments. For instance, they satisfy the sequential relation

$$U_I(t, \pm\infty) = U_I(t, t') U_I(t', \pm\infty) \quad (2.17)$$

and the adjoint relation

$$U_I(t, \pm\infty)^\dagger = U_I(\pm\infty, t) = U_I^{-1}(t, \pm\infty). \quad (2.18)$$

An important limiting case is obtained by letting $t \to \infty$ in Eq. (2.15) to give the state vector at very late times. This limit exists because at late times the projectile is again a free particle and is described by a constant state vector. One can follow essentially the same procedure as that used to derive Eqs. (2.14) and (2.16) to show that the proper limit is

$$U_I(\infty, -\infty) = 1 - (i/\hbar) \lim_{\varepsilon \to 0} \int_{-\infty}^{\infty} dt' \exp[-(\varepsilon/\hbar)|t'|] U_I(\infty, t') V(t'), \quad (2.19)$$

where $U_I(\infty, t')$ is defined by Eq. (2.16). Alternatively, one could consider the limit of Eq. (2.16) as $t_0 \to -\infty$. Here one finds

$$U_I(\infty, -\infty) = 1 - (i/\hbar) \lim_{\varepsilon \to 0} \int_{-\infty}^{\infty} dt' \exp[-(\varepsilon/\hbar)|t'|] V(t') U_I(t', -\infty), \quad (2.20)$$

with $U_I(t', -\infty)$ defined by Eq. (2.14). If Eq. (2.14) is inserted explicitly into Eq. (2.20), and Eq. (2.16) is inserted into Eq. (2.19), we can see that these two results are identical, as they must be.

As an additional consistency check, we may show that these definitions imply the relation
$$U_I(\infty, -\infty) = U_I(\infty, 0) U_I(0, -\infty), \qquad (2.21)$$
which is the analog of a similar relation for finite times. Using Eqs. (2.18) and (2.14), we have

$$\begin{aligned}
1 &= U_I(\infty, 0) U_I(0, \infty) \\
&= U_I(\infty, 0) \left\{ 1 - (i/\hbar) \lim_{\varepsilon \to 0} \int_\infty^0 dt' \exp[-(\varepsilon/\hbar)|t'|] U_I(0, t') V(t') \right\} \\
&= U_I(\infty, 0) + (i/\hbar) \lim_{\varepsilon \to 0} \int_0^\infty dt' \exp[-(\varepsilon/\hbar)|t'|] U_I(\infty, t') V(t'). \quad (2.22)
\end{aligned}$$

Inserting this into Eq. (2.19) yields

$$\begin{aligned}
U_I(\infty, -\infty) &= U_I(\infty, 0) - \frac{i}{\hbar} \lim_{\varepsilon \to 0} \int_{-\infty}^0 dt' \exp\left(-\frac{\varepsilon}{\hbar}|t'|\right) U_I(\infty, t') V(t') \\
&= U_I(\infty, 0) U_I(0, -\infty). \qquad (2.23)
\end{aligned}$$

3. Relation to the Time-Independent Theory

Because of the sequential relation, Eq. (2.17), it is only necessary to consider the state vector at $t = 0$; the vector at any other time can be obtained by a *finite* time translation. Using Eqs. (2.1) and (2.13) and recognizing that $\psi_I(t = 0) = \psi(t = 0)$, we find that the state vector at $t = 0$ is

$$\begin{aligned}
\psi(t = 0) &= U_I(0, -\infty) \phi_i \\
&= \phi_i - \frac{i}{\hbar} \lim_{\varepsilon \to 0} \int_{-\infty}^0 dt' \exp\left(-\frac{\varepsilon}{\hbar}|t'|\right) U_I(0, t') V(t') \phi_i \\
&= \phi_i - \frac{i}{\hbar} \lim_{\varepsilon \to 0} \int_{-\infty}^0 dt' \exp\left(-\frac{\varepsilon}{\hbar}|t'|\right) \exp\left(+\frac{i}{\hbar} H t'\right) \\
&\quad \times V \exp\left(-\frac{i}{\hbar} H_0 t'\right) \phi_i \\
&= \phi_i - \frac{i}{\hbar} \lim_{\varepsilon \to 0} \int_{-\infty}^0 dt' \exp\left(-\frac{\varepsilon}{\hbar}|t'|\right) \exp\left(\frac{i}{\hbar}(H - E_i) t'\right) V \phi_i \\
&= \phi_i + \lim_{\varepsilon \to 0} \frac{1}{E_i - H + i\varepsilon} V \phi_i. \qquad (3.1)
\end{aligned}$$

This is the explicit form of the state vector $\psi_i^{(+)}$, the outgoing-wave solution of the Schrödinger equation $(E_i - H)\psi_i^{(+)} = 0$. Equation (3.1) demonstrates

3. Relation to the Time-Independent Theory

that the limiting process defined by Eq. (2.14) enables us to generate an eigenfunction of the Schrödinger equation starting with the free-particle state ϕ_i. Thus we find

$$\psi_i^{(+)} = U_I(0, -\infty) \phi_i, \tag{3.2}$$

enabling us to identify $U_I(0, -\infty)$ with $\Omega^{(+)}$, the outgoing-wave matrix:

$$U_I(0, -\infty) = \Omega^{(+)} = 1 - \frac{i}{\hbar} \lim_{\varepsilon \to 0} \int_{-\infty}^{0} dt' \exp\left(-\frac{\varepsilon}{\hbar}|t'|\right) \exp\left(\frac{i}{\hbar} H t'\right)$$

$$\times V \exp\left(-\frac{i}{\hbar} H_0 t'\right). \tag{3.3}$$

Similarly, the ingoing-wave state is generated by

$$\psi_i^{(-)} = U_I(0, +\infty) \phi_i = \phi_i + \lim_{\varepsilon \to 0} \frac{1}{E_i - H - i\varepsilon} V \phi_i. \tag{3.4}$$

Since the free-particle eigenstates ϕ_i form a complete set ($\sum_i \phi_i \phi_i^\dagger = 1$), Eq. (3.2) can be inverted to give

$$U_I(0, -\infty) = \sum_i \psi_i^{(+)} \phi_i^\dagger = \sum_i |\psi_i^{(+)}\rangle \langle \phi_i |, \tag{3.5}$$

where the sum runs over all continuum states ($E_i > 0$). Similarly, we have

$$U_I(0, +\infty) = \sum_i |\psi_i^{(-)}\rangle \langle \phi_i |. \tag{3.6}$$

We can use these relations to show that these operators are not in general unitary. Thus, one product is unity, since the state $\psi_i^{(+)}$ is normalized:

$$U_I(0, -\infty)^\dagger U_I(0, -\infty) = \sum_{g,h} |\phi_g\rangle \langle \psi_g^{(+)} | \psi_h^{(+)} \rangle \langle \phi_h |$$

$$= \sum_g |\phi_g\rangle \langle \phi_g | = 1; \tag{3.7}$$

the other, giving the completeness property, need not be unity:

$$U_I(0, -\infty) U_I(0, -\infty)^\dagger = \sum_{g,h} |\psi_g^{(+)}\rangle \langle \phi_g | \phi_h \rangle \langle \psi_h^{(+)} |$$

$$= \sum_g |\psi_g^{(+)}\rangle \langle \psi_g^{(+)} |. \tag{3.8}$$

If the eigenstates of H include bound states, this sum is not unity, and $U_I(0, -\infty)$ is not unitary.

We can use these definitions of the wave operators to derive the implicit form of $\psi_i^{(+)}$ without using algebraic manipulations; this can then be taken as the justification for the manipulations performed in Chapters 6 and 7.

Using Eqs. (2.17) and (2.18), we find

$$U_I(0, -\infty) = 1 - U_I(-\infty, 0) U_I(0, -\infty) + U_I(0, -\infty)$$
$$= 1 - \{U_I(-\infty, 0) - 1\} U_I(0, -\infty)$$
$$= 1 - \frac{i}{\hbar} \lim_{\varepsilon \to 0} \int_{-\infty}^{0} dt' \exp\left(-\frac{\varepsilon}{\hbar}|t'|\right) V(t') U_I(t', 0) U_I(0, -\infty)$$
$$= 1 - \frac{i}{\hbar} \lim_{\varepsilon \to 0} \int_{-\infty}^{0} dt' \exp\left(-\frac{\varepsilon}{\hbar}|t'|\right) V(t') U_I(t', -\infty). \quad (3.9)$$

Allowing $U_I(0, -\infty)$ to act upon the state ϕ_i, we then obtain

$$\psi_i^{(+)} = U_I(0, -\infty) \phi_i$$
$$= \phi_i - \frac{i}{\hbar} \lim_{\varepsilon \to 0} \int_{-\infty}^{0} dt' \exp\left(\frac{\varepsilon}{\hbar}|t'|\right) V(t') U_I(t', 0) \psi_i^{(+)}$$
$$= \phi_i + \lim_{\varepsilon \to 0} \frac{1}{E_i - H_0 + i\varepsilon} V \psi_i^{(+)}, \quad (3.10)$$

using the definitions of $V(t)$ and $U_I(t, 0)$ and the Schrödinger equation for $\psi_i^{(+)}$.

The state vector at any time is given by

$$\psi_I(t) = U_I(t, 0) \psi_i^{(+)} = U_I(t, -\infty) \phi_i. \quad (3.11)$$

At very late times this becomes

$$\psi_I(t \to \infty) = U_I(\infty, -\infty) \phi_i, \quad (3.12)$$

where $U_I(\infty, -\infty)$ is given by Eq. (2.19) or (2.20). We will see in the next chapter that $U_I(\infty, -\infty)$ is the scattering operator S which gives the late time behavior of the state vector. Its matrix elements yield the S matrix:

$$S_{fi} = \langle \phi_f | U_I(\infty, -\infty) | \phi_i \rangle$$
$$= \delta_{fi} - \frac{i}{\hbar} \lim_{\varepsilon \to \infty} \int_{-\infty}^{\infty} dt' \exp\left(-\frac{\varepsilon}{\hbar}|t'|\right)$$
$$\times \langle \phi_f | V(t') U_I(t', -\infty) | \phi_i \rangle$$
$$= \delta_{fi} - \frac{i}{\hbar} \lim_{\varepsilon \to 0} \int_{-\infty}^{\infty} dt' \exp\left(-\frac{\varepsilon}{\hbar}|t'|\right)$$
$$\times \langle \phi_f | \exp\left(\frac{i}{\hbar} H_0 t'\right) V \exp\left(-\frac{i}{\hbar} H t'\right) | \psi_i^{(+)} \rangle$$
$$= \delta_{fi} - 2\pi i\, \delta(E_i - E_f) \langle \phi_f | V | \psi_i^{(+)} \rangle. \quad (3.13)$$

4. The Cross Section

The central result of the time-dependent theory is the transition rate, defined as the rate of change of the probability of finding the system in a particular final state. If the system is in the state ϕ_i at very early times, then at any time t the system is described by a state vector in the interaction picture as

$$\psi_I(t) = U_I(t, 0) \psi_I(t = 0) = \exp[(i/\hbar)H_0 t] \exp[-(i/\hbar)Ht] \psi_i^{(+)}$$
$$= \exp[(i/\hbar)(H_0 - E_i)t] \psi_i^{(+)}. \qquad (4.1)$$

The probability amplitude for finding the system in the free-particle state ϕ_f at any time t is $\langle \phi_f | \psi_I(t) \rangle$, and the transition rate is

$$w_{fi} = \frac{d}{dt} |\langle \phi_f | \psi_I(t) \rangle|^2. \qquad (4.2)$$

Since $(d/dt) \psi_I(t) = (1/i\hbar) V(t) \psi_I(t)$, for ordinary scattering this becomes

$$w_{fi} = (1/i\hbar) \exp[(i\hbar)(E_f - E_i)t] \langle \phi_f | V | \psi_i^{(+)} \rangle \langle \phi_f | \psi_I(t) \rangle^*$$
$$+ \text{complex conjugate.} \qquad (4.3)$$

Here we have used the fact that $\psi_i^{(+)}$ is an eigenstate of H and that H_0 is Hermitian when taken between ϕ_f and $V\psi_i^{(+)}$, if V is a finite-range potential. Use of Eqs. (3.10) and (4.1) shows the probability amplitude to be

$$\langle \phi_f | \psi_I(t) \rangle = \delta_{fi} - \lim_{\varepsilon \to 0} \frac{\exp[(i/\hbar)(E_f - E_i)t]}{E_f - E_i - i\varepsilon} \langle \phi_f | V | \psi_i^{(+)} \rangle. \qquad (4.4)$$

Thus the transition rate is

$$w_{fi} = \frac{1}{i\hbar} \langle \phi_i | V | \psi_i^{(+)} \rangle \delta_{fi} - \frac{1}{i\hbar} \lim_{\varepsilon \to 0} \frac{1}{E_f - E_i + i\varepsilon}$$
$$\times |\langle \phi_f | V | \psi_i^{(+)} \rangle|^2 + \text{complex conjugate}$$
$$= \frac{2}{\hbar} \text{Im} \langle \phi_i | V | \psi_i^{(+)} \rangle \delta_{fi} + \frac{2\pi}{\hbar} |\langle \phi_f | V | \psi_i^{(+)} \rangle|^2 \delta(E_f - E_i), \qquad (4.5)$$

independent of the time. Usually ϕ_f is a continuum state, and one must in practice sum over the range of momenta accepted by an experimental detector; this causes the delta function to be replaced by the density of states $\rho_f(E_i)$, as in Eq. (1.31) of Chapter 7.

In the case of a rearrangement collision, one can follow the same derivation for $w_{f'i}$ except that the interaction picture that is most convenient is that defined by $\psi_I(t) = \exp[(i/\hbar)H_0't] \psi(t)$, when the final state is an eigenfunction of H_0'. The result is identical with Eq. (4.5), but with ϕ_f replaced by $\phi_{f'}$ and V by V'.

Since the normalization $\langle \psi_I(t) | \psi_I(t) \rangle$ is constant, the *total* transition probability $\sum_f |\langle \phi_f | \psi_I(t) \rangle|^2$, where $\sum_f$ runs over all states in the complete set ϕ_f, must be independent of the time. This leads to the relation

$$w_{ii} = -\sum_{f \neq i} w_{fi}. \tag{4.6}$$

This relation expressing the conservation of probability, that is, the requirement that whatever leaves state i must appear in some state f. Taken together with Eq. (4.5), this implies the condition

$$\operatorname{Im} \langle \phi_i | V | \psi_i^{(+)} \rangle = -\pi \sum_f |\langle \phi_f | V | \psi_i^{(+)} \rangle|^2 \, \delta(E_f - E_i). \tag{4.7}$$

This is the familiar optical theorem.

The differential cross section is equal to the transition rate divided by the incident flux. If the incident state is normalized so that there is one particle per unit volume ($\phi_i(\mathbf{r}) = e^{i\mathbf{k}\cdot\mathbf{r}}$), the incident flux is simply v_i. Then the cross section is given by

$$\sigma_{fi} = \frac{w_{fi}}{v_i} = \frac{2\pi}{\hbar v_i} |\langle \phi_f | V | \psi_i^{(+)} \rangle|^2 \, \delta(E_f - E_i). \tag{4.8}$$

Summing over a finite set of states, this becomes

$$\frac{d\sigma_{fi}}{d\Omega} = \frac{2\pi}{\hbar v_i} |\langle \phi_f | V | \psi_i^{(+)} \rangle|^2 \, \rho_f(E_i), \tag{4.9}$$

which is the result obtained via wave packet methods in the previous chapter.

Chapter

9

The S Matrix and the K Matrix

Matrices that yield the full set of transition probabilities play a fundamental role in the general theory of collision processes. In previous chapters we examined the T matrix, which directly gives the scattering cross section. In this chapter we shall investigate the S matrix and the K matrix, two matrices that convey the same basic information but have different properties.

1. The S Matrix

In Chapter 7 we found that a scattering operator S arose in deriving the late-time behavior of the state vector $\psi(t)$. Let us examine further the meaning of the scattering operator, or S matrix.

The S matrix relates the scattering state for very late times to the initial state. Because of the intrinsic time dependence of this relation, it can be properly obtained only with the use of wave packets that are able to reproduce the time dependence of the collision process. From Eqs. (1.14) and (1.15) of Chapter 7 we have the asymptotic form for the scattering state vector

$$\psi(t) \xrightarrow[t \to \infty]{} \int \frac{d^3k}{(2\pi)^3} A(\mathbf{k}) S\phi_{\mathbf{k},\alpha} \exp[-(i/\hbar)E_k t]$$

$$= S \int \frac{d^3k}{(2\pi)^3} A(\mathbf{k}) \phi_{\mathbf{k},\alpha} \exp[-(i/\hbar)E_k t]. \tag{1.1}$$

The second equality in Eq. (1.1) follows from the fact that $A(\mathbf{k})$ is a number which therefore commutes with the operator S. The integral that remains is just $\phi(t)$, the free-particle packet, so that we have the result

$$\psi(t) \xrightarrow[t \to \infty]{} S\,\phi(t). \tag{1.2}$$

At very late times the S matrix transforms the free-particle packet into the scattering state. This is a formal restatement of the results obtained in Section 2 of Chapter 2, where we used the spatial representation of these packets. There we found that, a long time after the scattering has taken place, the wave function is again simply related to the initial wave packet. Equation (1.2) expresses that same relation formally.

SYMMETRIC FORM OF THE S MATRIX

In Section 1 of Chapter 7 we also computed the overlap of the final state $\phi_{\mathbf{k}',\alpha'} \exp[-(i/\hbar)E_{k'}t]$ with the wave packet $\psi(t)$. We found that in the limit of very late times this scalar product approached a constant value, viz.,

$$\exp[(i/\hbar)E_{k'}t] \langle \phi_{\mathbf{k}',\alpha'} | \psi(t) \rangle \xrightarrow[t \to \infty]{} \int \frac{d^3k}{(2\pi)^3} A(\mathbf{k}) [\langle \phi_{\mathbf{k}',\alpha'} | \phi_{\mathbf{k},\alpha} \rangle$$
$$- 2\pi i\, \delta(E_k - E_{k'}) \langle \phi_{\mathbf{k}',\alpha'} | V | \psi_{\mathbf{k},\alpha}^{(+)} \rangle]. \quad (1.3)$$

The expression in brackets is identified as the S-matrix element

$$\langle \phi_{\mathbf{k}',\alpha'} | S | \phi_{\mathbf{k},\alpha} \rangle = \langle \phi_{\mathbf{k}',\alpha'} | \phi_{\mathbf{k},\alpha} \rangle - 2\pi i\, \delta(E_k - E_{k'})$$
$$\times \langle \phi_{\mathbf{k}',\alpha'} | V | \psi_{\mathbf{k},\alpha}^{(+)} \rangle. \quad (1.4)$$

The S matrix then gives the overlap of $\psi(t)$ with each final state, in the limit as $t \to \infty$.

We can obtain another convenient expression for the S matrix by introducing the ingoing-wave state vector $\psi_{\mathbf{k}',\alpha'}^{(-)} \exp[-(i/\hbar)E_{k'}t]$ and recognizing that the overlap of this vector with $\psi(t)$ is equal to the right-hand side of Eq. (1.3) for all times. We shall now demonstrate this equality.

We first observe that the scalar product $\exp[(i/\hbar)E_{k'}t] \langle \psi_{\mathbf{k}',\alpha'}^{(-)} | \psi(t) \rangle$ is independent of time. This is because $\psi_{\mathbf{k}',\alpha'}^{(-)} \exp[-(i/\hbar)E_{k'}t]$ and $\psi(t)$ are solutions of the time-dependent Schrödinger equation $i\hbar \partial \psi / \partial t = H\psi$. If H is Hermitian and time independent, the time derivative of this scalar product is zero. The time independence also follows from the observation that the scalar product may be expressed as

$$\exp[(i/\hbar)E_{k'}t] \langle \psi_{\mathbf{k}',\alpha'}^{(-)} | \psi(t) \rangle = \int \frac{d^3k}{(2\pi)^3} A(\mathbf{k}) \langle \psi_{\mathbf{k}',\alpha'}^{(-)} | \psi_{\mathbf{k},\alpha}^{(+)} \rangle$$
$$\times \exp[-(i/\hbar)(E_k - E_{k'})t], \quad (1.5)$$

and, as we shall see below, the overlap of $\psi_{\mathbf{k},\alpha}^{(+)}$ with $\psi_{\mathbf{k}',\alpha'}^{(-)}$ is zero unless $E_k = E_{k'}$.

1. The S Matrix

While Eq. (1.5) is a constant independent of time, for very late times it can be expressed as $\exp[(i/\hbar)E_{k'}t]\langle\phi_{k',\alpha'}|\psi(t)\rangle$, the scalar product of $\psi(t)$ with the free-particle state. That is, the contribution of the ingoing spherical wave vanishes for very late times. This intuitively satisfactory result is demonstrated explicitly as follows:

The ingoing-wave state vector is

$$\psi_{k',\alpha'}^{(-)} = \phi_{k',\alpha'} + \frac{1}{E_{k'} - H - i\varepsilon} V\phi_{k',\alpha'}. \tag{1.6}$$

The overlap of the second, ingoing-wave term with $\psi(t)$ is

$$\int \frac{d^3k}{(2\pi)^3} A(\mathbf{k}) \exp[-(i/\hbar)E_k t] \langle \phi_{k',\alpha'} | V \frac{1}{E_{k'} - H + i\varepsilon} | \psi_{k,\alpha}^{(+)} \rangle$$

$$= \int \frac{d^3k}{(2\pi)^3} A(\mathbf{k}) \frac{\exp[-(i/\hbar)E_k t]}{E_{k'} - E_k + i\varepsilon} \langle \phi_{k',\alpha'} | V | \psi_{k,\alpha}^{(+)} \rangle, \tag{1.7}$$

since $H\psi_{k,\alpha}^{(+)} = E_k \psi_{k,\alpha}^{(+)}$. As $t \to \infty$ the integral can be performed by closing the contour in the lower half of the complex energy plane. As discussed in Section 1 of Chapter 7, the vector $V\psi_{k,\alpha}^{(+)}$ will not have any singularities, so long as V has a finite range. The only singularity in the integral then lies in the upper half plane at $E_k = E_{k'} + i\varepsilon$, so that Eq. (1.7), the contribution of the ingoing wave, is zero for very late times.

Equation (1.5) is then equal to $\exp[(i/\hbar)E_{k'}t]\langle\phi_{k',\alpha'}|\psi(t)\rangle$ in the limit as $t \to \infty$, and this is the right-hand side of Eq. (1.3). Comparing these two equivalent results, we find an alternative expression for the S matrix:

$$\langle \phi_{k',\alpha'} | S | \phi_{k,\alpha} \rangle = \langle \psi_{k',\alpha'}^{(-)} | \psi_{k,\alpha}^{(+)} \rangle, \tag{1.8}$$

or, in our simpler notation,

$$S_{fi} = \langle \psi_f^{(-)} | \psi_i^{(+)} \rangle. \tag{1.9}$$

A more succinct but less revealing derivation of this is obtained from Chapter 8 by recognizing that $U_I(\infty, -\infty)$ is just S, the scattering operator; Eq. (1.9) follows then directly from Eq. (2.21) of Chapter 8.

The S-matrix element S_{fi} is the amplitude for a system initially in a state ϕ_i to go into a final state ϕ_f, and Eq. (1.9) expresses this in terms of the overlap of the exact scattering states that asymptotically go into these free-particle states. This expression for the S matrix has many virtues. Unlike our previous expressions, it is symmetric between initial and final states, the only distinction between them being the necessary time ordering shown in the specification of outgoing and ingoing waves. In addition, this expression does not depend on any particular separation of the Hamiltonian into an unperturbed Hamiltonian H_0 and an interaction V, and it has the same form for both ordinary and rearrangement collisions.

Let us use Eq. (1.9) to obtain our previous expressions for the S matrix. Let us first consider ordinary collisions. From the integral equation for the incoming wave,

$$\psi_f^{(-)} = \phi_f + \frac{1}{E_f - H - i\varepsilon} V\phi_f \qquad (1.10)$$

and the identity

$$\frac{1}{E_f - H - i\varepsilon} = \frac{1}{E_f - H + i\varepsilon} + 2\pi i\, \delta(E_f - H), \qquad (1.11)$$

we obtain the relation between the incoming and outgoing wave functions

$$\psi_f^{(-)} = \psi_f^{(+)} + 2\pi i\, \delta(E_f - H) V\phi_f. \qquad (1.12)$$

Inserting this into Eq. (1.9) gives

$$S_{fi} = \langle \psi_f^{(+)} | \psi_i^{(+)} \rangle - 2\pi i \langle \phi_f | V\, \delta(E_f - H) | \psi_i^{(+)} \rangle. \qquad (1.13)$$

We could show using the same technique as in Section 2 of Chapter 6 that $\psi_f^{(+)}$ and $\psi_i^{(+)}$ are orthonormal. Then, using this and the fact that $H\psi_i^{(+)} = E_i\psi_i^{(+)}$, we have

$$S_{fi} = \delta_{fi} - 2\pi i\, \delta(E_f - E_i) \langle \phi_f | V | \psi_i^{(+)} \rangle. \qquad (1.14)$$

This will be recognized as our previous expression for the S matrix, Eq. (1.4). If we were instead to substitute for $\psi_i^{(+)}$, we would obtain the reciprocal form

$$S_{fi} = \delta_{fi} - 2\pi i\, \delta(E_f - E_i) \langle \psi_f^{(-)} | V | \phi_i \rangle. \qquad (1.15)$$

As we can see, S_{fi} is zero when the initial and final states do not have the same energy.

For rearrangement collisions we are interested in the S-matrix element $\langle \psi_{f'}^{(-)} | \psi_i^{(+)} \rangle$, where

$$\psi_{f'}^{(\pm)} = \phi_{f'} + \frac{1}{E_{f'} - H \pm i\varepsilon} V'\phi_{f'}$$

$$= \psi_{f'}^{(\mp)} \mp 2\pi i\, \delta(E_{f'} - H) V'\phi_{f'}. \qquad (1.16)$$

We can use the same derivation as in the ordinary scattering case, except that now $\psi_{f'}^{(+)}$ is orthogonal to $\psi_i^{(+)}$. To show this orthogonality, we follow

Eqs. (2.44)–(2.49) of Chapter 6. The first expression in Eq. (1.16) leads to

$$\langle \psi_{f'}^{(+)} | \psi_i^{(+)} \rangle = \langle \phi_{f'} | \psi_i^{(+)} \rangle + \langle \phi_{f'} | V' \frac{1}{E_{f'} - H - i\varepsilon} | \psi_i^{(+)} \rangle$$

$$= \langle \phi_{f'} | \psi_i^{(+)} \rangle + \frac{1}{E_{f'} - E_i - i\varepsilon} \langle \phi_{f'} | V' | \psi_i^{(+)} \rangle. \quad (1.17)$$

From Eq. (2.19) of Chapter 7,

$$\psi_i^{(+)} = \frac{1}{E_i - H_0' + i\varepsilon} V' \psi_i^{(+)}. \quad (1.18)$$

We see that the first term in Eq. (1.17) is

$$\langle \phi_{f'} | \psi_i^{(+)} \rangle = \langle \phi_{f'} | \frac{1}{E_i - H_0' + i\varepsilon} V' | \psi_i^{(+)} \rangle$$

$$= \frac{1}{E_i - E_{f'} + i\varepsilon} \langle \phi_{f'} | V' | \psi_i^{(+)} \rangle. \quad (1.19)$$

This cancels the second term of Eq. (1.17), so that

$$\langle \psi_{f'}^{(+)} | \psi_i^{(+)} \rangle = 0. \quad (1.20)$$

The analog of Eq. (1.13) for rearrangement collisions is then

$$S_{f'i} = \langle \psi_{f'}^{(+)} | \psi_i^{(+)} \rangle - 2\pi i \langle \phi_{f'} | V' \delta(E_{f'} - H) | \psi_i^{(+)} \rangle$$
$$= -2\pi i \, \delta(E_{f'} - E_i) \langle \phi_{f'} | V' | \psi_i^{(+)} \rangle, \quad (1.21)$$

in agreement with the results of Section 2 of Chapter 7.

EXISTENCE OF A SCATTERING OPERATOR

Let us consider further the implications of the orthogonality of $\psi_i^{(+)}$ and $\psi_{f'}^{(+)}$. A complete set of outgoing-wave eigenstates of the Hamiltonian must then be composed not only of the states $\psi_f^{(+)}$ corresponding to all the final states ϕ_f, but it must also include the states $\phi_{f'}^{(+)}$ corresponding to all possible *rearranged* final states $\phi_{f'}$. The full S matrix includes matrix elements joining all of these eigenstates, both the rearranged and the original ones.

One must also conclude that the S matrix cannot be obtainable from an operator acting upon the complete set of eigenstates for any one Hamiltonian H_0. If we were to introduce wave operators for the outgoing and ingoing states, we might think we could use Eq. (1.9) to obtain an operator expression for

9. THE S MATRIX AND THE K MATRIX

the scattering matrix, viz.,

$$S_{fi} = \langle \psi_f^{(-)} | \psi_i^{(+)} \rangle = \langle \phi_f | \Omega^{(-)\dagger} \Omega^{(+)} | \phi_i \rangle \qquad (1.22)$$

or

$$S = \Omega^{(-)\dagger} \Omega^{(+)}. \qquad (1.23)$$

However, this expression cannot be taken literally. In the case of ordinary scattering, $\Omega^{(-)}$ will satisfy

$$\Omega^{(-)} = 1 + \frac{1}{E - H_0 - i\varepsilon} V \Omega^{(-)}, \qquad (1.24)$$

while for rearrangement collisions $\Omega^{(-)}$ must be replaced by $\Omega'^{(-1)}$, which satisfies

$$\Omega'^{(-)} = 1 + \frac{1}{E - H_0' - i\varepsilon} V' \Omega'^{(-)}. \qquad (1.25)$$

This means that in Eqs. (1.22) and (1.23) either $\Omega^{(-)}$ or $\Omega'^{(-)}$ must be used, depending upon which matrix element is being considered. Equation (1.23) is then not a proper expression for the S operator when rearrangement collisions are possible; it must be viewed only as a symbolic statement that must be properly interpreted for each particular reaction. This must be remembered in later applications, for example in Section 3.

Suppose a legitimate scattering operator that could act upon a particular set of free-particle eigenstates were actually to exist. Then the S matrix could be obtained form its representation either in the set ϕ_f or in the set $\phi_{f'}$, which *separately* form complete sets of eigenstates for the Hamiltonians H_0 and H_0', respectively. This would mean that the full S matrix, including rearrangement matrix elements, could be obtained from the elements $\langle \phi_f | S | \phi_i \rangle$, corresponding to ordinary scattering alone. Physically, this would imply that differential cross sections for reaction processes could be obtained from measurements of elastic scattering alone. Such a result is, of course, not possible.

The free-particle states ϕ_f are in a sense artificial constructs that help us to visualize the asymptotic behavior of the scattering process, but they, in fact, do not *ever* actually describe the system. The only time-independent vectors that can properly describe the system are the full scattering state vectors $\psi_f^{(+)}$. State vectors of this kind corresponding to different initial configurations are mutually orthogonal and, with original and rearranged configurations included together, form a single complete set.

We can still define an operator whose matrix elements in the *proper* representation give the S matrix. The operator

$$S = \sum_i |\psi_i^{(+)}\rangle \langle \psi_i^{(-)}| \qquad (1.26)$$

gives the S matrix in the representation of outgoing-wave eigenstates of the full Hamiltonian, i.e., in the set $\psi_i^{(+)}$. This operator also gives, interestingly,

$$\psi_i^{(+)} = S\psi_i^{(-)}; \qquad (1.27)$$

one may easily check that this satisfies the asymptotic relation, Eq. (1.2).

UNITARITY OF THE S MATRIX

Equation (1.9) is convenient for demonstrating the unitarity of the S matrix. We must first note that, in general, the sets of ingoing- or outgoing-wave states must be supplemented by whatever bound states exist in order to form complete sets of states. The bound states are unique and contribute alike to either set. At the same time, we have seen that the S matrix couples only states of equal energy, so that the scattering states are not coupled to these bound states.

We show first that $S^\dagger S = 1$. The adjoint matrix element is

$$S^\dagger_{fg} = S^*_{gf} = \langle \psi_g^{(-)}|\psi_f^{(+)}\rangle^* = \langle \psi_f^{(+)}|\psi_g^{(-)}\rangle, \qquad (1.28)$$

so that

$$(S^\dagger S)_{fi} = \sum_g S^\dagger_{fg} S_{gi} = \sum_g \langle \psi_f^{(+)}|\psi_g^{(-)}\rangle \langle \psi_g^{(-)}|\psi_i^{(+)}\rangle. \qquad (1.29)$$

The sum here is over all the scattering states including all possible configurations and rearrangements. However, as we noted, the S-matrix elements joining scattering and bound states are zero ($\psi_i^{(+)}$ and the bound states ψ_b are orthogonal), so that the sum can be extended to include the bound states. Using the completeness relation

$$\sum_g |\psi_g^{(-)}\rangle \langle \psi_g^{(-)}| + \sum_b |\psi_b\rangle \langle \psi_b| = 1, \qquad (1.30)$$

we have

$$(S^\dagger S)_{fi} = \langle \psi_f^{(+)}|\psi_i^{(+)}\rangle = \delta_{fi}, \qquad (1.31)$$

or $S^\dagger S = 1$.

This result expresses the conservation of probability. We interpret S_{fi} as the probability amplitude for a state ϕ_i to go into some final state ϕ_f. The total probability that ϕ_i will go to *some* state is $\sum_f |S_{fi}|^2$, which is unity according to Eq. (1.31). For normalizable initial states, the probability amplitude for passage into the state $\phi_{\mathbf{k}',\alpha'}$ is given by Eq. (1.3) or Eq. (1.5). Using Eq. (1.31) in the form

$$\sum_{\alpha'} \int \frac{d^3k'}{(2\pi)^3} \langle \psi_{\mathbf{k}',\alpha'}^{(-)} | \psi_{\mathbf{k}_1,\alpha}^{(+)} \rangle^* \langle \psi_{\mathbf{k}',\alpha'}^{(-)} | \psi_{\mathbf{k}_2,\alpha}^{(+)} \rangle = (2\pi)^3 \, \delta(\mathbf{k}_1 - \mathbf{k}_2), \qquad (1.32)$$

the total probability of finding the system in some final state is found to be

$$\sum_{\alpha'} \int \frac{d^3k'}{(2\pi)^3} |\langle \psi_{\mathbf{k}',\alpha'}^{(-)} | \psi(t) \rangle|^2$$

$$= \sum_{\alpha'} \int \frac{d^3k'}{(2\pi)^3} \left| \int \frac{d^3k}{(2\pi)^3} A(\mathbf{k}) \langle \psi_{\mathbf{k}',\alpha'}^{(-)} | \psi_{\mathbf{k},\alpha}^{(+)} \rangle \exp[-(i/\hbar)E_k t] \right|^2$$

$$= \sum_{\alpha'} \int \frac{d^3k'}{(2\pi)^3} \frac{d^3k_1}{(2\pi)^3} \frac{d^3k_2}{(2\pi)^3} A^*(\mathbf{k}_1) A(\mathbf{k}_2) \langle \psi_{\mathbf{k}',\alpha'}^{(-)} | \psi_{\mathbf{k}_1,\alpha}^{(+)} \rangle^*$$

$$\times \langle \psi_{\mathbf{k}',\alpha'}^{(-)} | \psi_{\mathbf{k}_2,\alpha}^{(+)} \rangle \exp[-(i/\hbar)(E_1 - E_2)t]$$

$$= \int \frac{d^3k}{(2\pi)^3} |A(\mathbf{k})|^2 = \langle \phi(t) | \phi(t) \rangle. \qquad (1.33)$$

Then the total probability of finding the system in some final state is equal to the probability of finding it in the initial state.

If S is to be unitary, it must also satisfy $SS^\dagger = 1$. Again extending the sum over scattering states to be a sum over all states, we have

$$(SS^\dagger)_{fi} = \sum_g S_{fg} S_{gi}^\dagger = \sum_g \langle \psi_f^{(-)} | \psi_g^{(+)} \rangle \langle \psi_g^{(+)} | \psi_i^{(-)} \rangle$$

$$= \langle \psi_f^{(-)} | \psi_i^{(-)} \rangle = \delta_{fi} \qquad (1.34)$$

or $SS^\dagger = 1$.

This result is a statement of the completeness of the set of free-particle states $S | \phi_g \rangle$ and does not follow from Eq. (1.31). In fact, as we saw in Chapter 6, the wave operator satisfies the analog of Eq. (1.31), $\Omega^\dagger \Omega = 1$, but $\Omega \Omega^\dagger \neq 1$ if H has bound states. The reason is simply that the set of states $\Omega | \phi_g \rangle$ is not complete, since it includes the scattering eigenfunctions of H but not the bound states. In contrast to this, the scattering operator S creates a set of free-particle states, and Eq. (1.34) implies the completeness relations

$$\sum_g | S\phi_g \rangle \langle S\phi_g | = \sum_g S | \phi_g \rangle \langle \phi_g | S^\dagger = SS^\dagger = 1. \qquad (1.35)$$

Thus, unlike Ω, the scattering operator S is always unitary.

2. The Optical Theorem and the K Matrix

In this section we will investigate the consequences of the unitarity of S for the transition operator T. It is this operator that is most closely related to the physically measurable scattering cross section, and the restrictions that unitarity places on it are of great importance in interpreting experimental results.

DERIVATION OF THE OPTICAL THEOREM

Equation (1.14) for the S-matrix element implies that the scattering operator can be written either as

$$S = 1 - 2\pi i\, \delta(E - H_0)\, T, \qquad (2.1)$$

with E defined as the energy of the initial free-particle state, or as

$$S = 1 - 2\pi i T\, \delta(E - H_0), \qquad (2.2)$$

with E defined as the energy of the final state. These two forms are completely equivalent; they give the same matrix elements, and their formal equivalence can be shown using the time-dependent formulas of Chapter 8.

Let us then write the scattering operator in the form

$$S = 1 - i\tau, \qquad (2.3)$$

with

$$\tau = 2\pi\, \delta(E - H_0)\, T = 2\pi T\, \delta(E - H_0). \qquad (2.4)$$

Since $S^\dagger S = SS^\dagger = 1$, the operator τ satisfies

$$\tau - \tau^\dagger = -i\tau^\dagger \tau = -i\tau\tau^\dagger. \qquad (2.5)$$

The generalized optical theorem follows from this equation. The matrix element between two arbitrary states yields

$$\tau_{fi} - \tau^*_{if} = -i \sum_g \tau^*_{gf} \tau_{gi}. \qquad (2.6)$$

Inserting Eq. (2.4) into Eq. (2.6), we have

$$2\pi\, \delta(E_f - E_i)(T_{fi} - T^*_{if}) = -(2\pi)^2 i \sum_g \delta(E_f - E_g)\, T^*_{gf}\, \delta(E_g - E_i)\, T_{gi}. \qquad (2.7)$$

One of the δ functions can be canceled if we note that

$$\delta(E_f - E_g)\, \delta(E_g - E_i) = \delta(E_f - E_i)\, \delta(E_i - E_g), \qquad (2.8)$$

and if we remain on the energy shell so that $E_f = E_i$ in all matrix elements T_{fi}.

The result is the unitarity theorem for the T matrix on the energy shell:

$$\frac{1}{2i}(T_{fi} - T^*_{if}) = -\pi \sum_g T^*_{gf}\, \delta(E_i - E_g)\, T_{gi}. \tag{2.9}$$

In terms of the operator T, this is

$$\frac{1}{2i}(T - T^\dagger) = -\pi T^\dagger\, \delta(E - H_0)\, T. \tag{2.10}$$

The corresponding relation for the case of spinless elastic scattering was derived in Chapter 6. The present result shows that this relation follows in general from the unitarity of the S matrix.

The left-hand side of Eq. (2.9) is equal to the imaginary part of T_{fi} if the T matrix is symmetric, i.e., if $T_{fi} = T_{if}$. This symmetry under exchange of the initial and final states holds only in special cases. (For instance, in Chapter 10 we shall show that time-reversal invariance implies that the T matrix is symmetric in the angular-momentum representation.) However, for the *diagonal* element of Eq. (2.9), we find, even without this assumption,

$$\operatorname{Im} T_{ii} = -\pi \sum_g |T_{gi}|^2\, \delta(E_i - E_g). \tag{2.11}$$

This important result is completely general and does not depend upon the symmetry of the S matrix or upon time-reversal invariance.

The states ϕ_g may be identified by the energy E_g and by the variable α_g, which specifies the particles in the state and the direction of their momentum and spin components. The sum over states then becomes

$$\sum_g \to \int d\alpha_g\, dE_g\, \rho_g(E_g), \tag{2.12}$$

where $\rho_g(E_g)$ is the density of states and the integral over α_g includes a sum over all possible configurations or rearrangements. Equation (2.11) is then

$$\operatorname{Im} T_{ii} = -\pi \int d\alpha_g\, |T_{gi}|^2_{E_g = E_i}\, \rho_g(E_i). \tag{2.13}$$

If we compare this with Eqs. (1.32) and (2.21) of Chapter 7, we have

$$\operatorname{Im} T_{ii} = -\frac{\hbar v_i}{2}\, \sigma_{\text{TOT}}, \tag{2.14}$$

where σ_{TOT} is the total cross section for scattering from the initial system in the state ϕ_i. This is the optical theorem relating the imaginary part of the forward-scattering element of the T matrix (or the forward scattering amplitude $f(\theta = 0) = -(2m/4\pi\hbar^2)T_{ii}$) to the total cross section.

THE K MATRIX

The optical theorem and its generalization, Eq. (2.9), place severe restrictions upon the T matrix. Thus, Eq. (2.13) shows that the magnitude of each element of the T matrix must be limited, since the total number of particles removed from the incident beam by scattering or reactions must not exceed the number in the incident beam. As a result of such requirements, all elements of the T matrix are related through complicated nonlinear equations. These relations are useful in analyzing experimental data, but for the theoretical physicist, interested in calculating cross sections from assumed two-particle interactions, they present grave difficulties.

These problems can be illustrated by the common method of solution using a power series or Born expansion. In this method the transition operator is approximated by a finite number of terms in the series

$$T = V + V \frac{1}{E - H_0 + i\varepsilon} V + \cdots. \tag{2.15}$$

To any finite order in this series, the unitarity condition will be violated since, if T is approximated by a finite series of order n, the left-hand side of Eq. (2.13) is of order n while the right-hand side will be of order $2n$. Only for the complete infinite series will these relations be satisfied.

If the power series is rapidly convergent, the violation of these relations at a finite order may not seriously affect the numerical results. For instance, a weak inelastic process may not seriously deplete the incident state, and a first-order Born approximation could give reliable results. When the incident state is seriously depleted or "damped" by such a process, the restrictions of unitarity must be taken into account to obtain accurate results. Hence, for practical applications one often wants a formulation that permits approximate calculations incorporating the conditions of unitarity.

Such a formulation can be achieved if we use an expression for S different from our previous starting point, Eq. (2.3). This new expression gives S in terms of an inverse operator. In general, these are difficult to handle, but in this case we will be able to express the inverse operator in terms of a manageable integral equation. Therefore, let us define an operator κ by

$$S = \frac{1 - (i/2)\kappa}{1 + (i/2)\kappa}. \tag{2.16}$$

If κ is Hermitian, S will be unitary. The properties of κ can be immediately obtained by relating it to τ and thus to the transition operator T.

From Eqs. (2.3) and (2.16), we have

$$\tau = \frac{\kappa}{1 + (i/2)\,\kappa} \qquad (2.17)$$

or

$$\tau = \kappa - (i/2)\,\kappa\tau. \qquad (2.18)$$

By analogy with Eq. (2.4), we define a Hermitian operator K by

$$\kappa = 2\pi\,\delta(E - H_0)\,K. \qquad (2.19)$$

Proceeding as in Eqs. (2.5)–(2.9) and canceling the delta function, Eq. (2.18) becomes

$$T = K - i\pi K\,\delta(E - H_0)\,T. \qquad (2.20)$$

This equation is usually termed the "Heitler damping equation." It limits the possible magnitude of T, for arbitrary values of K, and thus ensures that the matrix elements of T will not exceed values permitted by probability conservation. To use the damping equation, one usually assumes that K is known from some previous approximate calculation; then T is determined from this equation. As we shall see below, so long as the approximation to K is Hermitian, the T matrix obtained from this equation will satisfy the unitarity condition.

The matrix elements of T are obtained from the equation

$$T_{fi} = K_{fi} - i\pi \sum_g K_{fg}\,\delta(E_i - E_g)\,T_{gi}. \qquad (2.21)$$

An interesting property of this equation is that, unlike the integral equations for T_{fi} obtained in Section 4 of Chapter 5, this formula involves only matrix elements on the energy shell, i.e., $E_i = E_f = E_g$. Since these are the only matrix elements of interest in obtaining cross sections, the computational problem is greatly simplified.

We have then a possible way of obtaining unitary T matrices, provided we can find a prescription for calculating approximate values of the K matrix. Such a rule can be derived if for the moment we treat Eq. (2.20) as an equation for K in terms of T. We must, however, recognize that the operator K is not unique. Equation (2.20) was derived as a relation between on-the-energy-shell matrix elements. Such matrix elements are not sufficient by themselves to determine the operator K, just as a subset of the Fourier coefficients of a function is not sufficient to determine the function. We shall fix the off-the-energy-shell matrix elements by assuming that Eq. (2.20) holds both on and off the energy shell, that is, that it is a generally valid operator relation.

We recognize from the derivation of Eq. (2.20) that this equation could equally well be expressed as

$$K = T + i\pi T\, \delta(E - H_0)\, K. \tag{2.22}$$

Inserting into this the integral equation for T in terms of V, we have

$$K = V + V \frac{1}{E - H_0 + i\varepsilon} T + i\pi V\, \delta(E - H_0)\, K$$

$$+ i\pi V \frac{1}{E - H_0 + i\varepsilon} T\, \delta(E - H_0)\, K$$

$$= V + V \frac{1}{E - H_0 + i\varepsilon} T + i\pi V\, \delta(E - H_0)\, K$$

$$+ V \frac{1}{E - H_0 + i\varepsilon} (K - T)$$

$$= V + V \frac{P}{E - H_0}\, K. \tag{2.23}$$

In obtaining Eq. (2.23), we have used the relation

$$\frac{1}{E - H_0 + i\varepsilon} = \frac{P}{E - H_0} - i\pi\, \delta(E - H_0). \tag{2.24}$$

Equation (2.23) has the same structure as the integral equation for T; only the definition of the Green's function is changed, with the singularity treated now by the principal value method. The Green's function is real, and on the energy shell the K matrix defined by this equation is Hermitian, as we expect.

Just as $\psi_i^{(+)}$ was related to T by $T\phi_i = V\psi_i^{(+)}$, so we can associate a new type of state with K. The principal-value Green's function defining this state is the sum of the ingoing- and outgoing-wave Green's functions:

$$\frac{P}{E - H_0} = \frac{1}{2}\left(\frac{1}{E - H_0 + i\varepsilon} + \frac{1}{E - H_0 - i\varepsilon}\right). \tag{2.25}$$

Hence this new state will contain standing waves, in addition to the usual incident plane wave. If we define a standing-wave state by

$$\psi_i^{(s)} = \phi_i + \frac{P}{E - H_0} V \psi_i^{(s)}, \tag{2.26}$$

the K matrix satisfies the relation

$$K\phi_i = V\psi_i^{(s)}. \tag{2.27}$$

Let us summarize what has been accomplished. The single integral equation for the non-Hermitian transition operator

$$T = V + V \frac{1}{E - H_0 + i\varepsilon} T \tag{2.28}$$

has been replaced by two equations involving the Hermitian operator K:

$$K = V + V \frac{P}{E - H_0} K \tag{2.29}$$

and

$$T = K - i\pi K \, \delta(E - H_0) \, T. \tag{2.30}$$

Only the energy-shell matrix elements are involved in the new equation for T. Furthermore, K need not be obtained exactly to maintain unitarity, since any Hermitian operator $\tilde{K}$, when inserted into the damping equation, will yield a T matrix satisfying the unitarity condition.

To see this, introduce

$$T^\dagger = \tilde{K} + i\pi T^\dagger \, \delta(E - H_0) \, \tilde{K} \tag{2.31}$$

and use this together with Eq. (2.30) to obtain

$$T = \tilde{K} - i\pi T^\dagger \, \delta(E - H_0) \, T - \pi^2 T^\dagger \, \delta(E - H_0) \, \tilde{K} \, \delta(E - H_0) \, T,$$
$$T^\dagger = \tilde{K} + i\pi T^\dagger \, \delta(E - H_0) \, T - \pi^2 T^\dagger \, \delta(E - H_0) \, \tilde{K} \, \delta(E - H_0) \, T,$$

and

$$T - T^\dagger = -2\pi i T^\dagger \, \delta(E - H_0) \, T. \tag{2.32}$$

Thus, any Hermitian K matrix leads to a T matrix satisfying the proper unitarity condition. In particular, a power-series approximation to K yields a Hermitian result and can be used for approximate calculations. The formal complication of having two equations is balanced in practice by the advantages of such a convenient approximation scheme.

APPLICATION OF THE DAMPING EQUATION

To see how this works in practice, let us consider the scattering of a spinless particle by a central potential as an example. In this case we know that

2. The Optical Theorem and the K Matrix

the orbital angular momentum will be conserved. This will enable us to solve the damping equation.

For this purpose it is convenient to use a representation in which the states are eigenstates of the orbital angular momentum. A solution of the free-particle Schrödinger equation that is simultaneously an eigenfunction of the orbital angular momentum is

$$\phi_{klm}(\mathbf{r}) = \langle \mathbf{r} | klm \rangle = N_k i^l \frac{F_l(kr)}{kr} Y_{lm}(\theta, \phi), \qquad (2.33)$$

with N_k an arbitrary normalization constant. In this representation the potential matrix on the energy shell is

$$\langle kl'm' | V | klm \rangle = N_k^2 \frac{1}{k^2} \int_0^\infty dr\, F_l^2(kr)\, V(r)\, \delta_{ll'} \delta_{mm'}, \qquad (2.34)$$

because of the orthonormality of the spherical harmonics. This matrix is thus diagonal in the orbital angular momentum and its magnetic quantum number, reflecting the conservation of these quantities in the scattering process. (This is a special case of the general conservation law discussed in Section 4 of Chapter 10.) The K matrix is likewise diagonal in these quantities, as one can see explicitly from Eq. (2.29). Equation (2.30) implies that the T matrix has a similar property.

The damping equation will be simplified if the delta function $\delta(E - H_0)$ is expressed in the angular-momentum representation. The plane-wave expansion can be written as

$$e^{i\mathbf{k}\cdot\mathbf{r}} = \langle \mathbf{r} | \mathbf{k} \rangle = \frac{4\pi}{N_k} \sum_{lm} Y_{lm}^*(\theta_k, \phi_k)\, \phi_{klm}(\mathbf{r}), \qquad (2.35)$$

which gives the state-vector relation

$$|\mathbf{k}\rangle = \frac{4\pi}{N_k} \sum_{ml} Y_{lm}^*(\theta_k, \phi_k) | klm \rangle. \qquad (2.36)$$

Using the completeness relation for linear-momentum eigenstates, we then have

$$\int \frac{d^3k}{(2\pi)^3} |\mathbf{k}\rangle \langle \mathbf{k}| = \frac{2}{\pi} \sum_{\substack{lm \\ l'm'}} \int \frac{k^2}{N_k^2} dk\, d\Omega_k\, Y_{lm}^*(\theta_k, \phi_k)\, Y_{l'm'}(\theta_k, \phi_k)$$
$$\times | klm \rangle \langle kl'm' |$$
$$= \frac{2}{\pi} \sum_{lm} \int \frac{k^2}{N_k^2} dk\, | klm \rangle \langle klm | = 1. \qquad (2.37)$$

From this, the delta function can be expressed as

$$\delta(E - H_0) = \frac{2}{\pi} \sum_{lm} \int \frac{k^2}{N_k^2} \, dk \, \delta(E - E_k) | klm \rangle \langle klm |$$

$$= \frac{2mk}{\pi \hbar^2 N_k^2} \sum_{lm} | klm \rangle \langle klm |, \qquad (2.38)$$

where $E = \hbar^2 k^2 / 2m$.

The fact that the T matrix will be diagonal in the angular momentum enables us to solve the damping equation, Eq. (2.30). We observe in Eq. (2.34) that the potential matrix is independent of m, so that we can write the K and T matrices in the angular-momentum representation as

$$\langle kl'm' | K | klm \rangle \equiv -K_l \delta_{ll'} \delta_{mm'}$$

and

$$\langle kl'm' | T | klm \rangle \equiv -T_l \delta_{ll'} \delta_{mm'}. \qquad (2.39)$$

(The minus sign is conventional and causes the amplitude for a weak attractive force to be positive.) The choice $N_k = (2mk/\hbar^2)^{1/2}$ and use of Eqs. (2.38) and (2.39) enable us to write the damping equation as

$$T_l = K_l + iK_l T_l \qquad (2.40)$$

or

$$T_l = \frac{K_l}{1 - iK_l}. \qquad (2.41)$$

If these same definitions are introduced into the optical theorem, Eq. (2.10), we find

$$\operatorname{Im} T_l = | T_l |^2. \qquad (2.42)$$

The effect of the damping is to enforce a limitation on the magnitude of T_l, which can never be larger than 1, regardless of the choice of K_l. Furthermore, since the K matrix is Hermitian, any choice for the K-matrix elements will lead to real values for the numbers K_l, and the solution for T_l given by Eq. (2.41) will automatically satisfy the optical theorem, Eq. (2.42).

Any solution of Eq. (2.42) can always be expressed in terms of a real phase shift δ_l, viz.,

$$T_l = e^{i\delta_l} \sin \delta_l. \qquad (2.43)$$

Inverting Eq. (2.41), we see that the K-matrix element is given by

$$K_l = \tan \delta_l. \qquad (2.44)$$

2. The Optical Theorem and the K Matrix

These relations are identical with those derived previously in Section 1 of Chapter 3.

Let us now extend this result to a situation in which a reaction can take place. Suppose that the incident particle is electrically neutral, but that it can pick up a positive electric charge during the scattering process. (For instance, these considerations might apply to the reactions $n + \text{He}^3 \to p + \text{H}^3$ or $\pi^0 + p \to \pi^+ + n$, under circumstances where the difference in mass between the charged and neutral particles may be neglected.) The interaction will then have a component that can transfer electric charge from the target to the projectile. With the original system denoted by a state vector $|klm; 0\rangle$ and the final charged system denoted by $|klm; +\rangle$, we can define a charge exchange operator P_{ex} having the properties

$$P_{ex} |klm; 0\rangle = |klm; +\rangle$$

and (2.45)

$$P_{ex}^2 = 1.$$

If the nonexchange or direct part of the interaction is the same for both charge states, the full interaction can be expressed as the sum of a direct part and an exchange part:

$$V = V_d + V_{ex} P_{ex}. \tag{2.46}$$

Provided that these interactions are both central potentials, orbital angular momentum will still be conserved, and we can write the elastic T-matrix elements and the charge-exchange element as

$$\langle kl'm'; 0 | T | klm; 0 \rangle = \langle kl'm'; + | T | klm; + \rangle$$
$$\equiv -T_l^d \delta_{ll'} \delta_{mm'}$$

and

$$\langle kl'm'; + | T | klm; 0 \rangle \equiv -T_l^{ex} \delta_{ll'} \delta_{mm'}. \tag{2.47}$$

Similar definitions will hold for the K matrix. It is important to note that a complete set of states now includes both charged and neutral states. With this fact taken into account, the damping equation for the elastic scattering amplitude will now become

$$\langle klm; 0 | T | klm; 0 \rangle = \langle klm; 0 | K | klm; 0 \rangle$$
$$+ i \langle klm; 0 | K | klm; 0 \rangle \langle klm; 0 | T | klm; 0 \rangle$$
$$+ i \langle klm; 0 | K | klm; + \rangle \langle klm; + | T | klm; 0 \rangle,$$

or
$$T_l^d = K_l^d + iK_l^d T_l^d + iK_l^{ex} T_l^{ex}, \qquad (2.48)$$
and the corresponding equation for the charge-exchange amplitude is
$$T_l^{ex} = K_l^{ex} + iK_l^{ex} T_l^d + iK_l^d T_l^{ex}. \qquad (2.49)$$
These can be solved to give
$$T_l^d = \frac{K_l^d(1 - iK_l^d) + i(K_l^{ex})^2}{(1 - iK_l^d)^2 + (K_l^{ex})^2}$$
and
$$T_l^{ex} = \frac{K_l^{ex}}{(1 - iK_l^d)^2 + (K_l^{ex})^2}. \qquad (2.50)$$

These results show several interesting features, when compared with the result for pure elastic scattering, Eq. (2.41). The elastic scattering amplitude is modified in two ways by the presence of the additional final state. A positive term representing depletion from the elastic scattering process has been added to the denominator. In addition, an "attractive" term has been added to the numerator. This represents another way, in addition to the direct scattering by V_d, that elastic scattering can take place: the incident particle can undergo a charge-exchange scattering, and can then exchange back to become a neutral particle again.

The solution for the charge-exchange amplitude shows the first effect, depletion due to the existence of a competing process, and shows the "self-damping" seen in pure elastic scattering as well. One may easily check that these amplitudes satisfy the unitarity relations, as they must, for any real values of K_l^d and K_l^{ex}.

3. Diagonalization of the S Matrix

It is of some interest to investigate the diagonalization of the S matrix. The definitions of κ and τ show that these operators commute with each other and with S (in fact, each can be expressed as a function of S alone). As we shall see, this implies that the S matrix and the energy-shell parts of the K and T matrices can be simultaneously diagonalized.

Let us suppose that there is a set of free-particle states that diagonalizes the S matrix. A system originally in one of these states will then not undergo a transition during the collision process, and the characteristic properties of

3. Diagonalization of the S Matrix

the initial state will be conserved throughout the reaction. We shall discuss conservation principles in the next chapter, but it is well known that such quantities as the total angular momentum are conserved in most reaction processes. Hence these eigenstates of S will also be eigenstates of these conserved quantities.

We can immediately identify one of these conserved quantities, since we know that the S matrix contains nonzero matrix elements only between states of equal energy. Then the total energy is conserved, and we can denote the eigenstates of S by $\phi_{E\lambda}$, where λ is the set of eigenvalues for the remaining conserved quantities. The S matrix is thus

$$\langle \phi_{E'\lambda'} | S | \phi_{E\lambda} \rangle = S_\lambda(E)\, \delta(E - E')\, \delta_{\lambda\lambda'}, \tag{3.1}$$

where $S_\lambda(E)$ is the eigenvalue of the scattering operator. Since the S matrix is unitary, the eigenstates form an orthogonal set and may be normalized according to

$$\langle \phi_{E'\lambda'} | \phi_{E\lambda} \rangle = \delta(E - E')\, \delta_{\lambda\lambda'}. \tag{3.2}$$

As an example, in the case of the scattering of spinless particles by a central potential, the states $\phi_{E\lambda}$ are eigenstates of orbital angular momentum, and properly normalized wave functions are

$$\phi_{Elm}(\mathbf{r}) = \left(\frac{2}{\pi}\frac{mk}{\hbar^2}\right)^{1/2} i^l \frac{F_l(kr)}{kr} Y_{lm}(\theta, \phi). \tag{3.3}$$

In deriving this normalization constant, we have used

$$\int_0^\infty dr\, F_l(kr)\, F_l(k'r) = \frac{\pi}{2} \delta(k - k') = \frac{\pi}{2} \frac{dE}{dk} \delta(E - E'). \tag{3.4}$$

With this normalization, the completeness relation is

$$\sum_\lambda \int dE\, | \phi_{E\lambda} \rangle \langle \phi_{E\lambda} | = 1. \tag{3.5}$$

An arbitrary S-matrix element S_{fi} is then expressed in terms of the eigenstates $\phi_{E\lambda}$ and eigenvalues $S_\lambda(E)$ by the relation

$$S_{fi} = \sum_\lambda \int dE\, \langle \phi_f | \phi_{E\lambda} \rangle\, S_\lambda(E)\, \langle \phi_{E\lambda} | \phi_i \rangle. \tag{3.6}$$

The scalar products $\langle \phi_{E\lambda} | \phi_i \rangle$ describe the initial state ϕ_i in the λ representation:

$$\phi_i = \sum_\lambda \int dE \, \langle \phi_{E\lambda} | \phi_i \rangle \, \phi_{E\lambda}. \tag{3.7}$$

If both $\phi_{E\lambda}$ and ϕ_i are eigenstates of the same free-particle Hamiltonian H_0, these coefficients are zero unless $E = E_i$. In this case they are proportional to an energy-conserving delta function and can be expressed as

$$\langle \phi_{E\lambda} | \phi_i \rangle = C_{\lambda,i}(E) \, \delta(E - E_i). \tag{3.8}$$

Then Eq. (3.7) becomes

$$\phi_i = \sum_\lambda C_{\lambda,i}(E_i) \, \phi_{E_i\lambda}. \tag{3.9}$$

For elastic scattering of spinless particles, this is just the partial-wave expansion. If rearrangement processes can occur, the eigenstates $\phi_{E\lambda}$ will be linear combinations of different configurations of free particles, corresponding to different free-particle Hamiltonians. In this case Eq. (3.7) will include contributions from all energies.

Since the S matrix is unitary, its eigenvalues $S_\lambda(E)$ must have a unit modulus; we therefore write

$$S_\lambda(E) = \exp[2i\delta_\lambda(E)], \tag{3.10}$$

where the real number $\delta_\lambda(E)$ is the "eigen-phase shift" for the state $\phi_{E\lambda}$. Again, for the spinless elastic case it is the ordinary scattering phase shift, and $S_\lambda(E)$ is the amplitude S_l. In more complicated problems for which $\phi_{E\lambda}$ is a mixture of observable states, the eigen-phase shift does not have a simple physical significance. For instance, if reactions are possible, the S matrix will be nondiagonal in the natural physical representation. The eigenfunctions $\phi_{E\lambda}$ will then be linear combinations of the "physical" states ϕ_i and will therefore not be directly observable.

The states $\phi_{E\lambda}$ or, equivalently, the mixing coefficients $\langle \phi_{E\lambda} | \phi_i \rangle$ are determined by the solution to the full scattering problem, that is, by the S matrix itself. If the S matrix couples N physical states, the N eigen-phase shifts, together with the $\frac{1}{2}N(N-1)$ mixing coefficients, completely determine the scattering cross sections and all observable spin-dependent parameters.

3. Diagonalization of the S Matrix

We can now express the eigenvalues of T and K in terms of the eigen-phase shifts. Since $S = 1 - 2\pi i \, \delta(E - H_0) \, T$, Eq. (3.1) implies

$$\langle \phi_{E'\lambda'} | S | \phi_{E\lambda} \rangle = \delta(E - E') [\delta_{\lambda\lambda'} - 2\pi i \langle \phi_{E\lambda'} | T | \phi_{E\lambda} \rangle]$$
$$= S_\lambda(E) \, \delta(E - E') \, \delta_{\lambda\lambda'}. \quad (3.11)$$

Thus, on the energy shell, the T matrix is diagonal,

$$\langle \phi_{E\lambda'} | T | \phi_{E\lambda} \rangle = T_\lambda(E) \, \delta_{\lambda\lambda'}, \quad (3.12)$$

and the diagonal element is given by

$$S_\lambda(E) = 1 - 2\pi i \, T_\lambda(E) \quad (3.13)$$

or

$$T_\lambda(E) = \frac{1 - \exp[2i\,\delta_\lambda(E)]}{2\pi i} = -\frac{\exp[i\,\delta_\lambda(E)]\sin\delta_\lambda(E)}{\pi}. \quad (3.14)$$

The physically observable matrix elements are

$$T_{fi} = \langle \phi_f | T | \phi_i \rangle = \sum_{\lambda\lambda'} \int dE\, dE' \, \langle \phi_f | \phi_{E'\lambda'} \rangle$$
$$\times \langle \phi_{E'\lambda'} | T | \phi_{E\lambda} \rangle \langle \phi_{E\lambda} | \phi_i \rangle. \quad (3.15)$$

On the energy shell $E_i = E_f$ and, in the absence of rearrangements, the mixing coefficients are given by Eq. (3.8). We can then express the physical T-matrix element in terms of the eigen-phase shifts through

$$T_{fi} = \frac{-1}{\pi} \sum_\lambda C^*_{\lambda,f}(E) \, C_{\lambda,i}(E) \exp[i\,\delta_\lambda(E)] \sin\delta_\lambda(E). \quad (3.16)$$

The damping equation may be used to find the eigenvalues of K on the energy shell. Proceeding as with T, we obtain

$$\langle \phi_{E\lambda'} | K | \phi_{E\lambda} \rangle = K_\lambda(E) \, \delta_{\lambda\lambda'} \quad (3.17)$$

and

$$T_\lambda(E) = K_\lambda(E) - i\pi \, K_\lambda(E) \, T_\lambda(E) \quad (3.18)$$

or

$$K_\lambda(E) = \frac{T_\lambda(E)}{1 - i\pi \, T_\lambda(E)} = -\frac{\tan\delta_\lambda(E)}{\pi}. \quad (3.19)$$

Thus the relations between the general S, T, and K matrices are identical in form with those for the simple case treated in Chapter 3. These general results must necessarily reduce to the formulas of that chapter for the case of elastic scattering by a central potential.

In that case ϕ_i is a momentum eigenstate and the states $\phi_{E\lambda}$ are eigenstates of orbital angular momentum. Then, using Eq. (3.3), the mixing coefficient is

$$C_{lm,i}(E) = 4\pi \left(\frac{\pi}{2} \frac{\hbar^2}{mk}\right)^{1/2} Y_{lm}^*(\theta_i, \phi_i) \tag{3.20}$$

and $\delta_\lambda(E) = \delta_l$. Equation (3.16) becomes

$$T_{fi} = -\frac{1}{\pi}(4\pi)^2 \left(\frac{\pi}{2}\frac{\hbar^2}{mk}\right) \sum_{lm} Y_{lm}(\theta_f, \phi_f) Y_{lm}^*(\theta_i, \phi_i) e^{i\delta_l} \sin \delta_l. \tag{3.21}$$

Using the addition theorem for spherical harmonics,

$$\sum_{m=-l}^{l} Y_{lm}(\theta_f, \phi_f) Y_{lm}^*(\theta_i, \phi_i) = \frac{2l+1}{4\pi} P_l(\cos \theta), \tag{3.22}$$

with θ defined by $\mathbf{k}_i \cdot \mathbf{k}_f = k^2 \cos \theta$, this is

$$T_{fi} = -2\pi \frac{\hbar^2}{mk} \sum_l (2l+1) e^{i\delta_l} \sin \delta_l P_l(\cos \theta). \tag{3.23}$$

The scattering amplitude is then

$$f(\theta) = -\frac{2m}{4\pi\hbar^2} T_{fi} = \frac{1}{k} \sum_l (2l+1) e^{i\delta_l} \sin \delta_l P_l(\cos \theta), \tag{3.24}$$

which is the result of Chapter 3.

Chapter
10

Invariance Principles and Conservation Laws

In the last chapter we discussed the diagonalization of the S matrix and pointed out that the possibility of such diagonalization is associated with the existence of certain conserved physical variables. If the S matrix is diagonal in some representation, then the eigenvalues that characterize that representation must be unchanged by the scattering process. In this chapter we want to survey in more detail these conserved quantities and to indicate the connection between these conservation laws and the invariance properties of the Hamiltonian.

1. Invariance under Space Translations

As an example of the general principles involved, we shall consider first the conservation of momentum and the associated invariance property, invariance under space translations.

FORMAL CONSEQUENCES OF TRANSLATION INVARIANCE

In all of our previous work we have assumed that we were examining the scattering process in the center-of-mass system. We now want to drop this constraint, so that we are free to use any coordinate system we wish. In an arbitrary coordinate system the initial state is described by *two* momenta rather than just the center-of-mass momentum we have used until now. Likewise, in the coordinate representation the wave function will be a function of *two* coordinates $(\mathbf{r}_1, \mathbf{r}_2)$ rather than just the relative coordinate $\mathbf{r}$.

Let us now imagine that at the beginning of the scattering process the entire two-body system is displaced by $\mathbf{r}_0$. (Throughout this chapter we shall speak of displacing the *particles* rather than transforming the *coordinate system*; either viewpoint is correct, however.) Under the displacement, the wave function is transformed so that it becomes

$$\phi_i(\mathbf{r}_1, \mathbf{r}_2) \rightarrow \bar{\phi}_i(\mathbf{r}_1, \mathbf{r}_2) = \phi_i(\mathbf{r}_1 - \mathbf{r}_0, \mathbf{r}_2 - \mathbf{r}_0), \tag{1.1}$$

where $\bar{\phi}_i(\mathbf{r}_1, \mathbf{r}_2)$ is the wave function of the displaced system of particles.

Suppose the Hamiltonian is invariant under this transformation. Since this transformation leaves momenta and momentum operators unchanged, the transformed Hamiltonian is then

$$\bar{H}(\mathbf{r}_1, \mathbf{r}_2, \mathbf{p}_1, \mathbf{p}_2) = H(\mathbf{r}_1 - \mathbf{r}_0, \mathbf{r}_2 - \mathbf{r}_0, \mathbf{p}_1, \mathbf{p}_2) = H(\mathbf{r}_1, \mathbf{r}_2, \mathbf{p}_1, \mathbf{p}_2). \tag{1.2}$$

We have here written H as a function of the vectors $\mathbf{r}_1$ and $\mathbf{r}_2$; we could equally well have expressed it as a function of the center-of-mass coordinate $\mathbf{R} = (m_1\mathbf{r}_1 + m_2\mathbf{r}_2)/(m_1 + m_2)$ and the relative coordinate $\mathbf{r} = \mathbf{r}_1 - \mathbf{r}_2$. Under the translation only the center-of-mass coordinate is affected:

$$\begin{aligned} \mathbf{R} &\rightarrow \mathbf{R} - \mathbf{r}_0 \\ \mathbf{r} &\rightarrow \mathbf{r}. \end{aligned} \tag{1.3}$$

Hence the translation invariance expressed by Eq. (1.2) implies that H is independent of $\mathbf{R}$, the location of the center of mass of the two particles. This invariance argument, a simple example of others we shall make, has therefore shown that translational invariance implies that in its coordinate dependence the Hamiltonian is a function only of the relative separation of the two particles.

If H is translation invariant, the S matrix is also. To see this, we need only observe that the free-particle Hamiltonian and thus V, the potential, are each translation-invariant, while S depends only on these operators, via the relation

$$S = \Omega^{(-)\dagger}\Omega^{(+)} = \left(1 + V\frac{1}{E - H + i\varepsilon}\right)\left(1 + \frac{1}{E - H + i\varepsilon}V\right). \tag{1.4}$$

In general, any invariance property of H implies the corresponding invariance property for S.

Momentum Conservation Using the Coordinate Representation

We must now discover the effect of translation invariance upon an S-matrix element S_{fi}. This can be done quite explicitly if we recognize that in the spatial representation S becomes a nonlocal operator which we can write

1. Invariance under Space Translations

as $S(\mathbf{r}_1', \mathbf{r}_2'; \mathbf{r}_1, \mathbf{r}_2)$. (This representation for S is seldom used in practice, but it is convenient in this application.) We can equally well consider this to be a function of four other variables: the two relative coordinates $\mathbf{r}$ and $\mathbf{r}'$ and the difference and sum of the two center-of-mass coordinates, $\mathbf{R} - \mathbf{R}'$ and $\mathbf{R} + \mathbf{R}'$. The first three vectors do not change under translation, while the last one *is* modified. Hence translation invariance implies that S can depend only on the two relative coordinates and on the *difference* of the center-of-mass coordinates. We will see below that invariance under velocity transformations implies that $\mathbf{R} = \mathbf{R}'$, but we can assume here that the dependence on $\mathbf{R} - \mathbf{R}'$ is arbitrary. Therefore we have

$$S(\mathbf{r}_1', \mathbf{r}_2'; \mathbf{r}_1, \mathbf{r}_2) = S(\mathbf{r}', \mathbf{r}, \mathbf{R} - \mathbf{R}'). \tag{1.5}$$

In spatial representation the S-matrix element may be written as

$$S_{fi} = \langle \phi_f | S | \phi_i \rangle = \int \phi_f^*(\mathbf{r}_1', \mathbf{r}_2') \, d^3r_1' \, d^3r_2' \\ \times S(\mathbf{r}_1', \mathbf{r}_2'; \mathbf{r}_1, \mathbf{r}_2) \, d^3r_1 \, d^3r_2 \, \phi_i(\mathbf{r}_1, \mathbf{r}_2). \tag{1.6}$$

If we introduce center-of-mass and relative coordinates and use

$$\phi_i(\mathbf{r}_1, \mathbf{r}_2) = e^{i\mathbf{k}_1 \cdot \mathbf{r}_1} e^{i\mathbf{k}_2 \cdot \mathbf{r}_2} = e^{i\mathbf{K}_i \cdot \mathbf{R}} e^{i\mathbf{k}_i \cdot \mathbf{r}}, \tag{1.7}$$

where the total momentum is $\mathbf{K}_i = \mathbf{k}_1 + \mathbf{k}_2$ and the relative momentum is $\mathbf{k}_i = (m_2 \mathbf{k}_1 - m_1 \mathbf{k}_2) / (m_1 + m_2)$, Eq. (1.6) becomes

$$S_{fi} = \int d^3R \, d^3R' \, d^3r \, d^3r' \, e^{-i\mathbf{K}_f \cdot \mathbf{R}'} e^{-i\mathbf{k}_f \cdot \mathbf{r}'} \\ \times S(\mathbf{r}', \mathbf{r}, \mathbf{R} - \mathbf{R}') \, e^{i\mathbf{K}_i \cdot \mathbf{R}} e^{i\mathbf{k}_i \cdot \mathbf{r}}. \tag{1.8}$$

Replacing $\mathbf{R}'$ with the coordinate $\mathbf{X} = \mathbf{R} - \mathbf{R}'$, the scattering operator no longer depends on $\mathbf{R}$ and we can immediately perform the integral over that variable. The result is

$$S_{fi} = (2\pi)^3 \, \delta(\mathbf{K}_f - \mathbf{K}_i) \int d^3X \, d^3r \, d^3r' \, e^{-i\mathbf{k}_f \cdot \mathbf{r}'} S(\mathbf{r}', \mathbf{r}, \mathbf{X}) \, e^{i\mathbf{k}_i \cdot \mathbf{r}}. \tag{1.9}$$

The delta function in Eq. (1.9) tells us that the scattering process does not take place unless $\mathbf{K}_f = \mathbf{K}_i$. Thus we find the important result that translational invariance implies that the total momentum is conserved. This illustrates how an invariance property leads directly to a conservation law.

The Transformation Operator

We shall now derive this result using a more general operator formalism that is applicable to all systems and to other invariance properties as well. Let us postulate a set of states ϕ_i and a set $\bar{\phi}_i$, each of which describes the same particles, and each of which forms a complete set. We know that there must then be a unitary operator U that relates them:

$$\bar{\phi}_i = U\phi_i. \tag{1.10}$$

In the case presently being considered U is called the translation operator. The operator U transforms the state ϕ_i into the translated state $\bar{\phi}_i$. For the case of a single structureless particle U can be obtained by the following procedure which, with only slight generalization, applies to many other transformations as well.

The result of a displacement can be obtained by observing that the transformed single-particle wave function is simply

$$\bar{\phi}(\mathbf{r}) = \phi(\mathbf{r} - \mathbf{r}_0). \tag{1.11}$$

(The minus sign appears because we are displacing the particle, not the coordinate system; see Fig. 10.1.)

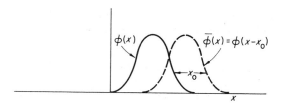

Fig. 10.1

This function may be expanded in a Taylor series

$$\phi(\mathbf{r} - \mathbf{r}_0) = \phi(\mathbf{r}) - \mathbf{r}_0 \cdot \nabla \phi(\mathbf{r}) + \frac{1}{2!} (\mathbf{r}_0 \cdot \nabla)(\mathbf{r}_0 \cdot \nabla) \phi(\mathbf{r}) - \cdots. \tag{1.12}$$

A formal expression for this Taylor series is

$$\bar{\phi}(\mathbf{r}) = e^{-\mathbf{r}_0 \cdot \nabla} \phi(\mathbf{r}). \tag{1.13}$$

If the exponential is expanded in a power series in $\mathbf{r}_0$, it reproduces the Taylor

1. Invariance under Space Translations

series. When we express Eq. (1.13) in terms of the momentum operator

$$\mathbf{p} = -i\hbar \nabla, \tag{1.14}$$

we have the fundamental result

$$\bar{\phi}(\mathbf{r}) = \exp\left(\frac{-i}{\hbar} \mathbf{p} \cdot \mathbf{r}_0\right) \phi(\mathbf{r}). \tag{1.15}$$

The transformation operator is then given by

$$U = \exp\left(\frac{-i}{\hbar} \mathbf{p} \cdot \mathbf{r}_0\right). \tag{1.16}$$

This final form is independent of the particular representation being used.

In general, an operator for a continuous transformation can be expressed as an exponential operator in which the exponent is the product of the coordinate displacement and the operator canonically conjugate to this coordinate. In this context the conjugate operator is the "generator" of the transformation.

One can easily check that the operator U induces the proper transformations in other operators. Thus, the coordinate, momentum, and intrinsic spin operators transform as

$$\begin{aligned} \bar{\mathbf{r}} &= U\mathbf{r}U^{-1} = \mathbf{r} - \mathbf{r}_0 \\ \bar{\mathbf{p}} &= U\mathbf{p}U^{-1} = \mathbf{p} \\ \bar{\mathbf{S}} &= U\mathbf{S}U^{-1} = \mathbf{S}. \end{aligned} \tag{1.17}$$

These relations can be demonstrated by treating the magnitude of the displacement r_0 as a continuous variable and integrating the differential relation $r_0 \, d\bar{\mathcal{O}}/dr_0 = (i/\hbar) [\bar{\mathcal{O}}, \mathbf{p} \cdot \mathbf{r}_0]$ with $\mathcal{O} = \mathbf{r}, \mathbf{p}, \mathbf{S}$, or any other operator.

GENERAL DERIVATION OF MOMENTUM CONSERVATION

We now find the conservation law by using the unitary translation operator. If we insert $U^{-1}U = 1$ into the S matrix, we obtain

$$\begin{aligned} S_{fi} &= \langle \phi_f | S | \phi_i \rangle \\ &= \langle \phi_f | U^{-1}USU^{-1}U | \phi_i \rangle \\ &= \langle \bar{\phi}_f | \bar{S} | \bar{\phi}_i \rangle. \end{aligned} \tag{1.18}$$

This form-independence can be given physical content by recalling that S is invariant under the transformation, i.e., $\bar{S} = S$. Thus we have

$$S_{fi} = \langle \phi_f | S | \phi_i \rangle = \langle \bar{\phi}_f | S | \bar{\phi}_i \rangle. \tag{1.19}$$

If we use a bar to denote the eigenvalues of the transformed state, so that

$\bar{\phi}_i = \eta_i \phi_i$, where η_i is a phase factor, we can write this result in concise form as

$$S_{fi} = \eta_f^* \eta_i S_{fi}. \qquad (1.20)$$

The conservation law can be obtained from Eq. (1.19) by introducing the explicit form of U, given in Eq. (1.16). The two-particle state ϕ_i given by Eq. (1.7) is an eigenstate of the momentum operator for each particle, with eigenvalues $\hbar \mathbf{k}_1$ and $\hbar \mathbf{k}_2$, respectively. Hence,

$$\begin{aligned} \bar{\phi}_i = U\phi_i &= \exp\left(-\frac{i}{\hbar}\mathbf{p}_1\cdot\mathbf{r}_0\right)\exp\left(-\frac{i}{\hbar}\mathbf{p}_2\cdot\mathbf{r}_0\right)\phi_i \\ &= \exp\left(-\frac{i}{\hbar}\mathbf{P}\cdot\mathbf{r}_0\right)\phi_i \\ &= \exp(-i\mathbf{K}_i\cdot\mathbf{r}_0)\phi_i, \end{aligned} \qquad (1.21)$$

where $\mathbf{P} = \mathbf{p}_1 + \mathbf{p}_2$ is the operator for the total momentum. Thus

$$\langle \bar{\phi}_f | S | \bar{\phi}_i \rangle = \langle \phi_f | S | \phi_i \rangle \exp[i(\mathbf{K}_f - \mathbf{K}_i)\cdot\mathbf{r}_0]. \qquad (1.22)$$

Since this must equal $\langle \phi_f | S | \phi_i \rangle$ for arbitrary values of $\mathbf{r}_0$, we must have conservation of momentum: $\mathbf{K}_f = \mathbf{K}_i$. Thus we find that translation invariance has led to conservation of the conjugate variable, the total momentum.

This result can be stated in another general form that will permit a concise application of these principles to other cases. If we use an infinitesimal translation $\delta \mathbf{r}_0$ and introduce the total momentum operator $\mathbf{P}$ with $U = \exp[(-i/\hbar)\mathbf{P}\cdot\delta\mathbf{r}_0]$, the transformed scattering operator is

$$\begin{aligned} \bar{S} = USU^{-1} &= (1 - \frac{i}{\hbar}\mathbf{P}\cdot\delta\mathbf{r}_0) S (1 + \frac{i}{\hbar}\mathbf{P}\cdot\delta\mathbf{r}_0) \\ &= S + \frac{i}{\hbar}\delta\mathbf{r}_0\cdot[S, \mathbf{P}], \end{aligned} \qquad (1.23)$$

to first order in $\delta\mathbf{r}_0$. If the invariance condition $\bar{S} = S$ is to hold, the commutator must vanish:

$$[S, \mathbf{P}] = 0. \qquad (1.24)$$

This implies that the eigenvalues of $\mathbf{P}$ are conserved in any collision. Denoting $|\phi_i\rangle$ by $|\mathbf{P}_i\rangle$ and suppressing the other variables, we have

$$0 = \langle \mathbf{P}_f | [S, \mathbf{P}] | \mathbf{P}_i \rangle = (\mathbf{P}_i - \mathbf{P}_f) \langle \mathbf{P}_f | S | \mathbf{P}_i \rangle. \qquad (1.25)$$

Thus $\mathbf{P}_i = \mathbf{P}_f$ whenever $\langle \mathbf{P}_f | S | \mathbf{P}_i \rangle \neq 0$.

We can now apply this general procedure to other invariance properties.

2. Invariance under Time Translations

If the Hamiltonian is independent of time, the Schrödinger equation $i\hbar\, \partial\psi(t)/\partial t = H\psi(t)$ is invariant under translations along the time axis, that is, under the transformation $t \to t + t_0$. Using the reasoning of the previous section, we can see from the Schrödinger equation that H is the generator of time translations. Thus we have the immediate result

$$U = \exp\left(-\frac{i}{\hbar} H t_0\right) \tag{2.1}$$

and

$$\bar{\psi}(t) = \psi(t + t_0) = \exp\left(-\frac{i}{\hbar} H t_0\right) \psi(t). \tag{2.2}$$

This in fact gives us a formal solution to the Schrödinger equation, since it describes the behavior of the state as t_0 increases. We have exploited this in Chapter 8 where a formal time-dependent scattering theory was developed.

The conservation law that follows from invariance under time translations has been discussed many times previously in this volume. The eigenstates of the generator H are simply the states $\psi_i^{(\pm)}$. Using the form of the S matrix developed in Chapter 9, we have

$$\begin{aligned} S_{fi} &= \langle \psi_f^{(-)} | \psi_i^{(+)} \rangle = \langle \bar{\psi}_f^{(-)} | \bar{\psi}_i^{(+)} \rangle \\ &= \langle \psi_f^{(-)} | \exp\left(\frac{i}{\hbar} H t_0\right) \exp\left(-\frac{i}{\hbar} H t_0\right) | \psi_i^{(+)} \rangle \\ &= S_{fi} \exp\left(\frac{i}{\hbar} (E_f - E_i) t_0\right). \end{aligned} \tag{2.3}$$

If this is to hold for all t_0, we must have $E_f = E_i$ when S_{fi} is nonzero; this is conservation of energy.

3. Galilean Invariance

The properties of many systems are invariant under a change in the velocity of their center of mass. This may be considered to be an actual change in the motion of the system or, equivalently, a change in the motion of the observer. Galilean invariance refers to the identical results obtained by a stationary observer or by an observer moving with a constant velocity relative to the center of mass of the system. If the interactions do not depend on the velocities of the particles relative to any fixed coordinate system, that is, if no external velocity-dependent forces such as magnetic fields are present, the system will be Galilean invariant.

The Transformation Operator

Under a Galilean or velocity transformation the system is caused to move with a velocity $\mathbf{v}_0$ relative to its original velocity. At time t the coordinate of each particle is shifted by $\mathbf{v}_0 t$ and the momentum by $\mathbf{p}_0 = m\mathbf{v}_0$. The transformation operator U that accomplishes this is therefore a product of the coordinate-translation operator $\exp[-(i/\hbar)\,\mathbf{p}\cdot\mathbf{v}_0 t]$ and the momentum-translation operator $\exp[(i/\hbar)\,\mathbf{p}_0\cdot\mathbf{r}]$. (The latter can be derived by the same methods as those used to derive U in Section 1, except that the momentum representation of the state is used.) Then we have

$$U = \exp\left(\frac{i}{\hbar}\mathbf{p}_0\cdot\mathbf{r}\right)\exp\left(-\frac{i}{\hbar}\mathbf{p}\cdot\mathbf{v}_0 t\right). \tag{3.1}$$

The coordinate, momentum, and spin operators transform as

$$\begin{aligned}\bar{\mathbf{r}} &= U\mathbf{r}U^{-1} = \mathbf{r} - \mathbf{v}_0 t \\ \bar{\mathbf{p}} &= U\mathbf{p}U^{-1} = \mathbf{p} - m\mathbf{v}_0 \\ \bar{\mathbf{S}} &= U\mathbf{S}U^{-1} = \mathbf{S}.\end{aligned} \tag{3.2}$$

The Galilean transformation is unique among coordinate transformations in that it does not leave the Hamiltonian unchanged. Instead, it changes the kinetic energy by modifying the velocity of each particle. The change in the Hamiltonian can be traced to the inherent time dependence of the transformation defined by Eq. (3.1). The modified time dependence must arise from a modification in the Hamiltonian, which generates the motion in time. Using Eq. (3.2), the free-particle Hamiltonian for a single particle is indeed transformed to

$$\bar{H}_0 = UH_0U^{-1} = (\mathbf{p} - m\mathbf{v}_0)^2/2m. \tag{3.3}$$

Let us consider the time-dependent Schrödinger equation for a single free particle:

$$i\hbar\frac{\partial\phi(t)}{\partial t} = H_0\,\phi(t). \tag{3.4}$$

3. Galilean Invariance

The principle of Galilean invariance states that, if this equation is satisfied in one inertial frame, it will be satisfied in all inertial frames. Equivalently, if it is satisfied by a given state vector $\phi(t)$, it will be satisfied by the state vector $\bar{\phi}(t) = U \phi(t)$, where U generates the transformations of Eq. (3.2). If we use the form of U given by Eq. (3.1), we find that $\phi(t)$ satisfies the new equation

$$i\hbar \frac{\partial \phi(t)}{\partial t} = (H_0 - \tfrac{1}{2} m v_0^2) \bar{\phi}(t),$$

rather than the original Schrödinger equation. The invariance of the Schrödinger equation can be restored if we instead use for the transformation operator

$$U = \exp\left[-\frac{i}{\hbar}(\tfrac{1}{2}mv_0^2)t\right] \exp\left(\frac{i}{\hbar} \mathbf{p}_0 \cdot \mathbf{r}\right) \exp\left(-\frac{i}{\hbar} \mathbf{p} \cdot \mathbf{v}_0 t\right). \qquad (3.5)$$

This differs from the previous choice for U by a phase factor of unit modulus, so that Eqs. (3.2) and (3.3) are unchanged. However, we now find that the Schrödinger equation

$$i\hbar \frac{\partial \bar{\phi}(t)}{\partial t} = H_0 \bar{\phi}(t) \qquad (3.6)$$

is satisfied.

The necessity for the additional phase factor is of course a result of the change of kinetic energy. The final form for U transforms a free-particle state as we would expect intuitively. If $\phi(t)$ is a one-particle state having momentum $\mathbf{p}_1$, the wave function is

$$\phi(\mathbf{r}, t) = \langle \mathbf{r} | \phi(t) \rangle = \exp\left(\frac{i}{\hbar} \mathbf{p}_1 \cdot \mathbf{r}\right) \exp\left(-\frac{i}{\hbar} \frac{p_1^2}{2m} t\right). \qquad (3.7)$$

The transformed wave function is

$$\bar{\phi}(\mathbf{r}, t) = \langle \mathbf{r} | \bar{\phi}(t) \rangle = \exp\left[\frac{i}{\hbar}(\mathbf{p}_1 + \mathbf{p}_0) \cdot \mathbf{r}\right] \exp\left[-\frac{i}{\hbar} \frac{(p_1 + p_0)^2}{2m} t\right], \qquad (3.8)$$

which is just a state having momentum $\mathbf{p}_1 + \mathbf{p}_0$.

General Consequences of Galilean Invariance

When interactions are present, the condition of Galilean invariance is that the interaction V be invariant under the transformation of Eq. (3.2). This gives the condition

$$V(\mathbf{r}_1 - \mathbf{v}_0 t, \mathbf{r}_2 - \mathbf{v}_0 t, \mathbf{p}_1 - m_1 \mathbf{v}_0, \mathbf{p}_2 - m_2 \mathbf{v}_0) = V(\mathbf{r}_1, \mathbf{r}_2, \mathbf{p}_1, \mathbf{p}_2), \quad (3.9)$$

where the spin variable has been suppressed. This result is simpler when expressed in terms of center-of-mass and relative coordinates and momenta. These transform as

$$\begin{array}{ll} \mathbf{R} \to \mathbf{R} - \mathbf{v}_0 t, & \mathbf{P} \to \mathbf{P} - (m_1 + m_2)\mathbf{v}_0, \\ \mathbf{r} \to \mathbf{r}, & \mathbf{p} \to \mathbf{p}, \end{array} \quad (3.10)$$

where the relative momentum is

$$\mathbf{p} = \frac{m_2 \mathbf{p}_1 - m_1 \mathbf{p}_2}{m_1 + m_2} = \frac{m_1 m_2}{m_1 + m_2}(\mathbf{v}_1 - \mathbf{v}_2).$$

We now see that Eq. (3.9) implies that a Galilean-invariant interaction must be independent of the center-of-mass position and momentum and can depend only on the relative coordinate and momentum. Thus we have

$$V(\mathbf{r}_1, \mathbf{r}_2, \mathbf{p}_1, \mathbf{p}_2) = V(\mathbf{r}, \mathbf{p}). \quad (3.11)$$

Equation (3.10) also implies that the free-particle Hamiltonian for a pair of particles,

$$H_0 = \frac{p_1^2}{2m_1} + \frac{p_2^2}{2m_2} = \frac{P^2}{2M} + \frac{p^2}{2m}, \quad (3.12)$$

transforms as

$$\frac{P^2}{2M} + \frac{p^2}{2m} \to \frac{(\mathbf{P} - M\mathbf{v}_0)^2}{2M} + \frac{p^2}{2m}. \quad (3.13)$$

Here $M = m_1 + m_2$ is the total mass and $m = m_1 m_2/(m_1 + m_2)$ is the reduced mass. We can use this to show that S is invariant under a Galilean transformation.

From Eq. (1.4) we see that the invariance of the S matrix hinges on the invariance of the Green's function $1/(E - H + i\varepsilon)$. This invariance follows from the observation that the energy E is the energy of the initial state which, under a Galilean transformation, undergoes the change

$$\frac{P_i^2}{2M} + \frac{p_i^2}{2m} \to \frac{(\mathbf{P}_i - M\mathbf{v}_0)^2}{2M} + \frac{p_i^2}{2m}. \quad (3.14)$$

Since V does not depend on the position of the center of mass, the total momentum operator $\mathbf{P}$ commutes with it and can act "through" V onto the initial state, giving just the total momentum $\mathbf{P}_i$. Hence, using Eqs. (3.12)

3. Galilean Invariance

and (3.14), the Green's function becomes effectively

$$\frac{1}{E - H + i\varepsilon} = \frac{1}{(p_i^2/2m) - (p^2/2m) - V + i\varepsilon}, \quad (3.15)$$

independent of the center-of-mass variables and thus invariant under a Galilean transformation.

The scattering operator is thus Galilean invariant, and the S matrix has the property given in Eq. (1.19):

$$\langle \phi_f | S | \phi_i \rangle = \langle \bar{\phi}_f | S | \bar{\phi}_i \rangle. \quad (3.16)$$

Suppose, for instance, that ϕ_i and ϕ_f contain two particles each and that the momentum of each particle is specified. (Since the spin is not affected by a velocity transformation, it is immaterial for our present purposes whether or not the particles have intrinsic spins.) The initial state is then

$$\phi_i = | \mathbf{p}_1, \mathbf{p}_2 \rangle = | \mathbf{P}_i, \mathbf{p}_i \rangle.$$

Under the Galilean transformation the momenta transform as

$$\begin{aligned} \mathbf{P}_i &\to \mathbf{P}_i + M\mathbf{v}_0 \\ \mathbf{p}_i &\to \mathbf{p}_i. \end{aligned} \quad (3.17)$$

Thus Eq. (3.16) implies

$$\langle \mathbf{P}_f, \mathbf{p}_f | S | \mathbf{P}_i, \mathbf{p}_i \rangle = \langle \mathbf{P}_f + M\mathbf{v}_0, \mathbf{p}_f | S | \mathbf{P}_i + M\mathbf{v}_0, \mathbf{p}_i \rangle, \quad (3.18)$$

and this S-matrix element, as a function of the total momenta, can depend only on the difference $\mathbf{P}_i - \mathbf{P}_f$. However, we have seen that translation invariance makes this difference zero, so that the combination of invariance under translations of position and of velocity implies that the S matrix cannot depend at all on the total momentum.

The conservation law that results from Galilean invariance depends on the observation that, if S is independent of the total momentum $\mathbf{P}$, it will commute with the canonically conjugate variable, the position of the center of mass. As a result, the position of the center of mass will be conserved and, in the coordinate representation, the S matrix will be a local operator in the center-of-mass coordinate. Together with Eq. (1.5), this shows that the spatial representation of S has the form

$$S(\mathbf{R}', \mathbf{r}'; \mathbf{R}, \mathbf{r}) = \delta(\mathbf{R} - \mathbf{R}') S(\mathbf{r}', \mathbf{r}). \quad (3.19)$$

Thus the combination of translation invariance and Galilean invariance requires that the S matrix be independent of the center-of-mass coordinate.

We may note the conjugate results we have obtained: Translation invariance implies conservation of total momentum; Galilean invariance implies conservation of the center-of-mass position. The latter is not a conservation law in the same sense as is momentum conservation, since the center-of-mass coordinate is not fixed in the initial state of a scattering experiment. This result does, however, place restrictions on models such as "pseudopotentials," which one may wish to use to obtain simple approximations to S or T: they may depend only on *relative* coordinates and momenta.

SEPARATION OF CENTER-OF-MASS MOTION

This combination of invariance properties also allows us to describe the scattering process using the most convenient inertial frame available. Galilean invariance allows us, for instance, to move to the center-of-mass system for the initial state, where $\mathbf{P}_i = 0$, and translation invariance ensures that this will also be the center-of-mass system for the final state (since $\mathbf{P}_f = \mathbf{P}_i$). This provides a justification for our use of the center-of-mass system throughout the earlier portion of this book.

The S matrix we have been using in this chapter describes collisions in an arbitrary reference frame. To relate this general S matrix to the center-of-mass description, we can introduce the complete set of coordinate eigenstates into the S matrix describing a two-particle collision:

$$\langle \mathbf{p}_1', \mathbf{p}_2' | S | \mathbf{p}_1, \mathbf{p}_2 \rangle = \langle \mathbf{P}_f, \mathbf{p}_f | S | \mathbf{P}_i, \mathbf{p}_i \rangle$$
$$= \int d^3R\, d^3R'\, d^3r\, d^3r' \langle \mathbf{P}_f, \mathbf{p}_f | \mathbf{R}', \mathbf{r}' \rangle$$
$$\times S(\mathbf{R}', \mathbf{r}'; \mathbf{R}, \mathbf{r}) \langle \mathbf{R}, \mathbf{r} | \mathbf{P}_i, \mathbf{p}_i \rangle. \tag{3.20}$$

The two-particle wave function is

$$\langle \mathbf{R}, \mathbf{r} | \mathbf{P}_i, \mathbf{p}_i \rangle = e^{i\mathbf{K}_i \cdot \mathbf{R}} e^{i\mathbf{k}_i \cdot \mathbf{r}}. \tag{3.21}$$

Introducing Eq. (3.19), we have

$$\langle \mathbf{p}_1', \mathbf{p}_2' | S | \mathbf{p}_1, \mathbf{p}_2 \rangle = \int d^3R\, d^3r\, d^3r' \exp[-i(\mathbf{K}_f \cdot \mathbf{R} + \mathbf{k}_f \cdot \mathbf{r}')]$$
$$\times S(\mathbf{r}', \mathbf{r}) \exp[i(\mathbf{K}_i \cdot \mathbf{R} + \mathbf{k}_i \cdot \mathbf{r})]$$
$$= (2\pi)^3 \delta(\mathbf{K}_i - \mathbf{K}_f) \langle \mathbf{k}_f | S | \mathbf{k}_i \rangle, \tag{3.22}$$

where

$$\langle \mathbf{k}_f | S | \mathbf{k}_i \rangle = \int d^3r\, d^3r'\, e^{-i\mathbf{k}_f \cdot \mathbf{r}'} S(\mathbf{r}', \mathbf{r}) e^{i\mathbf{k}_i \cdot \mathbf{r}}. \tag{3.23}$$

This is the S-matrix element we have used in our earlier work. It is equivalent to the S matrix one would obtain in the center-of-mass system, where the relative momentum is the only relevant quantity. Since the scattering operator does not act upon the center-of-mass coordinate and momentum, it is effectively a unit operator with respect to these variables. One can then check that all the formal relations we have discussed in earlier chapters hold both for the full S matrix, Eq. (3.22), and for the center-of-mass S matrix, Eq. (3.23). Furthermore, the latter can be used in all practical computations.

4. Rotation Invariance

The interactions between particles in an isolated system are generally invariant under rotations about any axis. For instance, the central potentials we dealt with earlier are obviously rotation invariant. In discussing this invariance, we shall content ourselves with noting the conservation laws and the implications for the dependence of the S matrix upon observables. The detailed properties of rotations and the transformation properties of angular-momentum eigenstates have been thoroughly discussed in several recent books.

Under an infinitesimal rotation through an angle $\delta\theta$ and about a direction $\hat{n}$, scalars are unchanged while a vector $\mathbf{V}$ transforms according to

$$\mathbf{V} \to \overline{\mathbf{V}} = \mathbf{V} + \delta\theta\, \hat{n} \times \mathbf{V}. \tag{4.1}$$

For an arbitrary rotation, the ith component of a vector will transform as

$$V_i \to \overline{V}_i = \sum_{j=1}^{3} R_{ij} V_j, \tag{4.2}$$

where R_{ij} is an element of the 3×3 rotation matrix. For simplicity, we shall denote this transformation by

$$\mathbf{V} \to \overline{\mathbf{V}} = R\mathbf{V}, \tag{4.3}$$

where R is an operator on the vector space satisfying $R^T R = 1$. Thus, under a rotation

$$\begin{aligned} \mathbf{r} &\to R\mathbf{r} \\ \mathbf{p} &\to R\mathbf{p} \\ \mathbf{S} &\to R\mathbf{S}. \end{aligned} \tag{4.4}$$

If the Hamiltonian is invariant under such a transformation, that is, if there is no preferred direction in space, the S matrix will likewise be rotation invariant. The corresponding conservation law is obtained in the representation in which the canonically conjugate operator is diagonal. If the rotation is about the z axis, this operator is J_z, the z component of the total angular momentum. If the rotation is about an arbitrary axis, the generator of the transformation is the operator corresponding to the angular momentum about that axis. Thus, if $\boldsymbol{\theta}_0$ is a vector having a magnitude equal to the angle of rotation θ_0 and directed along the axis of rotation, defined conventionally by a right-hand rule, the transformation operator is

$$U = \exp\left(-\frac{i}{\hbar} \mathbf{J} \cdot \boldsymbol{\theta}_0\right). \tag{4.5}$$

The appropriate representation is that in which J^2 and J_z (for some arbitrary z axis) are diagonal. If S is invariant under rotations, it will commute with both J^2 and J_z, and the matrix $\langle J_f, M_f | S | J_i, M_i \rangle$ will be diagonal, that is, this element will be zero unless $J_i = J_f$ and $M_i = M_f$. Here $J_i(J_i + 1)\hbar^2$ is the eigenvalue of J^2, $M_i\hbar$ is the eigenvalue of J_z, and we have suppressed the dependence of the S matrix on other variables such as the energy.

This matrix element must also be independent of M_i, since it cannot depend upon the choice of z axis or the orientation of the system. This can be seen formally by using the raising and lowering operators $J_{\pm} = J_x \pm iJ_y$, which commute with S and which raise or lower the value of M_i when acting on $|J_i, M_i\rangle$. (This property follows from the commutation relation $[J_z, J_{\pm}] = \pm J_{\pm}$.) Suppose $M_i < J_i$. Then

$$J_+ | J_i, M_i \rangle = \hbar [J_i(J_i + 1) - M_i(M_i + 1)]^{\frac{1}{2}} | J_i, M_i + 1 \rangle. \tag{4.6}$$

The normalization is obtained from the relation $J_-J_+ = J^2 - J_z(J_z + 1)$, together with the fact that $(J_-)^\dagger = J_+$. These imply that

$$\langle J_i, M_i | J_-J_+ | J_i, M_i \rangle = [J_i(J_i + 1) - M_i(M_i + 1)]\hbar^2. \tag{4.7}$$

Now, using $[S, J_-] = 0$ and Eqs. (4.6) and (4.7), we have the identity

$$\langle J_i, M_i | S | J_i, M_i \rangle = \langle J_i, M_i \left| \frac{S}{\hbar^2} \left[\frac{J_-J_+}{J_i(J_i + 1) - M_i(M_i+1)} \right] \right| J_i, M_i \rangle$$

$$= [J_i(J_i + 1) - M_i(M_i + 1)]^{-1}$$

$$\times \hbar^{-2} \langle J_i, M_i | J_-SJ_+ | J_i, M_i \rangle$$

$$= \langle J_i, M_i + 1 | S | J_i, M_i + 1 \rangle. \tag{4.8}$$

Hence all elements $\langle J_i, M_i | S | J_i, M_i \rangle$ having a given value of J_i are equal.

We find then that the S matrix is given by

$$\langle J_f, M_f | S | J_i, M_i \rangle = S_{J_i} \delta_{J_i, J_f} \delta_{M_i, M_f}, \tag{4.9}$$

showing explicitly the conservation of total angular momentum and of its component along any axis. The number S_{J_i} is the eigenvalue of S in the J representation; if reactions are possible, it will be a matrix joining all possible states having the same total angular momentum.

5. Reflection Invariance

Under a complete reflection of coordinates, right-hand coordinate systems are transformed into left-hand coordinate systems. The concept of reflection invariance implies that nature does not distinguish between these two systems.

Description of Reflections

A space reflection distinguishes polar vectors such as **r** and **p** from axial vectors such as $\mathbf{L} = \mathbf{r} \times \mathbf{p}$ and **S**. Under this operation polar vectors change sign, while axial vectors remain unchanged:

$$\begin{aligned} \mathbf{r} &\to -\mathbf{r} \\ \mathbf{p} &\to -\mathbf{p} \\ \mathbf{L} &\to \mathbf{L} \\ \mathbf{S} &\to \mathbf{S}. \end{aligned} \tag{5.1}$$

Reflection is a discontinuous operation and cannot be obtained by a succession of infinitesimal transformations, as were the continuous operations discussed in previous sections. We expect therefore that the operator that generates space reflections cannot be expressed in a convenient exponential form, as were the previous transformation operators. Nevertheless, we can define a unitary space reflection operator $\mathscr{P}$ by its action on a complete set of states.

Consider, for example, the complete set of one-particle states $|\mathbf{p}, m_s\rangle$ having a well-defined momentum and, for particles with intrinsic spin, a well-defined z component of the spin. The transformations of Eq. (5.1) imply that, in general, these states behave under a reflection as

$$\mathscr{P} | \mathbf{p}, m_s \rangle = \Pi_\alpha | -\mathbf{p}, m_s \rangle, \tag{5.2}$$

where the phase factor Π_α introduced by the reflection is characteristic of the particle being described (α is the particle label). If we assume that two successive reflections lead back to the original state, then $\mathscr{P}^2 = 1$ and $\Pi_\alpha^2 = 1$. Hence $\Pi_\alpha = \pm 1$. The number Π_α is called the intrinsic parity of the particle. If $\Pi_\alpha = +1$, the particle is said to be a scalar; if $\Pi_\alpha = -1$, the particle is said to be a pseudoscalar. The intrinsic parity of a many-particle state is the product of the intrinsic parities of each particle.

In general, the absolute parity of a particle can be determined only if the particle is destroyed or created singly. It is conventional to define the intrinsic parities of the electron and of the nucleon, which must be produced in association with antiparticles, as positive. (They do not transform into each other, so that their relative parity cannot be experimentally measured.) The parity of the positron is found, by observing the annihilation of positronium, to be negative, in agreement with the prediction of the Dirac equation; thus the positron is pseudoscalar relative to the electron. Likewise, the antinucleon is believed to be pseudoscalar.

GENERAL CONSEQUENCES OF REFLECTION INVARIANCE

Since all observables are expressed in terms of matrix elements involving the overlap of *two* states, only the *relative* parity of two different systems can be determined experimentally. For, if the Hamiltonian and the S matrix are reflection invariant, we have the relation

$$\langle \mathbf{p}_f, m_s' | S | \mathbf{p}_i, m_s \rangle = \langle \mathbf{p}_f, m_s' | \mathscr{P}^{-1}\mathscr{P} S \mathscr{P}^{-1}\mathscr{P} | \mathbf{p}_i, m_s \rangle$$
$$= \Pi_f \Pi_i \langle -\mathbf{p}_f, m_s' | S | -\mathbf{p}_i, m_s \rangle. \qquad (5.3)$$

Hence, if the initial and final systems are composed of identical particles, only $\Pi_i^2 = 1$ arises in the reflection operation. Meaningful information can be obtained only if the initial and final states are composed of different particles.

If the systems are unchanged in the collision, or if both systems have the same intrinsic parity, Eq. (5.3) tells us that the corresponding S-matrix element is a scalar function of the initial and final momenta and spin directions. In a reaction in which the initial and final states have opposite intrinsic parities (such as one in which a π-meson is produced), $\Pi_f \Pi_i = -1$. According to Eq. (5.3), the corresponding S-matrix element must be odd under reflection.

The conservation law that follows from reflection invariance is, as usual, obtained in a representation in which the transformation operator is diagonal. Under a reflection the spherical harmonics behave as

$$Y_{lm}(\theta, \phi) \to Y_{lm}(\pi - \theta, \pi + \phi) = (-1)^l Y_{lm}(\theta, \phi). \qquad (5.4)$$

5. Reflection Invariance

Thus the spherical harmonics are eigenfunctions of the reflection operator with eigenvalues $(-1)^l$. This implies that angular momentum eigenstates transform as

$$\mathscr{P} \, | \, klmm_s \rangle = \Pi_\alpha (-1)^l \, | \, klmm_s \rangle. \tag{5.5}$$

We can derive this from Eq. (5.2). Suppressing the intrinsic spin, which does not affect these considerations, the coordinate representation of an angular momentum eigenstate is

$$\langle \mathbf{r} \, | \, klm \rangle = N i^l \frac{F_l(kr)}{kr} Y_{lm}(\theta, \phi). \tag{5.6}$$

The Legendre expansion of the plane wave leads to the expression

$$\int d\Omega_k \, Y_{lm}(\theta_k, \phi_k) \, e^{i \mathbf{k} \cdot \mathbf{r}} = 4\pi i^l \frac{F_l(kr)}{kr} Y_{lm}(\theta, \phi)$$

$$= \frac{4\pi}{N} \langle \mathbf{r} \, | \, klm \rangle. \tag{5.7}$$

This is the coordinate representation of the state-vector expansion

$$\int d\Omega_k \, Y_{lm}(\theta_k, \phi_k) \, | \, \mathbf{k} \rangle = \frac{4\pi}{N} \, | \, klm \rangle, \tag{5.8}$$

which expresses the angular momentum eigenstate in terms of the linear momentum eigenstates. Using Eqs. (5.2) and (5.4), we then have the relations

$$\mathscr{P} \, | \, klm \rangle = \Pi_\alpha \frac{N}{4\pi} \int d\Omega_k \, Y_{lm}(\theta_k, \phi_k) \, | -\mathbf{k} \rangle$$

$$= \Pi_\alpha \frac{N}{4\pi} \int d\Omega_{k'} \, Y_{lm}(\pi - \theta_{k'}, \pi + \phi_{k'}) \, | \, \mathbf{k}' \rangle$$

$$= \Pi_\alpha (-1)^l \, | \, klm \rangle. \tag{5.9}$$

Introducing this result, the S matrix satisfies

$$\langle k_f l_f m_f m_s' \, | \, S \, | \, k_i l_i m_i m_s \rangle = \langle k_f l_f m_f m_s' \, | \, \mathscr{P}^{-1} \mathscr{P} S \mathscr{P}^{-1} \mathscr{P} \, | \, k_i l_i m_i m_s \rangle$$
$$= \Pi_f \Pi_i (-1)^{l_i + l_f} \langle k_f l_f m_f m_s' \, | \, S \, | \, k_i l_i m_i m_s \rangle, \tag{5.10}$$

which implies the conservation law

$$(-1)^{l_i + l_f} = \Pi_f \Pi_i. \tag{5.11}$$

Thus the quantity $\Pi_i (-1)^{l_i}$ is conserved, and it is the evenness or oddness of l that is affected in a collision by a change in the intrinsic parity. For elastic scattering $\Pi_f = \Pi_i$, and even-l states can only lead to other even-l states, and similarly for odd-l states.

6. Time-Reversal Invariance

The time-reversal transformation is also discontinuous, but again we can define an operator that induces this transformation. However, it must be quite different from the operators we have discussed earlier. All of the previous transformations were useful because they left the Schrödinger equation unchanged; the time-reversal operator is useful precisely because it *does* modify the Schrödinger equation.

Description of Time-Reversal

Under the time-reversal transformation $t \to -t$ and the Schrödinger equation for the time-dependent wave function becomes

$$i\hbar \frac{\partial \psi(\mathbf{r}, t)}{\partial t} = H \psi(\mathbf{r}, t) \to -i\hbar \frac{\partial \psi(\mathbf{r}, -t)}{\partial t} = H \psi(\mathbf{r}, -t). \tag{6.1}$$

(We assume throughout that H is independent of time; however, as we shall see, this does not guarantee time-reversal invariance.) One might expect time-reversal invariance to imply that, if $\psi(\mathbf{r}, t)$ is a solution of the Schrödinger equation, then $\psi(\mathbf{r}, -t)$ is also a solution. But Eq. (6.1) shows that this is not the case; the Schrödinger equation itself has been transformed and $\psi(\mathbf{r}, -t)$ is a solution of this *transformed* equation.

One way to recover the original Schrödinger equation is to take the complex conjugate of Eq. (6.1), giving

$$i\hbar \frac{\partial \psi^*(\mathbf{r}, -t)}{\partial t} = H^* \psi^*(\mathbf{r}, -t). \tag{6.2}$$

If there exists a unitary operator U having the property

$$UH^*U^{-1} = H, \tag{6.3}$$

then the wave function

$$\bar{\psi}(\mathbf{r}, t) = U \psi^*(\mathbf{r}, -t) \tag{6.4}$$

6. Time-Reversal Invariance

does satisfy the Schrödinger equation:

$$i\hbar \frac{\partial \bar{\psi}(\mathbf{r}, t)}{\partial t} = H \, \bar{\psi}(\mathbf{r}, t). \tag{6.5}$$

Although this wave function may differ somewhat from what we might have expected for the time-reversed wave function, we can see readily that it does indeed have the correct properties.

Let us first consider what is meant by the time-reversal transformation. We do *not* mean that time will run backwards. Besides leading to states that would not be observable, this would not be in accord with the basic postulates of quantum mechanics. These require that a measurement on a system destroy all information about the previous states of the system, so that it is impossible from a knowledge of the state at time t to know the behavior of the system at earlier times. This is reflected in the one-sidedness of the time-dependent Schrödinger equation which, like a diffusion equation, is linear in the time.

What is required can be seen by considering the classical case. Suppose a classical system moves along a certain path determined by the laws of motion (Fig. 10.2). If we perform a time inversion on this system, the system will

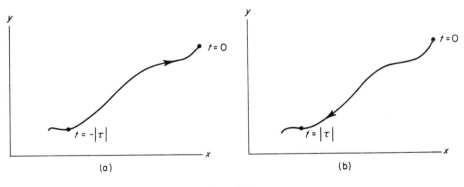

Fig. 10.2

retrace this path, returning ultimately to its starting point with the reversed momentum and spin. This concept of "time reversal" is properly thought of as "reversal of the motion"; it is accomplished by reversing the directions of the momenta and spins, but permitting time to continue to run forward. Such a reversal of the motion is connected with the transformation $t \to -t$ through the fact that observables that are odd in the time change their sign, while functions that are even in the time do not. For instance, the momentum changes sign, while the kinetic energy does not.

Returning now to our transformed wave function, we can examine, for instance, the expectation value of the momentum. We will see below that U commutes with the momentum operator in the spatial representation. Then

$$\begin{aligned}
\langle \bar{\psi}(t) | \mathbf{p} | \bar{\psi}(t) \rangle &= \int d^3r \, \bar{\psi}^*(\mathbf{r}, t)(-i\hbar \nabla) \bar{\psi}(\mathbf{r}, t) \\
&= \int d^3r \, \psi(\mathbf{r}, -t)(-i\hbar \nabla) \psi^*(\mathbf{r}, -t) \\
&= \int d^3r \, \psi^*(\mathbf{r}, -t)(i\hbar \nabla) \psi(\mathbf{r}, -t) \\
&= -\langle \psi(-t) | \mathbf{p} | \psi(-t) \rangle,
\end{aligned} \quad (6.6)$$

where a partial integration has been performed. This is the negative of the momentum of the original state at the time $(-t)$, which is what we expect from Fig. 10.2. One can check that other quantities as well behave as expected for a time-reversed state.

By comparison with the classical situation, we see that reversal of the motion requires that representative operators transform as follows:

$$\begin{aligned} \mathbf{r} &\to \mathbf{r} \\ \mathbf{p} &\to -\mathbf{p} \\ \mathbf{J} &\to -\mathbf{J}. \end{aligned} \quad (6.7)$$

Since the energy of the system is invariant under this operation, the Hamiltonian must remain unchanged.

The Transformation Operator

We now want to pass from the wave function description to the abstract operator formalism. In this case we must introduce an operator, conventionally denoted by K, which generates the time-reversal transformation. According to Eq. (6.4), this operator must transform the state vector $\psi(-t)$ into a state vector $\bar{\psi}(t)$, via

$$\bar{\psi}(t) = K \psi(-t), \quad (6.8)$$

such that the new wave function is proportional to the *complex conjugate* of the original wave function. Such an operator is called an "antilinear" operator.

An antilinear operator satisfies the relation

$$K(c_1 \psi_1 + c_2 \psi_2) = c_1^* K \psi_1 + c_2^* K \psi_2; \quad (6.9)$$

this differs from ordinary linearity through the presence of the complex conjugate of the numerical coefficients. An operator that is antilinear and that preserves the normalization of all state vectors is antiunitary:

$$\langle K\psi | K\psi \rangle = \langle \psi | K^\dagger K | \psi \rangle = \langle \psi | \psi \rangle. \tag{6.10}$$

It follows from Eqs. (6.9) and (6.10), using $\psi = \psi_1 + \psi_2$ and $\psi = \psi_1 + i\psi_2$ successively in Eq. (6.10), that

$$\langle K\psi_1 | K\psi_2 \rangle = \langle \psi_1 | K^\dagger K | \psi_2 \rangle = \langle \psi_2 | \psi_1 \rangle. \tag{6.11}$$

Hence $K^\dagger K$ maintains orthonormality, as it would for a unitary operator, but in addition it reverses the order of the two state vectors. This property is perhaps not surprising, since in an S-matrix element, for instance, this just corresponds to the interchange of initial and final states. This is what we expect of a time-reversal operator. If K is to accomplish the desired reversal of motion, it must also induce the transformations of Eq. (6.7):

$$\begin{aligned} K\mathbf{r}K^{-1} &= \mathbf{r} \\ K\mathbf{p}K^{-1} &= -\mathbf{p} \\ K\mathbf{J}K^{-1} &= -\mathbf{J}. \end{aligned} \tag{6.12}$$

These conditions characterize completely the antilinear, antiunitary operator K.

To see explicitly how the properties of K lead to a time-reversed state, let us consider as an example a free-particle wave packet for a spinless particle. Such a packet is a superposition of momentum eigenstates $\phi_\mathbf{k}$. These states transform into states having the opposite momentum,

$$\bar{\phi}_\mathbf{k} = K\phi_\mathbf{k} = \phi_{-\mathbf{k}}, \tag{6.13}$$

since, according to Eq. (6.12),

$$\mathbf{p}K\phi_\mathbf{k} = KK^{-1}\mathbf{p}K\phi_\mathbf{k} = -K\mathbf{p}\phi_\mathbf{k} = -\hbar\mathbf{k}K\phi_\mathbf{k}. \tag{6.14}$$

(Any arbitrary phase factor can be absorbed into K as is done below.) The wave packet

$$\phi(t) = \int \frac{d^3k}{(2\pi)^3} A(\mathbf{k}) \exp\left(-\frac{i}{\hbar} E_k t\right) \phi_\mathbf{k} \tag{6.15}$$

is then transformed according to Eq. (6.8) into

$$\bar{\phi}(t) = K\phi(-t) = \int \frac{d^3k}{(2\pi)^3} A^*(\mathbf{k}) \exp\left(-\frac{i}{\hbar} E_k t\right) \phi_{-\mathbf{k}}, \tag{6.16}$$

through the use of Eq. (6.13) and the antilinearity of K. The transformed wave function is then

$$\bar{\phi}(\mathbf{r}, t) = \langle \mathbf{r} | \bar{\phi}(t) \rangle$$
$$= \int \frac{d^3k}{(2\pi)^3} A^*(\mathbf{k}) \exp\left(-\frac{i}{\hbar} E_k t\right) e^{-i\mathbf{k}\cdot\mathbf{r}} = \phi^*(\mathbf{r}, -t). \quad (6.17)$$

We can see also, as in Eq. (6.6), that the mean momentum in $\bar{\phi}(\mathbf{r}, t)$ is the negative of that in $\phi(\mathbf{r}, -t)$.

It is useful to define a complex conjugation operator K_0 which is simply an instruction to take the complex conjugate of any numerical coefficients or functions to its right. Since two complex conjugations nullify each other, $K_0^2 = 1$ or $K_0^{-1} = K_0$. This operator is an antiunitary operator, but one may verify that the product KK_0 is a unitary operator. Then one may always write K as the product of a unitary operator U and the complex conjugation operator K_0:

$$K = UK_0. \quad (6.18)$$

If we wish to use complex conjugation as an operation in the Hilbert space of state vectors, we must recognize that its effect depends upon the representation we are using. For instance, in the coordinate representation the momentum operator is $\mathbf{p} = -i\hbar \nabla$ and $K_0 \mathbf{p} K_0^{-1} = \mathbf{p}^* = -\mathbf{p}$; in the momentum representation $K_0 \mathbf{p} K_0^{-1} = \mathbf{p}^* = \mathbf{p}$. Thus one cannot say unambiguously whether an operator is "real" or "imaginary." This may be contrasted with the operation of Hermitian conjugation; a Hermitian operator satisfying $\mathcal{O}^\dagger = \mathcal{O}$ is Hermitian in all representations that are related by a unitary transformation. As a consequence, the form of the unitary operator U in Eq. (6.18) depends upon the representation we are using.

The time-reversal operator K must lead to the operator transformations of Eq. (6.12). Using Eq. (6.18), the transformation of an arbitrary operator is

$$K\mathcal{O}K^{-1} = UK_0\mathcal{O}K_0^{-1}U^{-1} = U\mathcal{O}^*U^{-1}. \quad (6.19)$$

In the spatial representation complex conjugation is sufficient to accomplish the transformations of $\mathbf{r}$ and $\mathbf{p}$ to $\mathbf{r}$ and $-\mathbf{p}$, respectively. The operator U must then act only on the spin components. For spin-$\frac{1}{2}$ particles the requirement is that $K\boldsymbol{\sigma}K^{-1} = -\boldsymbol{\sigma}$. The usual representation of the Pauli matrices is

$$\sigma_x = \begin{pmatrix} 0 & 1 \\ 1 & 0 \end{pmatrix}, \quad \sigma_y = \begin{pmatrix} 0 & -i \\ i & 0 \end{pmatrix}, \quad \sigma_z = \begin{pmatrix} 1 & 0 \\ 0 & -1 \end{pmatrix}. \quad (6.20)$$

6. Time-Reversal Invariance

Complex conjugation changes σ_y to $-\sigma_y$ but leaves σ_x and σ_z unaffected. However, σ_y anticommutes with these latter two (and of course commutes with itself) so U must be proportional to σ_y. For reasons that will become apparent shortly, we let

$$U = i\sigma_y. \tag{6.21}$$

Then we find

$$K\sigma K^{-1} = U\sigma^* U^{-1} = \sigma_y^* \sigma_y = -\sigma. \tag{6.22}$$

Then for a single spin-$\frac{1}{2}$ particle $K = i\sigma_y K_0$ in the coordinate representation. For instance, for a free particle having a momentum $\mathbf{k}$ and $m_z = +\frac{1}{2}$, the wave function is

$$\phi(\mathbf{r}, t) = \exp\left(i\mathbf{k}\cdot\mathbf{r} - \frac{i}{\hbar} E_k t\right) \begin{pmatrix} 1 \\ 0 \end{pmatrix} \tag{6.23}$$

while the transformed wave function is

$$\bar{\phi}(\mathbf{r}, t) = K\phi(\mathbf{r}, -t) = i\sigma_y K_0 \exp\left(i\mathbf{k}\cdot\mathbf{r} + \frac{i}{\hbar} E_k t\right) \begin{pmatrix} 1 \\ 0 \end{pmatrix}$$

$$= -\exp\left(-i\mathbf{k}\cdot\mathbf{r} - \frac{i}{\hbar} E_k t\right) \begin{pmatrix} 0 \\ 1 \end{pmatrix}. \tag{6.24}$$

As expected, this function has momentum and spin components opposite to those of the original function. (Because K has no effect upon space coordinates, we have taken some liberties with the notation in Eq. (6.24). If we were to follow rigidly the rules of Chapter 6, we would let $\langle \mathbf{r} | K | \mathbf{r}' \rangle = i\sigma_y K_0 \delta(\mathbf{r} - \mathbf{r}')$ and Eq. (6.24) would be

$$\bar{\phi}(\mathbf{r}, t) = \int d^3 r' \langle \mathbf{r} | K | r' \rangle \phi(\mathbf{r}', -t) = i\sigma_y K_0 \phi(\mathbf{r}, -t),$$

the same result as above.)

The transformation operator K may then be obtained in general by requiring that the operators for all observables transform properly. The condition of time-reversal invariance is that $\bar{\psi}(t) = K\psi(-t)$ must be a solution of the Schrödinger equation. Acting on the operator form of the Schrödinger equation for $\psi(-t)$, we have

$$K\left(-i\hbar \frac{\partial \psi(-t)}{\partial t}\right) = i\hbar \frac{\partial \bar{\psi}(t)}{\partial t} = KH\psi(-t) = \bar{H}\bar{\psi}(t) \tag{6.25}$$

and $\bar{H} = KHK^{-1}$. If the Hamiltonian is invariant under this transformation, that is, if $\bar{H} = H$, we will obtain the desired result. This condition for time-reversal invariance can also be expressed in terms of the unitary operators as

$$\bar{H} = KHK^{-1} = UH^*U^{-1} = H. \tag{6.26}$$

Let us consider some examples of this invariance condition. Together with Eq. (6.21), it implies that any spin-independent Hamiltonian must be real. In particular, a potential $V(r)$ must be real if time-reversal invariance is to hold. In Chapter 12 we will discuss situations where it is useful to introduce potentials having imaginary components; these describe processes where particles can be absorbed out of the incident beam. The time-reversed process, involving the creation of particles, is a very different process, and the potential is manifestly not time-reversal invariant.

If H contains a spin-orbit term $V_{so}(r)(\boldsymbol{\sigma} \cdot \mathbf{L})$ acting on a single spin-$\frac{1}{2}$ particle, the invariance condition is

$$U V_{so}^*(r)(\boldsymbol{\sigma}^* \cdot \mathbf{L}^*)U^{-1} = V_{so}^*(r)(\boldsymbol{\sigma} \cdot \mathbf{L}) = V_{so}(r)(\boldsymbol{\sigma} \cdot \mathbf{L}), \tag{6.27}$$

using Eq. (6.22) and $\mathbf{L} = \mathbf{r} \times \mathbf{p} = -i\hbar \mathbf{r} \times \boldsymbol{\nabla} = -\mathbf{L}^*$. Then, again, $V_{so}(r)$ must be real. On the other hand, if H were to contain a term of the form $V_p(r)(\boldsymbol{\sigma} \cdot \mathbf{r})$, the invariance condition would be

$$U V_p^*(r)(\boldsymbol{\sigma}^* \cdot \mathbf{r})U^{-1} = -V_p^*(r)(\boldsymbol{\sigma} \cdot \mathbf{r}) = V_p(r)(\boldsymbol{\sigma} \cdot \mathbf{r}), \tag{6.28}$$

so that $V_p(r)$ would have to be imaginary. Note that such a term is not reflection invariant.

The Phase of the Time-Reversed State

We will henceforth assume that the Hamiltonian is invariant under time reversal so that a transformed state vector $\bar{\phi}_i$, as well as the original vector ϕ_i, are each solutions of the Schrödinger equation. We will also deal only with energy eigenstates, that is, with solutions of the time-independent Schrödinger equation. The relation of $\bar{\phi}_i$ to ϕ_i can be determined from the operator transformations, Eq. (6.12). If ϕ_a is an eigenstate of a set of observables A, $\bar{\phi}_a$ will be an eigenstate of the transformed observables with eigenvalue a or, equivalently, can be considered to be an eigenstate of the original set of operators with transformed eigenvalues. We adopt the latter viewpoint, in which the system of particles — and not the reference frame of the observer — is transformed. We then write

$$\bar{\phi}_a \equiv K\phi_a = \eta_a \phi_{\bar{a}}, \tag{6.29}$$

where $\bar{a}$ denotes the transformed eigenvalue and η_a is a phase factor. This relation represents a reversal of the motion; for instance, if a denotes the momentum and the z component of the spin, the transformed state will have its momentum and spin-component reversed.

The phase factor η_a cannot arbitrarily be set equal to 1. If ϕ_a is a spinless system in an eigenstate of energy and orbital angular momentum, its spatial representation is

$$\phi_{klm}(\mathbf{r}) = N i^l \frac{F_l(kr)}{kr} Y_{lm}(\theta, \phi) \tag{6.30}$$

(Compare Eq. (5.6).) In this case $K = K_0$ and the spatial representation of Eq. (6.29) is

$$\bar{\phi}_{klm}(\mathbf{r}) = \phi_{klm}^*(\mathbf{r}) = N(-i)^l \frac{F_l(kr)}{kr} (-1)^m Y_{l,-m}(\theta, \phi)$$

$$= (-1)^{l+m} \phi_{kl,-m}(\mathbf{r}). \tag{6.31}$$

Then, as expected, the transformed eigenvalues are $\bar{a} = (kl, -m)$. The phase factor introduced by the transformation depends on m, so that it cannot be removed by multiplying K by a fixed phase factor. On the other hand, the transformation of an eigenstate of linear momentum does not introduce any phase factor. In fact, the original and transformed wave functions in this case are

$$\phi_\mathbf{k}(\mathbf{r}) = e^{i\mathbf{k}\cdot\mathbf{r}} \quad \text{and} \quad \bar{\phi}_\mathbf{k}(\mathbf{r}) = e^{-i\mathbf{k}\cdot\mathbf{r}} = \phi_{-\mathbf{k}}(\mathbf{r}). \tag{6.32}$$

If ϕ_a describes a single spin-$\tfrac{1}{2}$ particle, $K = i\sigma_y K_0$. Considering only the spin variable, so that ϕ_a is one of the spinors $\phi_+ = \binom{1}{0}$ or $\phi_- = \binom{0}{1}$, we have

$$\bar{\phi}_+ = i\sigma_y \phi_+ = -\phi_-, \tag{6.33}$$

while

$$\bar{\phi}_- = i\sigma_y \phi_- = \phi_+.$$

Thus we find different phase factors for $m_z = \pm\tfrac{1}{2}$, and these cannot be removed simply by multiplying K by a single compensating factor. This is in fact required in order that the expectation value of σ_x and σ_y also change sign under time reversal.

One can show in general that η_a depends upon the magnetic quantum number in just the way illustrated by these examples. Suppose we have an eigenstate of J^2 and J_z, $\phi_{J,M,\alpha}$, where α denotes all other variables of the system, assumed to be unaffected by time reversal. Then Eq. (6.29) is

$$\bar{\phi}_{J,M,\alpha} = K\phi_{J,M,\alpha} = \eta_{J,M,\alpha}\phi_{J,-M,\alpha}. \tag{6.34}$$

Equation (6.12) implies that

$$KJ_- K^{-1} = -J_+. \tag{6.35}$$

If this is used together with Eq. (4.6) and the corresponding relation for J_-, we find

$$\eta_{J,M,\alpha} = -\eta_{J,M-1,\alpha}, \tag{6.36}$$

so that the phase can be expressed in general as

$$\eta_{J,M,\alpha} = (-1)^M \eta_{J,\alpha}. \tag{6.37}$$

Thus the dependence of the phase factor upon M is fixed; the remaining dependence on J and α is still arbitrary. It is convenient to let this be $\eta_{J,\alpha} = (-1)^J$ so that

$$\eta_{J,M,\alpha} = (-1)^{J+M}. \tag{6.38}$$

(We saw in the above examples that only the angular momentum variables contribute to these phase factors.) This choice makes $\eta_{J,M,\alpha}$ real for both integral and half-integral values of J and corresponds to the use of $U = i\sigma_y$ for a spin-$\tfrac{1}{2}$ particle.

CONSEQUENCES OF TIME-REVERSAL INVARIANCE

Knowing the effect of K on operators and state vectors, we can now determine the effects of time-reversal invariance on the elements of the S matrix. To do this we need only use the transformation property of the wave operator. If time-reversal invariance holds:

$$K\Omega^{(+)}K^{-1} = K\left(1 + \frac{1}{E - H + i\varepsilon}V\right)K^{-1}$$

$$= 1 + \frac{1}{E - H - i\varepsilon}V = \Omega^{(-)}. \tag{6.39}$$

Thus, as we might expect, outgoing scattered waves are transformed into ingoing waves by the operation of time reversal. Using this and Eq. (6.29) we have

$$\bar{\psi}_a^{(+)} = K\psi_a^{(+)} = \eta_a \psi_{\bar{a}}^{(-)}. \tag{6.40}$$

This relation, together with the antiunitarity of K, informs us that the S

6. Time-Reversal Invariance

matrix satisfies

$$S_{fi} = \langle \psi_f^{(-)} | \psi_i^{(+)} \rangle = \langle K\psi_i^{(+)} | K\psi_f^{(-)} \rangle$$
$$= \eta_i^* \eta_f \langle \psi_i^{(-)} | \psi_f^{(+)} \rangle = \eta_i^* \eta_f S_{\bar{i}\bar{f}}. \quad (6.41)$$

This implies that $|S_{fi}|^2 = |S_{\bar{i}\bar{f}}|^2$, which gives us a direct relation between the cross section for one process and that for its transformed inverse.

This identity between the forward and backward transition probabilities is the analog of the classical statement that the system will retrace its original path under time reversal. If the initial and final states each contain two particles, Eq. (6.41) can be written in the center-of-mass system as

$$\langle \mathbf{k}'; s_1's_2'm_1'm_2'; \alpha_f | S | \mathbf{k}; s_1s_2m_1m_2; \alpha_i \rangle$$
$$= (-1)^{s_1+m_1+s_2+m_2+s_1'+m_1'+s_2'+m_2'}$$
$$\times \langle -\mathbf{k}; s_1s_2 -m_1 -m_2; \alpha_i | S | -\mathbf{k}'; s_1's_2' -m_1' -m_2'; \alpha_f \rangle. \quad (6.42)$$

Here α_i and α_f denote all other variables for each state, such as type of particle, excitation energy, etc. An identical relation holds for the T and K matrices. Two processes which thus have identical transition probabilities are illustrated schematically in Figs. 10.3(a) and 10.3(b).

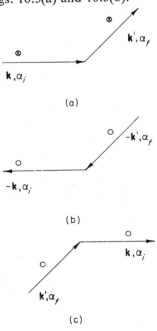

FIG. 10.3

A further relation can be obtained by performing a space reflection, which sends **k** into $-\mathbf{k}$ but leaves the spins unchanged. Then we find

$$\langle \mathbf{k}'; s_1's_2'm_1'm_2'; \alpha_f \,|\, S \,|\, \mathbf{k}; s_1 s_2 m_1 m_2; \alpha_i \rangle$$
$$= (-1)^{s_1 + m_1 + s_2 + m_2 + s_1' + m_1' + s_2' + m_2'}$$
$$\times \Pi_f \Pi_i \langle \mathbf{k}; s_1 s_2 - m_1 - m_2; \alpha_i \,|\, S \,|\, \mathbf{k}'; s_1's_2' - m_1' - m_2'; \alpha_f \rangle. \quad (6.43)$$

This third situation is illustrated in Fig. 10.3(c).

These relations imply special identities between collision cross sections in which all particle spins are measured. We can also use them to obtain a useful relation between the cross sections for these various processes when the spins are not measured. We first note that the T matrix satisfies an equation identical to Eq. (6.43). If we sum over the spins of the final state and average over those of the initial states, we obtain the differential cross section for process (a):

$$\frac{d\sigma(a)}{d\Omega} = \frac{2\pi}{\hbar v_i} \frac{1}{D_i} \sum |T_{fi}|^2 \rho_f(E), \quad (6.44)$$

where D_i is the spin-degeneracy of the initial state: $D_i = (2s_1 + 1)(2s_2 + 1)$ (see Section 4). Since we are summing over the spin-components, we can invert their signs in T_{fi}. Using Eq. (6.43), the net result is that $|T_{fi}|$ can be replaced by $|T_{if}|$. We then obtain the "detailed balance theorem":

$$\frac{d\sigma(a)}{d\Omega} = \frac{v_f}{v_i} \frac{D_f}{D_i} \frac{\rho_f(E)}{\rho_i(E)} \frac{d\sigma(c)}{d\Omega} \quad (6.45)$$

or

$$v_i D_i \rho_i(E) \frac{d\sigma(a)}{d\Omega} = v_f D_f \rho_f(E) \frac{d\sigma(c)}{d\Omega}. \quad (6.46)$$

This gives a direct relation between the cross sections for processes (a) and (c) in Fig. 10.3. It can be used to predict one from the other, to determine spins from the measured cross sections, and so on.

Equation (6.42) shows that time-reversal invariance also implies a relation between the phases of the T-matrix elements. These can have experimental consequences if interferences, such as between several partial waves, can be measured. These phase relations take their simplest form in the angular-momentum representation.

In that representation, if we use the results of Section 4, we have

$$\langle J_f, M_f, \alpha_f \,|\, S \,|\, J_i, M_i, \alpha_i \rangle = \langle \alpha_f \,|\, S_{J_i} \,|\, \alpha_i \rangle \delta_{J_i, J_f} \delta_{M_i, M_f}. \quad (6.47)$$

6. Time-Reversal Invariance

Equation (6.41) implies the relation

$$\langle J_f, M_f, \alpha_i | S | J_i, M_i, \alpha_i \rangle = (-1)^{J_i+M_i+J_f+M_f}$$
$$\times \langle J_i, -M_i, \alpha_i | S | J_f, -M_f, \alpha_f \rangle \quad (6.48)$$

or, with Eq. (6.47),

$$\langle \alpha_f | S_{J_i} | \alpha_i \rangle = \langle \alpha_i | S_{J_i} | \alpha_f \rangle. \quad (6.49)$$

Hence the S matrix is symmetric in the angular momentum representation.

This result can be used to obtain explicit parameterizations of the S matrix and to find relations between the phases of different S-matrix elements. For instance, a common case is that in which two states α_1 and α_2 are coupled, e.g., (pion + nucleon) to (photon + nucleon), (electron + positron) to two photons, etc. Then, in the angular-momentum representation the S matrix is a sequence of 2×2 matrices, one matrix for each value of J. The symmetry condition, Eq. (6.49), and the unitarity condition together imply that these 2×2 matrices can be expressed in terms of three parameters (rather than the four parameters required for an arbitrary unitary matrix). Thus, for example, each may be written in terms of the real numbers θ_J, $\delta_J^{(1)}$, and $\delta_J^{(2)}$:

$$\| S_J \| = \begin{pmatrix} \cos\theta_J \exp(2i\delta_J^{(1)}) & i\sin\theta_J \exp[i(\delta_J^{(1)} + \delta_J^{(2)})] \\ i\sin\theta_J \exp[i(\delta_J^{(1)} + \delta_J^{(2)})] & \cos\theta_J \exp(2i\delta_J^{(2)}) \end{pmatrix}. \quad (6.50)$$

The parameter θ_J measures the amplitude of transitions between the two states; $\delta^{(1)}$ and $\delta^{(2)}$ are the scattering phase shifts in the two states.

Chapter

11

Spin and Angular Momentum

When particles having intrinsic spin collide, the resulting angular distributions are much more complicated than for the case of the scattering of spinless particles, which we have considered until now. To see this clearly, consider Figs. 11.1 and 11.2. In Fig. 11.1(a) we illustrate schematically the

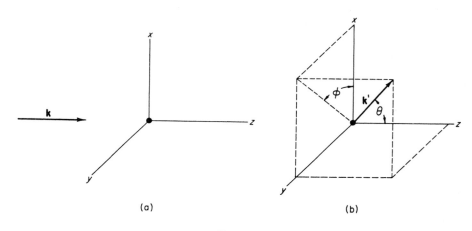

Fig. 11.1

incident state when both the projectile and the target are spinless and only the incident momentum is needed to identify the state. Since this state has cylindrical symmetry about the direction of propagation of the incident wave (the z axis), the final state will also have this symmetry. The scattering (Fig. 11.1(b)) must then be independent of the azimuthal angle ϕ, and we may write the cross section as

$$\sigma(\theta, \phi) = \sigma(\theta) = |f(\theta)|^2.$$

Now suppose a particle having intrinsic spin, with its spin vector initially perpendicular to the direction of propagation, is scattered by a spinless target. Then the initial state may be represented as in Fig. 11.2. In this case the

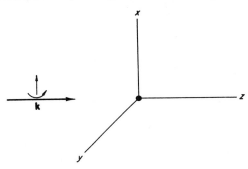

Fig. 11.2

initial state is not symmetric about the direction of propagation, and the differential cross section need not be independent of ϕ. One symmetry property which does remain can be inferred from Fig. 11.2; the initial state is unchanged if a reflection is performed in the y-z plane so that, if parity is conserved, the final state, and hence the cross section, must exhibit an up-down symmetry.

In addition to the greater complexity of the angular distribution, the direction of the spin in the final state need not be the same as in the initial state. There will be a distribution of spins as a function of the scattering angle, and the direction of the spin must be specified in describing the final state.

In this chapter we shall discuss techniques for describing the scattering of particles with spin. We shall develop the general angular-momentum expansions for the cross section and consider some methods for treating spin-dependent scattering. First, however, in order clearly to see the essential ingredients without undue algebraic complexity, we shall treat the case of a spin-$\frac{1}{2}$ particle scattering from a spin-zero target.

1. System of Spin $\frac{1}{2}$

For a system having intrinsic spin, the incident wave can no longer be represented simply as a plane wave but, in addition, must include a spin wave function χ. It is convenient to choose χ to be an eigenfunction of the square of the spin operator $\mathbf{S}$ and of one component of $\mathbf{S}$, as discussed in standard texts on quantum mechanics. For spin $\frac{1}{2}$ the usual representation of the spin matrices is $\mathbf{S} = (\hbar/2)\boldsymbol{\sigma}$, with the Pauli matrices

$$\sigma_x = \begin{pmatrix} 0 & 1 \\ 1 & 0 \end{pmatrix}, \quad \sigma_y = \begin{pmatrix} 0 & -i \\ i & 0 \end{pmatrix}, \quad \sigma_z = \begin{pmatrix} 1 & 0 \\ 0 & -1 \end{pmatrix}, \tag{1.1}$$

and the eigenfunctions of S^2 and S_z are

$$\chi_{\frac{1}{2}\frac{1}{2}} = \begin{pmatrix} 1 \\ 0 \end{pmatrix} \quad \text{and} \quad \chi_{\frac{1}{2}-\frac{1}{2}} = \begin{pmatrix} 0 \\ 1 \end{pmatrix}. \tag{1.2}$$

Thus if the incident beam is a plane wave of spin-$\frac{1}{2}$ particles moving along the z axis with spins polarized along the positive z axis, the incident wave is

$$\phi_{\frac{1}{2}}(\mathbf{r}) = e^{ikz} \chi_{\frac{1}{2}\frac{1}{2}}. \tag{1.3}$$

The polarization of a beam of spin-$\frac{1}{2}$ particles is defined as the average value of the spin vector $\boldsymbol{\sigma}$, so that, for the eigenfunctions of Eq. (1.2), the polarization is

$$\mathbf{P} = \langle \chi_{\frac{1}{2}\pm\frac{1}{2}} | \boldsymbol{\sigma} | \chi_{\frac{1}{2}\pm\frac{1}{2}} \rangle = \pm \hat{\mathbf{z}}. \tag{1.4}$$

That is to say, for $m_s = \pm \frac{1}{2}$ the beam is completely polarized along the z axis in either the positive or negative sense. For an arbitrary spin function that is a linear combination of $\chi_{\frac{1}{2}\frac{1}{2}}$ and $\chi_{\frac{1}{2}-\frac{1}{2}}$, i.e.,

$$\chi = \alpha \chi_{\frac{1}{2}\frac{1}{2}} + \beta \chi_{\frac{1}{2}-\frac{1}{2}} \tag{1.5}$$

with

$$|\alpha|^2 + |\beta|^2 = 1, \tag{1.6}$$

the polarization is

$$\mathbf{P} = \hat{\mathbf{x}}(\alpha^*\beta + \beta^*\alpha) - i\hat{\mathbf{y}}(\alpha^*\beta - \beta^*\alpha) + \hat{\mathbf{z}}(|\alpha|^2 - |\beta|^2). \tag{1.7}$$

in the general case of Eq. (1.7) also, the beam is completely polarized, that is, $P^2 = 1$, but the spin is oriented along the direction given by Eq. (1.7).

It might appear from this that any beam of spin-$\frac{1}{2}$ particles must be completely polarized in some direction. A spin-$\frac{1}{2}$ particle in a specific state such as Eq. (1.5) has its spin aligned along some one direction. However, a beam consists of many particles, and these may be in many different states. An unpolarized beam, for instance, has a vanishing polarization vector, when averaged over all the particles in the beam. This is the case when the spins of

1. System of Spin ½

the particles are randomly oriented, for example. Later we shall see how to describe a beam that has an arbitrary polarization. For the present, it suffices to remark that knowledge of the scattering of the two pure spin states $\chi_{\frac{1}{2}\frac{1}{2}}$ and $\chi_{\frac{1}{2}-\frac{1}{2}}$ will enable us to describe any physical situation.

Partial-Wave Expansion

Let us determine the scattered wave produced by such incident waves. If the interaction Hamiltonian is spin independent, the solution of the Schrödinger equation is the product

$$\psi^{(+)}_{\pm\frac{1}{2}}(\mathbf{r}) = \psi(\mathbf{r})\chi_{\frac{1}{2}\pm\frac{1}{2}}, \tag{1.8}$$

and the spin has no effect on the scattering.

Let us then consider a Hamiltonian that is spin dependent. Since higher powers of $\boldsymbol{\sigma}$ yield either 1 or $\boldsymbol{\sigma}$ itself, there are only two types of interactions, and we will restrict the Hamiltonian to be of the form

$$H = T + V_c(r) + V_s(r)(\mathbf{L}\cdot\mathbf{S}) = H_c + H_s, \tag{1.9}$$

where the subscripts c and s stand for "central," i.e., spin independent, and "spin dependent," respectively. Our problem is to solve the spin-dependent Schrödinger equation implied by Eq. (1.9), subject to the boundary condition

$$\psi^{(+)}_{\pm\frac{1}{2}}(\mathbf{r}) \xrightarrow[r\to\infty]{} \phi_{\pm\frac{1}{2}}(\mathbf{r}) + \psi_{sc}(\mathbf{r}), \tag{1.10}$$

where $\phi_{\pm\frac{1}{2}}(\mathbf{r})$ is the incident wave given by

$$\begin{aligned}\phi_{\pm\frac{1}{2}}(\mathbf{r}) &= e^{ikz}\chi_{\frac{1}{2}\pm\frac{1}{2}}\\ &= (kr)^{-1}\sum_l i^l[4\pi(2l+1)]^{1/2}F_l(kr)Y_{l,0}(\theta,\phi)\chi_{\frac{1}{2}\pm\frac{1}{2}}\end{aligned} \tag{1.11}$$

and $\psi_{sc}(\mathbf{r})$ is an outgoing spherical wave.

To accomplish this, we must effect a partial-wave expansion of the wave function $\psi^{(+)}_{\pm\frac{1}{2}}(\mathbf{r})$. If we were to adopt the approach of Chapter 3, we would attempt to expand the wave function as

$$\psi^{(+)}_{\pm\frac{1}{2}}(\mathbf{r}) = (kr)^{-1}\sum_l u_l(r) Y_{l,0}(\theta,\phi)\chi_{\frac{1}{2}\pm\frac{1}{2}}. \tag{1.12}$$

However, if we operate with $\mathbf{L}\cdot\mathbf{S}$ on the spin-angle function $Y_{l,0}(\theta,\phi)\chi_{\frac{1}{2}\pm\frac{1}{2}}$, we find

$$\begin{aligned}\mathbf{L}\cdot\mathbf{S}\, Y_{l,0}(\theta,\phi)\chi_{\frac{1}{2}\pm\frac{1}{2}} &= (\tfrac{1}{2}L_+S_- + \tfrac{1}{2}L_-S_+ + L_zS_z)Y_{l,0}(\theta,\phi)\chi_{\frac{1}{2}\pm\frac{1}{2}}\\ &= \tfrac{1}{2}\hbar^2[l(l+1)]^{1/2}Y_{l,\pm 1}(\theta,\phi)\chi_{\frac{1}{2}\mp\frac{1}{2}},\end{aligned} \tag{1.13}$$

where the angular momentum raising and lowering operators $L_\pm$ and $S_\pm$ are those described in the previous chapter. Thus this function is not an eigenfunction of $\mathbf{L}\cdot\mathbf{S}$, and the expansion in Eq. (1.12) cannot yield an eigenfunction of H.

Equation (1.13) implies that the expansion for $\psi_{\frac{1}{2}}^{(+)}(\mathbf{r})$, for example, must have the form

$$\psi_{\frac{1}{2}}^{(+)}(\mathbf{r}) = (kr)^{-1} \sum_l \{u_l(r)\, Y_{l,0}(\theta,\phi)\, \chi_{\frac{1}{2}\frac{1}{2}} + w_l(r)\, Y_{l,1}(\theta,\phi)\, \chi_{\frac{1}{2}-\frac{1}{2}}\}. \tag{1.14}$$

If we operate with $\mathbf{L}\cdot\mathbf{S}$ on $Y_{l,1}(\theta,\phi)\,\chi_{\frac{1}{2}-\frac{1}{2}}$, we obtain

$$\begin{aligned}\mathbf{L}\cdot\mathbf{S}\, Y_{l,1}(\theta,\phi)\,\chi_{\frac{1}{2}-\frac{1}{2}} &= \tfrac{1}{2}\hbar^2 [l(l+1)]^{1/2}\, Y_{l,0}(\theta,\phi)\,\chi_{\frac{1}{2}\frac{1}{2}} \\ &\quad - \tfrac{1}{2}\hbar^2\, Y_{l,1}(\theta,\phi)\,\chi_{\frac{1}{2}-\frac{1}{2}},\end{aligned} \tag{1.15}$$

which indicates that an expansion of the form of Eq. (1.14) will in fact produce a solution. Using the orthogonality of the spin functions, the radial equations are then found to be

$$\left\{-\frac{\hbar^2}{2m}\frac{d^2}{dr^2} + \frac{l(l+1)\hbar^2}{2mr^2} + V_c(r)\right\} u_l(r)$$
$$+ \tfrac{1}{2}\hbar^2 [l(l+1)]^{1/2}\, V_s(r)\, w_l(r) = 0 \tag{1.16}$$

and

$$\left\{-\frac{\hbar^2}{2m}\frac{d^2}{dr^2} + \frac{l(l+1)\hbar^2}{2mr^2} + V_c(r) - \tfrac{1}{2}\hbar^2\, V_s(r)\right\} w_l(r)$$
$$+ \tfrac{1}{2}\hbar^2 [l(l+1)]^{1/2}\, V_s(r)\, u_l(r) = 0. \tag{1.17}$$

Equations (1.16) and (1.17) are coupled second-order differential equations for two separate wave functions, and the requirement of regularity at the origin is not sufficient to specify the solution. However, if the boundary condition of Eq. (1.10) is invoked, the desired solution can be determined.

DIAGONALIZATION

The procedure we have just outlined, leading to coupled differential equations, is cumbersome because we have chosen to work in an inconvenient representation of the angular momentum operators. In particular, the operator $\mathbf{L}\cdot\mathbf{S}$ is not diagonal in the orbital and spin angular momentum eigenfunctions. If we can find a representation in which the operator $\mathbf{L}\cdot\mathbf{S}$ is diagonal, the problem will be very much simplified. Thus we will now seek a

1. System of Spin ½

linear combination of $Y_{l,0}(\theta, \phi) \chi_{\frac{1}{2}\frac{1}{2}}$ and $Y_{l,1}(\theta, \phi) \chi_{\frac{1}{2}-\frac{1}{2}}$ which is an eigenfunction of $\mathbf{L} \cdot \mathbf{S}$.

We write this eigenfunction as

$$\mathscr{Y}_l = Y_{l,0}(\theta, \phi) \chi_{\frac{1}{2}\frac{1}{2}} + A\, Y_{l,1}(\theta, \phi) \chi_{\frac{1}{2}-\frac{1}{2}} \tag{1.18}$$

and operate on it with $\mathbf{L} \cdot \mathbf{S}$ to obtain

$$\begin{aligned}\mathbf{L} \cdot \mathbf{S}\, \mathscr{Y}_l &= \tfrac{1}{2}\hbar^2 \{ A[l(l+1)]^{1/2}\, Y_{l,0}(\theta, \phi) \chi_{\frac{1}{2}\frac{1}{2}} \\ &\quad + [(l(l+1))^{1/2} - A]\, Y_{l,1}(\theta, \phi) \chi_{\frac{1}{2}-\frac{1}{2}} \} \\ &= C\{ Y_{l,0}(\theta, \phi) \chi_{\frac{1}{2}\frac{1}{2}} + A\, Y_{l,1}(\theta, \phi) \chi_{\frac{1}{2}-\frac{1}{2}} \}. \end{aligned} \tag{1.19}$$

To satisfy this equation, A must satisfy the relation

$$A^2 [l(l+1)]^{1/2} + A - [l(l+1)]^{1/2} = 0, \tag{1.20}$$

which implies the solutions

$$A = \left(\frac{l}{l+1}\right)^{1/2} \quad \text{and} \quad -\left(\frac{l+1}{l}\right)^{1/2}. \tag{1.21}$$

Thus there are two normalized eigenfunctions of $\mathbf{L} \cdot \mathbf{S}$,

$$\mathscr{Y}_l^{(a)} = [(l+1)/(2l+1)]^{1/2}\, Y_{l,0}(\theta, \phi) \chi_{\frac{1}{2}\frac{1}{2}} + [l/(2l+1)]^{1/2}\, Y_{l,1}(\theta, \phi) \chi_{\frac{1}{2}-\frac{1}{2}}$$

and

$$\mathscr{Y}_l^{(b)} = [l/(2l+1)]^{1/2}\, Y_{l,0}(\theta, \phi) \chi_{\frac{1}{2}\frac{1}{2}} - [(l+1)/(2l+1)]^{1/2}\, Y_{l,1}(\theta, \phi) \chi_{\frac{1}{2}-\frac{1}{2}}. \tag{1.22}$$

In terms of these eigenfunctions, the original spin-angle function is

$$Y_{l,0}(\theta, \phi) \chi_{\frac{1}{2}\frac{1}{2}} = [(l+1)/(2l+1)]^{1/2}\, \mathscr{Y}_l^{(a)} + [l/(2l+1)]^{1/2}\, \mathscr{Y}_l^{(b)}, \tag{1.23}$$

and the incident state is

$$\begin{aligned}\phi_{\frac{1}{2}}(\mathbf{r}) &= e^{ikz} \chi_{\frac{1}{2}\frac{1}{2}} \\ &= (4\pi)^{1/2}(kr)^{-1} \sum_l i^l F_l(kr) \{(l+1)^{1/2} \mathscr{Y}_l^{(a)} + l^{1/2} \mathscr{Y}_l^{(b)}\}. \end{aligned} \tag{1.24}$$

We may expand $\psi_{\frac{1}{2}}^{(+)}$ in terms of these eigenfunctions:

$$\psi_{\frac{1}{2}}^{(+)} = (4\pi)^{1/2}(kr)^{-1} \sum_l i^l \{(l+1)^{1/2} u_l^{(a)}(r)\, \mathscr{Y}_l^{(a)} + l^{1/2} u_l^{(b)}(r)\, \mathscr{Y}_l^{(b)}\}. \tag{1.25}$$

From the eigenvalue equations

$$\mathbf{L} \cdot \mathbf{S}\, \mathscr{Y}_l^{(a)} = \tfrac{1}{2}\hbar^2 l\, \mathscr{Y}_l^{(a)} \tag{1.26}$$

and
$$\mathbf{L} \cdot \mathbf{S} \mathcal{Y}_l^{(b)} = -\tfrac{1}{2}\hbar^2(l+1)\mathcal{Y}_l^{(b)}, \tag{1.27}$$

we obtain the uncoupled radial equations

$$\left\{-\frac{\hbar^2}{2m}\frac{d^2}{dr^2} + \frac{l(l+1)\hbar^2}{2mr^2} + V_c(r) + \tfrac{1}{2}\hbar^2 l\, V_s(r)\right\} u_l^{(a)}(r) = 0 \tag{1.28}$$

and

$$\left\{-\frac{\hbar^2}{2m}\frac{d^2}{dr^2} + \frac{l(l+1)\hbar^2}{2mr^2} + V_c(r) - \tfrac{1}{2}\hbar^2(l+1)\, V_s(r)\right\} u_l^{(b)}(r) = 0. \tag{1.29}$$

Their regular solutions are unique to within a normalization constant, which is fixed by requiring that Eq. (1.10) be satisfied. The asymptotic wave functions with the resulting normalization are

$$u_l^{(a,b)}(r) \xrightarrow[r \to \infty]{} \sin(kr - \tfrac{1}{2}\pi l)$$
$$+ \exp(i\delta_l^{(a,b)}) \sin \delta_l^{(a,b)} \exp[i(kr - \tfrac{1}{2}\pi l)], \tag{1.30}$$

and we find for the full wave function

$$\psi_{\frac{1}{2}}^{(+)}(\mathbf{r}) \xrightarrow[r \to \infty]{} \phi_{\frac{1}{2}}(\mathbf{r}) + \frac{e^{ikr}}{r}\frac{(4\pi)^{1/2}}{k}\sum_l \{(l+1)^{1/2} \exp(i\delta_l^{(a)}) \sin \delta_l^{(a)} \mathcal{Y}_l^{(a)}$$
$$+ l^{1/2} \exp(i\delta_l^{(b)}) \sin \delta_l^{(b)} \mathcal{Y}_l^{(b)}\}$$
$$= \phi_{\frac{1}{2}}(\mathbf{r}) + \frac{e^{ikr}}{r}\frac{1}{k}\sum_l \left(\frac{4\pi}{2l+1}\right)^{1/2}$$
$$\times \{[(l+1) \exp(i\delta_l^{(a)}) \sin \delta_l^{(a)} + l \exp(i\delta_l^{(b)}) \sin \delta_l^{(b)}] Y_{l,0}(\theta, \phi) \chi_{\frac{1}{2}\frac{1}{2}}$$
$$+ [l(l+1)]^{1/2} [\exp(i\delta_l^{(a)}) \sin \delta_l^{(a)} - \exp(i\delta_l^{(b)}) \sin \delta_l^{(b)}] Y_{l,1}(\theta, \phi) \chi_{\frac{1}{2}-\frac{1}{2}}\}$$
$$\equiv \phi_{\frac{1}{2}}(\mathbf{r}) + \frac{e^{ikr}}{r}\{f_{++}(\theta, \phi) \chi_{\frac{1}{2}\frac{1}{2}} + f_{-+}(\theta, \phi) \chi_{\frac{1}{2}-\frac{1}{2}}\}. \tag{1.31}$$

Here the spin-dependent scattering amplitudes are

$$f_{++}(\theta, \phi) = k^{-1} (4\pi)^{1/2} \sum_l (2l+1)^{-1/2}\{(l+1) \exp(i\delta_l^{(a)}) \sin \delta_l^{(a)}$$
$$+ l \exp(i\delta_l^{(b)}) \sin \delta_l^{(b)}\} Y_{l,0}(\theta) \tag{1.32}$$

and

$$f_{-+}(\theta, \phi) = (2ik)^{-1} (4\pi)^{1/2} \sum_l [l(l+1)/(2l+1)]^{1/2}\{\exp(2i\delta_l^{(a)})$$
$$- \exp(2i\delta_l^{(b)})\} Y_{l,1}(\theta, \phi). \tag{1.33}$$

1. System of Spin $\frac{1}{2}$

An analogous calculation for the negative spin polarization yields

$$\psi^{(+)}_{-\frac{1}{2}}(\mathbf{r}) \xrightarrow[r \to \infty]{} \phi_{-\frac{1}{2}}(\mathbf{r}) + (e^{ikr}/r)\{f_{+-}(\theta, \phi)\chi_{\frac{1}{2}\frac{1}{2}}$$
$$+ f_{--}(\theta, \phi)\chi_{\frac{1}{2}-\frac{1}{2}}\}, \tag{1.34}$$

where the scattering amplitudes are

$$f_{+-}(\theta, \phi) = -(2ik)^{-1}(4\pi)^{1/2}\sum_l [l(l+1)/(2l+1)]^{1/2}\{\exp(2i\delta_l^{(a)})$$
$$- \exp(2i\delta_l^{(b)})\} Y_{l,-1}(\theta, \phi) \tag{1.35}$$

and

$$f_{--}(\theta, \phi) = f_{++}(\theta, \phi). \tag{1.36}$$

Thus, to obtain a phase shift expansion and radial equations that permit the phase shifts to be computed, it is desirable to introduce appropriate spin-angle eigenfunctions that diagonalize the interaction. Once this is done, results analogous to those obtained in the spinless case are found.

We have had to go to some length to arrive by direct calculation at a result that is very much easier to derive by somewhat more general methods. The total angular momentum operator is $\mathbf{J} = \mathbf{L} + \mathbf{S}$ and, on very general grounds, we know that $\mathbf{J}$ commutes with the Hamiltonian. Furthermore, the Hamiltonian of Eq. (1.9) commutes with L^2 and S^2. Thus we may construct spin-angle functions that are eigenfunctions of J^2, J_z, L^2, and S^2, and J, M, l, and s will be conserved. Because of the identity

$$\mathbf{L} \cdot \mathbf{S} = \tfrac{1}{2}(\mathbf{J}^2 - \mathbf{L}^2 - \mathbf{S}^2), \tag{1.37}$$

the eigenfunctions of these operators, which we denote by $\mathscr{Y}_{JM}^{ls}$, satisfy the eigenvalue equation

$$\mathbf{L} \cdot \mathbf{S}\,\mathscr{Y}_{JM}^{ls} = \tfrac{1}{2}\hbar^2\{J(J+1) - l(l+1) - \tfrac{3}{4}\}\mathscr{Y}_{JM}^{ls}. \tag{1.38}$$

Comparing this with Eqs. (1.26) and (1.27), we see that the spin-angle functions $\mathscr{Y}_l^{(a)}$ and $\mathscr{Y}_l^{(b)}$ are, in fact, eigenfunctions of J^2, J_z, L^2, and S^2 with $M = \tfrac{1}{2}$, and the (a) solution corresponds to $J = l + \tfrac{1}{2}$, while the (b) solution corresponds to $J = l - \tfrac{1}{2}$. The previous argument was longer than necessary because we chose to derive the spin-angle eigenfunctions directly by diagonalizing $\mathbf{L} \cdot \mathbf{S}$. In the next section we shall outline the general method for obtaining such spin-angle eigenfunctions.

2. Addition of Spin and Orbital Angular Momentum

In earlier chapters we used orbital angular momentum expansions of the wave functions and scattering amplitudes in order to obtain differential equations in a single variable and to describe low-energy scattering cross sections in terms of a limited number of energy-dependent quantities, the phase shifts. The physical fact which permitted this simplification was that orbital angular momentum was conserved in the scattering process.

When particles having intrinsic spin collide, it is no longer generally true that orbital angular momentum is conserved. Instead, it is the total angular momentum, the vector sum of the orbital angular momentum and spin angular momentum, which is conserved, and the orbital angular momentum can change during the collision. For instance, a spin-one particle can approach a spinless target in an S state ($l = 0$) and emerge in a D state ($l = 2$). (Conservation of parity does not permit a change of one unit in the orbital angular momentum.) In both cases the total angular momentum would be $J = 1$. Thus, as we saw in the previous section, we must form eigenstates of the total angular momentum and expand the wave function in terms of them, if the differential equations are to be simplified and the T matrix diagonalized to obtain the phase shifts.

A particle having an intrinsic spin s can be characterized by a fourth degree of freedom, which for convenience we denote by the variable σ. This variable takes on the values attainable by the magnetic quantum number, namely, $s, s-1, \ldots, -s$, and should not be confused with the Pauli spin matrices $\boldsymbol{\sigma}$. As we will see in a later section, each value of σ is associated with an axis in spin space or, equivalently, an entry in the column vector that represents the wave function in spin space. We can write the eigenfunctions of S^2 and S_z as

$$\chi_{sm_s}(\sigma) = \begin{cases} 1, & \sigma = m_s, \\ 0, & \sigma \neq m_s, \end{cases} \quad (2.1)$$

and these functions satisfy the usual orthonormality relation

$$\sum_{\sigma=-s}^{s} \chi_{sm_s}^\dagger(\sigma) \chi_{sm_{s'}}(\sigma) = \delta_{m_s m_{s'}}. \quad (2.2)$$

Their use will facilitate later manipulations and permit spatial and spin variables to be treated on an equivalent footing.

We can now form eigenfunctions of the square of the total angular momentum J^2 and its z component J_z. It is useful first to note several facts concerning these operators:

2. Addition of Spin and Orbital Angular Momentum

(1) The operators J_z and $J^2 = (\mathbf{L} + \mathbf{S})^2$ commute with both L^2 and S^2, so that the new eigenfunctions can be simultaneous eigenfunctions of these four operators. States having a given spin and orbital angular momentum can also have a well-defined total angular momentum.

(2) The operators L_z and S_z do *not* commute with J^2, so that a function cannot in general be a simultaneous eigenfunction of J^2 and the pair (L_z, S_z). Only in the special case when the eigenvalues J and M take on their maximum values ($J = l + s$, $|M| = |l + s|$) does such an eigenfunction exist.

As a result one can construct eigenfunctions of the set of operators (L^2, S^2, L_z, S_z) or of the set (L^2, S^2, J^2, J_z), but a function cannot, in general, be a simultaneous eigenfunction of both sets. The eigenfunctions of the first set are simply the products $Y_{lm_l}(\theta, \phi)\, \chi_{sm_s}(\sigma)$; those of the second set will be denoted by $\mathscr{Y}_{JM}^{ls}(\theta, \phi, \sigma)$. Since, for a given l and s, either of these functions yields a complete set, we can expand the eigenfunctions of the second set of functions in terms of the first set. We then have

$$\mathscr{Y}_{JM}^{ls}(\theta, \phi, \sigma) = \sum_{m_l=-l}^{l} \sum_{m_s=-s}^{s} C_{ls}(JM; m_l m_s)\, Y_{lm_l}(\theta, \phi)\, \chi_{sm_s}(\sigma)$$

$$\equiv \sum_{m_l=-l}^{l} \sum_{m_s=-s}^{s} \langle l, s; m_l m_s | l, s; JM \rangle\, Y_{lm_l}(\theta, \phi)\, \chi_{sm_s}(\sigma), \quad (2.3)$$

where the numerical coefficients $C_{ls}(JM; m_l m_s)$ are given by

$$C_{ls}(JM; m_l m_s) = \sum_{\sigma=-s}^{s} \int d\Omega\, Y_{lm_l}^*(\theta, \phi)\, \chi_{sm_s}^\dagger(\sigma)\, \mathscr{Y}_{JM}^{ls}(\theta, \phi, \sigma)$$

$$\equiv \langle l, s; m_l m_s | l, s; JM \rangle. \quad (2.4)$$

They are the scalar products of these two sets of spin-angle eigenfunctions. The Dirac notation for the scalar product, which is here used in Eqs. (2.3) and (2.4), was discussed in Chapter 6.

These formulas and those that will be developed below are general formulas for the addition of angular momentum. While we have focused our attention on the addition of spin and orbital angular momentum to obtain a resultant total angular momentum, the results apply to the addition of any two angular momenta. All we need to do to make these formulas completely general is to interpret the vectors **L** and **S** as any two angular momenta that are added vectorially to form a resultant angular momentum **J**.

The numerical coefficients in Eq. (2.4) are called vector addition coefficients or Clebsch–Gordan coefficients. Their values follow from the general properties of angular momentum operators. They may be chosen to be real. Allowed eigenvalues J lie in the range $|l - s| \leq J \leq l + s$, and the Clebsch–Gordan coefficients are nonzero in this range. Since $J_z = L_z + S_z$, the Clebsch–Gordan coefficient $\langle l, s; m_l m_s | l, s; JM \rangle$ is zero unless $m_l + m_s = M$; nevertheless, it is convenient formally to write the sum in Eq. (2.3) as a double sum. The Clebsch–Gordan coefficients are discussed fully in all standard books on angular momentum.

The orthonormality properties of the Clebsch–Gordan coefficients are of special interest. These follow from the orthonormality and completeness of the eigenfunctions $\mathcal{Y}^{ls}_{JM}(\theta, \phi, \sigma)$ and of the functions $Y_{l m_l}(\theta, \phi) \chi_{s m_s}(\sigma)$ and are most easily seen in the Dirac notation. We may write Eq. (2.3) as a ket equation,

$$|l, s; JM\rangle = \sum_{m_l m_s} |l, s; m_l m_s\rangle \langle l, s; m_l m_s | l, s; JM\rangle, \qquad (2.5)$$

or, equivalently, as a bra equation,

$$\langle l, s; JM| = \sum_{m_l m_s} \langle l, s; JM | l, s; m_l m_s\rangle \langle l, s; m_l m_s |$$
$$= \sum_{m_l m_s} \langle l, s; m_l m_s | l, s; JM\rangle \langle l, s; m_l m_s |, \qquad (2.6)$$

since the Clebsch–Gordan coefficients can be chosen to be real. From the orthonormality of the state vectors we find

$$\langle l, s; JM | l, s; J'M'\rangle = \sum_{m_l m_s} \langle l, s; m_l m_s | l, s; JM\rangle$$
$$\times \langle l, s; m_l m_s | l, s; J'M'\rangle$$
$$= \delta_{JJ'}\delta_{MM'}. \qquad (2.7)$$

The inverse equation to Eq. (2.3) is

$$Y_{l m_l}(\theta, \phi) \chi_{s m_s}(\sigma) = \sum_{J,M} \langle l, s; JM | l, s; m_l m_s\rangle \mathcal{Y}^{ls}_{JM}(\theta, \phi, \sigma), \qquad (2.8)$$

or, equivalently, as a Dirac ket equation,

$$|l, s; m_l m_s\rangle = \sum_{J,M} |l, s; JM\rangle \langle l, s; JM | l, s; m_l m_s\rangle. \qquad (2.9)$$

Using the orthonormality property, we obtain from this

$$\langle l, s; m_l m_s | l, s; m_l' m_s'\rangle = \sum_{JM} \langle l, s; m_l m_s | l, s; JM\rangle$$
$$\times \langle l, s; m_l' m_s' | l, s; JM\rangle$$
$$= \delta_{m_l m_l'}\delta_{m_s m_s'}. \qquad (2.10)$$

These coefficients also satisfy the following useful symmetry relations:

$$\langle l, s; m_l m_s | l, s; JM \rangle = (-1)^{l+s-J} \langle l, s; -m_l, -m_s | l, s; J, -M \rangle$$

$$= (-1)^{l-m_l} \left(\frac{2J+1}{2s+1} \right)^{1/2}$$

$$\times \langle l, J; m_l, -M | l, J; s, -m_s \rangle. \quad (2.11)$$

Another useful relation,

$$\sum_{M, m_s} \langle l, s; m_l m_s | l, s; JM \rangle \langle l'; s; m_l' m_s | l', s; JM \rangle$$

$$= \frac{2J+1}{2l+1} \delta_{ll'} \delta_{m_l m_l'}, \quad (2.12)$$

follows from Eq. (2.7) upon application of these symmetry relations.

3. Radial Integral Equation for Scattering by a Spin-Zero Target

The formulas of the preceding section may be applied in a straightforward manner to compute the scattering of a particle with spin from a spinless target. The initial state may be written

$$\phi_{m_s}(\mathbf{r}, \sigma) = e^{ikz} \chi_{sm_s}(\sigma), \quad (3.1)$$

where we assume that the initial magnetic quantum number m_s is known. Use of the expansion of the plane wave and Eq. (2.8) leads to the expansion

$$\phi_{m_s}(\mathbf{r}, \sigma) = e^{ikz} \chi_{sm_s}(\sigma)$$

$$= \frac{1}{kr} \sum_{lJ} i^l [4\pi(2l+1)]^{1/2} F_l(kr) \langle l, s; 0 m_s | l, s; J m_s \rangle \mathcal{Y}_{J m_s}^{ls}(\theta, \phi, \sigma). \quad (3.2)$$

The scattering of this beam depends, of course, on the interaction with the target, which is a function of the position and spin variables of the projectile. Assuming this is a local potential denoted by $V(\mathbf{r}, \sigma', \sigma'')$, the outgoing-wave solution of the Schrödinger equation satisfies the integral equation

$$\psi_{m_s}^{(+)}(\mathbf{r}, \sigma) = \phi_{m_s}(\mathbf{r}, \sigma) + \sum_{\sigma', \sigma''=-s}^{s} \int d^3 r' \, G(\mathbf{r}, \sigma; \mathbf{r}', \sigma')$$

$$\times V(\mathbf{r}', \sigma', \sigma'') \psi_{m_s}(\mathbf{r}', \sigma''). \quad (3.3)$$

The Green's function for a free particle of spin s is

$$G(\mathbf{r}, \sigma; \mathbf{r}'; \sigma') = -\frac{2m}{4\pi \hbar^2} \frac{\exp(ik|\mathbf{r} - \mathbf{r}'|)}{|\mathbf{r} - \mathbf{r}'|} \sum_{m_s'} \chi_{s m_s'}(\sigma) \chi_{s m_s'}^\dagger(\sigma'), \quad (3.4)$$

where the spin sum may be recognized as simply $\delta_{\sigma\sigma'}$, using Eq. (2.1); it is thus just the unit operator in spin space. The relation

$$\frac{\exp(ik|\mathbf{r}-\mathbf{r}'|)}{|\mathbf{r}-\mathbf{r}'|} = \frac{1}{krr'} \sum_l (2l+1)$$

$$\times P_l(\cos)\theta_{\mathbf{rr}'} \begin{cases} F_l(kr) H_l^{(+)}(kr'), & r < r', \\ H_l^{(+)}(kr) F_l(kr'), & r > r', \end{cases} \quad (3.5)$$

which we developed in Chapter 4, may be used, together with the addition theorem for Legendre polynomials, to rewrite the Green's function in terms of the eigenfunctions of the total angular momentum:

$$G(\mathbf{r},\sigma;\mathbf{r}',\sigma') = \frac{-2m}{\hbar^2 krr'} \sum_{lJM} \mathscr{Y}_{JM}^{ls}(\theta,\phi,\sigma) \mathscr{Y}_{JM}^{ls*}(\theta',\phi',\sigma')$$

$$\times F_l(kr_<) H_l^{(+)}(kr_>). \quad (3.6)$$

Similarly, the wave function may be expanded in terms of these eigenfunctions:

$$\psi_{m_s}^{(+)}(\mathbf{r},\sigma) = \frac{1}{kr} \sum_{lJM} u_{lMm_s}^{Js}(r) \mathscr{Y}_{JM}^{ls}(\theta,\phi,\sigma). \quad (3.7)$$

This is the natural generalization of Eq. (1.10) of Chapter 3. Although $M = m_s$ because of the conservation of J_z, we keep these two quantum numbers formally independent of each other in order to simplify the manipulations with Clebsch–Gordan coefficients.

These expansions may be introduced into Eq. (3.3) to give the radial equation

$$u_{lMm_s}^{Js}(r) = i^l[4\pi(2l+1)]^{1/2} \langle l,s;0m_s | l,s;JM \rangle F_l(kr)$$

$$-\frac{2m}{\hbar^2 k} \sum_{l'} \int_0^\infty dr' \, F_l(kr_<) H_l^{(+)}(kr_>) V_{ll'}^{Js}(r') u_{lMm_s}^{Js}(r'). \quad (3.8)$$

The interaction $V_{ll'}^{Js}(r)$ is defined by the integral

$$\sum_{\sigma',\sigma''=-s}^{s} \int d\Omega' \, \mathscr{Y}_{JM}^{ls*}(\theta',\phi',\sigma') V(\mathbf{r}',\sigma',\sigma'') \mathscr{Y}_{J'M'}^{l's}(\theta',\phi',\sigma'')$$

$$\equiv \langle l,s;JM | V(\mathbf{r}',\mathbf{S}) | l',s;J'M' \rangle = V_{ll'}^{Js}(r') \delta_{JJ'} \delta_{MM'}. \quad (3.9)$$

The fact that $\langle l,s;JM | V(\mathbf{r}',\mathbf{S}) | l',s;J'M' \rangle$ is diagonal in J and M follows from the assumption that the interaction is a scalar and hence conserves the total angular momentum. We have also recognized in Eq. (3.9) that the matrix element is independent of M. (See Eqs. (4.6)–(4.8) of Chapter 9.) If

3. Radial Integral Equation

the interaction is reflection-invariant, parity is conserved and $V_{l'l}^{Js}(r) = 0$ if $(l - l')$ is odd for elastic scattering. (In general, as shown in Chapter 9, $(l - l')$ is even or odd, depending on whether or not the final state has the same intrinsic parity as the initial state.)

In general, dynamical equations such as Eq. (3.8) do not depend in any intrinsic way on the magnetic quantum number, provided the interaction is a scalar. Any apparent dependence on M involves only geometric effects and can always be eliminated. In this case the dependence on M can be eliminated by multiplying Eq. (3.8) by $\langle l', s; 0m_s | l', s; JM \rangle$ and summing over m_s and M. A new radial wave function, defined by the transformation

$$i^l [4\pi(2l+1)]^{1/2} u_{l'l}^{Js}(r) \equiv \frac{2l+1}{2J+1} \sum_{m_s, M} \langle l, s; 0m_s | l, s; JM \rangle u_{lMm_s}^{Js}(r), \quad (3.10)$$

is then found to satisfy the radial integral equation

$$u_{l'l}^{Js}(r) = F_l(kr) \delta_{l'l} - \frac{2m}{\hbar^2 k} \sum_{l''} \int_0^\infty dr' \, F_{l'}(kr_<) H_{l'}^{(+)}(kr_>) V_{l'l''}^{Js}(r') u_{l''l}^{Js}(r'). \quad (3.11)$$

(Note that, to simplify the notation later, we have exchanged l and l' in passing from Eq. (3.8) to (3.11).) There is now no remaining dependence on the magnetic quantum number, and this equation can be used as the basis for computing the scattering cross section.

The wave function defined by this integral equation has a simple interpretation: $u_{l'l}^{Js}(r)$ is the wave function for a state in which a projectile of total angular momentum J is incident on the target with orbital angular momentum l and emerges with orbital angular momentum l'. In lowest-order Born approximation it is

$$u_{l'l}^{Js}(r) \approx F_l(kr) \delta_{ll'} - \frac{2m}{\hbar^2 k} \int_0^\infty dr' \, F_{l'}(kr_<) H_{l'}^{(+)}(kr_>) V_{l'l}^{Js}(r') F_l(kr'). \quad (3.12)$$

The second term involves a transition from l to l'. This behavior persists to all orders in V, so that $u_{l'l}^{Js}(r)$ always involves a transition to the final orbital angular momentum l'. The total angular momentum is, of course, always conserved, so that l' lies in the range $|J - s| \leq l' \leq J + s$. Also conservation of parity requires l and l' to differ by an even integer for elastic scattering.

In the case studied in Section 1 the spin was $\frac{1}{2}$, so that l' might take on the values $l' = l, l \pm 1$. Invoking parity conservation, there can then be no change in the orbital angular momentum, and $u_{l'l}^{Js}(r)$ defined in Eq. (3.10) must be of the form

$$u_{l'l}^{J\frac{1}{2}}(r) = u_l^{J\frac{1}{2}}(r)\,\delta_{ll'}. \tag{3.13}$$

For $V(\mathbf{r}, \mathbf{S}) = V_c(r) + V_s(r)\,\mathbf{L}\cdot\mathbf{S}$, Eq. (1.38) implies that the potential matrix is

$$\langle l, s; JM \mid V(\mathbf{r}, \mathbf{S}) \mid l', s; J'M'\rangle = \{V_c(r) + \tfrac{1}{2}\hbar^2\,V_s(r)\,[J(J+1) \\ - l(l+1) - \tfrac{3}{4}]\}\,\delta_{ll'}\,\delta_{JJ'}\delta_{MM'}. \tag{3.14}$$

Using this and Eq. (3.13), the radial integral equation becomes

$$u_l^{J\frac{1}{2}}(r) = F_l(kr) + (2m/\hbar^2 k)\int_0^\infty dr'\,F_l(kr_<)\,H_l^{(+)}(kr_>)\,\{V_c(r') \\ + \tfrac{1}{2}\hbar^2\,V_s(r')\,[J(J+1) - l(l+1) - \tfrac{3}{4}]\}\,u_l^{J\frac{1}{2}}(r'). \tag{3.15}$$

This is just the integral equation corresponding to the earlier differential equations, Eqs. (1.28) and (1.29).

The number of functions $u_{l'l}^{Js}(r)$ for fixed J and l is identical with the number of functions $u_{lMm_s}^{Js}(r)$. The latter function is zero unless $M = m_s$, so that there are either $(2J+1)$ or $(2s+1)$ such functions, depending upon whether J is less than or greater than s. Likewise, since l' must lie in the range $|J - s| \le l' \le J + s$, there are either $(2J+1)$ or $(2s+1)$ functions of the first kind. The advantage of Eq. (3.11) over Eq. (3.8) is not that it involves fewer functions, but that the functions defined therein have a physical significance which depends in a simple way upon the properties of the Hamiltonian and, in particular, upon the ability of the interaction to induce transitions between different orbital angular momentum states. The effects of such features as spin dependence and tensor character also become far clearer in this form.

If these new functions are used, the expansion of $\psi_{m_s}^{(+)}(\mathbf{r}, \sigma)$, Eq. (3.7), takes the form

$$\psi_{m_s}^{(+)}(\mathbf{r}, \sigma) = \frac{1}{kr} \sum_{l,l'JM} i^l[4\pi(2l+1)]^{1/2}\,\langle l, s; 0m_s \mid l, s; JM\rangle \\ \times u_{l'l}^{Js}(r)\,\mathcal{Y}_{JM}^{l's}(\theta, \phi, \sigma), \tag{3.16}$$

which we will use in succeeding sections.

4. Scattering Amplitudes and T Matrices

Let us now consider the asymptotic form of $u_{l'l}^{Js}(r)$, assuming that the interaction with the target has a finite range. Applying to Eq. (3.11) the known asymptotic properties of the spherical Bessel functions, the asymptotic radial wave function is

$$u^{Js}_{l'l}(r) \xrightarrow[r\to\infty]{} F_l(kr)\,\delta_{ll'} + T^{Js}_{l'l}\exp[i(kr - \tfrac{1}{2}\pi l')], \qquad (4.1)$$

where the T-matrix element for scattering from a state $|l, s; JM\rangle$ to a state $|l', s; JM\rangle$ is

$$T^{Js}_{l'l} = \frac{-2m}{\hbar^2 k} \sum_{l''} \int_0^\infty dr\, F_{l'}(kr)\, V^{Js}_{l'l''}(r)\, u^{Js}_{l''l}(r). \qquad (4.2)$$

If we substitute the asymptotic form for $u^{Js}_{l'l}(r)$ into the wave function $\psi^{(+)}_{m_s}(\mathbf{r},\sigma)$ given in Eq. (3.16), we obtain for the asymptotic wave function

$$\begin{aligned}
\psi^{(+)}_{m_s}(\mathbf{r},\sigma) \xrightarrow[r\to\infty]{} \; & e^{ikz}\chi_{sm_s}(\sigma) \\
& + \frac{e^{ikr}}{r}\frac{1}{k}\sum_{l,l'JM}[4\pi(2l+1)]^{1/2}\langle l,s;0m_s\,|\,l,s;JM\rangle \\
& \times i^{(l-l')} T^{Js}_{l'l}\, \mathcal{Y}^{l's}_{JM}(\theta,\phi,\sigma) \\
= \; & \frac{1}{2ikr}\sum_{l,l'JM} i^{(l-l')}[4\pi(2l+1)]^{1/2} \\
& \times \langle l,s;0m_s\,|\,l,s;JM\rangle\{-(-1)^l \delta_{ll'}\, e^{-ikr} \\
& + (\delta_{ll'} + 2iT^{Js}_{l'l})\, e^{ikr}\}\, \mathcal{Y}^{l's}_{JM}(\theta,\phi,\sigma). \qquad (4.3)
\end{aligned}$$

From this asymptotic form the scattering amplitude, and hence the cross section, can be determined. First, however, we will examine some properties of the T matrices we have found.

Current Conservation and the Optical Theorem

Let us imagine an experiment in which the incident particle is prepared in an initial angular momentum eigenstate $|l, s; JM\rangle$, rather than in the usual plane wave state. From Eq. (4.3) we see that the asymptotic wave function for such a state would be proportional to

$$\begin{aligned}
\psi^{ls(+)}_{JM}(\mathbf{r},\sigma) \xrightarrow[r\to\infty]{} \; & \frac{1}{r}\sum_{l'} i^{-l'}\{-(-1)^l \delta_{l'l}\, e^{-ikr} \\
& + (\delta_{l'l} + 2iT^{Js}_{l'l})\, e^{ikr}\}\, \mathcal{Y}^{l's}_{JM}(\theta,\phi,\sigma). \qquad (4.4)
\end{aligned}$$

If we calculate the probability current for this wave function, we find the current through a spherical surface to be

$$\int d\mathbf{S}\cdot\mathbf{j} \xrightarrow[r\to\infty]{} \left(\frac{\hbar k}{m}\right)\sum_{l'}\{-\delta_{ll'} + |\delta_{l'l} + 2iT^{Js}_{l'l}|^2\}. \qquad (4.5)$$

The hermiticity of the Hamiltonian implies that no *net* current is carried by this wave function, and the current calculated in Eq. (4.5) must vanish. This implies that the outgoing and ingoing currents are equal,

$$\sum_{l'=|J-s|}^{J+s} |\delta_{l'l} + 2iT^{Js}_{l'l}|^2 = 1, \tag{4.6}$$

or, equivalently,

$$\operatorname{Im} T^{Js}_{ll} = \sum_{l'=|J-s|}^{J+s} |T^{Js}_{l'l}|^2. \tag{4.7}$$

This is the unitarity relation when one particle has spin s. It could also have been derived directly from the operator relations of Section 2, Chapter 8, by taking the matrix element of those relations between angular momentum eigenstates. Equation (4.7) can be generalized by using the relations developed in Section 2, Chapter 8, which imply the relation

$$T^{Js}_{l'l} - T^{Js*}_{ll''} = 2i \sum_{l'} T^{Js*}_{l'l''} T^{Js}_{l'l}. \tag{4.8}$$

Equation (4.7) is the diagonal element of this matrix relation.

Equation (4.7) implies that, for $s = \frac{1}{2}$, the T matrix is given by $T^{J\frac{1}{2}}_{l'l} = T^{J\frac{1}{2}}_l \delta_{ll'}$, with

$$T^{J=l\pm\frac{1}{2},\frac{1}{2}}_l \equiv T^{\pm}_l = \exp(i\delta_l^{\pm}) \sin \delta_l^{\pm}, \tag{4.9}$$

and $\delta_l^{\pm}$ are real phase shifts. Thus, for a fixed value of J, the T matrix is a 2×2 diagonal matrix,

$$\|T^{J\frac{1}{2}}\| = \begin{pmatrix} \exp(i\delta^+_{l=J-\frac{1}{2}}) \sin \delta^+_{l=J-\frac{1}{2}} & 0 \\ 0 & \exp(i\delta^-_{l=J+\frac{1}{2}}) \sin \delta^-_{l=J+\frac{1}{2}} \end{pmatrix}, \tag{4.10}$$

characterized by two real numbers $\delta^+_{l=J-\frac{1}{2}}$ and $\delta^-_{l=J+\frac{1}{2}}$. The S matrix is similarly diagonal and characterized by the same two phase shifts:

$$\|S^J\| = \begin{pmatrix} \exp(2i\delta^+_{l=J-\frac{1}{2}}) & 0 \\ 0 & \exp(2i\delta^-_{l=J+\frac{1}{2}}) \end{pmatrix}. \tag{4.11}$$

For higher values of the spin these matrices are more complicated. Let us briefly examine the spin-1 case. For a fixed value of J, l and l' may take on only the three values J, $J \pm 1$. For fixed J, $V^{J1}_{l'l}(r)$, $T^{J1}_{l'l}$, and $S^{J1}_{l'l}$ are then each 3×3 matrices. For $l = J$, we must have $l' = l$ since the values $l' = l \pm 1$,

4. Scattering Amplitudes and T Matrices

although allowed by conservation of angular momentum, are ruled out if parity is conserved. For $l = J$, the T matrix is then

$$T^{J1}_{l',l=J} = \exp(i\delta^{J1}_{l=J}) \sin \delta^{J1}_{l=J} \, \delta_{l'J}. \tag{4.12}$$

For $l = J \pm 1$, the appropriate matrix is a 2×2 matrix, since $l = J \pm 1$ can be coupled to $l' = J \mp 1$. Since the K matrix is a real symmetric matrix, only three real numbers are required to specify it completely, and similarly for the T and S matrices. We may think of these three numbers as two real phase shifts $\delta^{J1}_{l=J-1} \equiv \delta_-$ and $\delta^{J1}_{l=J+1} \equiv \delta_+$, plus a real mixing parameter ε. We might write a unitary S matrix so parameterized as

$$\| S^{J1}_{l \neq J} \| = U^+ \begin{pmatrix} e^{2i\delta_-} & 0 \\ 0 & e^{2i\delta_+} \end{pmatrix} U, \tag{4.13}$$

with

$$U = \begin{pmatrix} \cos \varepsilon & \sin \varepsilon \\ -\sin \varepsilon & \cos \varepsilon \end{pmatrix}, \tag{4.14}$$

to obtain the unitary S matrix

$$\| S^{J1}_{l \neq J} \| = \begin{pmatrix} \cos^2\varepsilon \, e^{2i\delta_-} + \sin^2\varepsilon \, e^{2i\delta_+} & \sin \varepsilon \cos \varepsilon \, (e^{2i\delta_-} - e^{2i\delta_+}) \\ \sin \varepsilon \cos \varepsilon \, (e^{2i\delta_-} - e^{2i\delta_+}) & \sin^2\varepsilon \, e^{2i\delta_-} + \cos^2\varepsilon \, e^{2i\delta_+} \end{pmatrix}. \tag{4.15}$$

In the limit $\varepsilon \to 0$, this has the standard uncoupled form.

Another possible parameterization is

$$\| S^{J1}_{l \neq J} \| = \begin{pmatrix} \cos \varepsilon' \exp(2i\delta_-') & i \sin \varepsilon' \exp[i(\delta_+' + \delta_-')] \\ i \sin \varepsilon' \exp[i(\delta_+' + \delta_-')] & \cos \varepsilon' \exp(2i\delta_+') \end{pmatrix}, \tag{4.16}$$

where $\delta_\pm'$ and ε' are real parameters. It is obvious that the parameterization in not unique. Howsoever we choose to express the three numbers needed to characterize the T matrix for $s = 1$ and $J \neq l$, the matrix elements $T^{J,s=1}_{l'l}$ are still, of course, given by Eq. (4.2).

The Spin-Dependent Scattering Amplitude

The asymptotic wave function, Eq. (4.3), is not in the most useful form for describing the usual scattering experiment. Spin projections are, in principle, directly measurable, whereas the total angular momentum cannot be directly measured by experimental apparatus (although it often can be inferred from the observed distributions). In an experiment we might have a plane wave in a state of complete polarization as our initial situation. Such an incident wave can be represented by the wave function $\phi_{m_s}(\mathbf{r}, \sigma)$ given in Eq. (3.2). We may observe the particles scattered outward along a ray (θ, ϕ) with a given spin polarization and may thus wish to know the component of the outgoing wave having a particular spin projection. To this end we rewrite the spin-angle eigenfunction in terms of the original eigenfunctions of m_l and m_s, according to the prescription of Eq. (2.3). We then find from Eq. (3.16)

$$\psi_{m_s}^{(+)}(\mathbf{r}, \sigma) = \frac{1}{kr} \sum_{l,l',J,m_l',m_s'} i^l [4\pi(2l+1)]^{1/2} \langle l, s; 0 m_s | l, s; J m_s \rangle$$

$$\times \langle l', s; m_l' m_s' | l', s; J m_s \rangle u_{l'l}^{Js}(r) Y_{l'm_l'}(\theta, \phi) \chi_{sm_s'}(\sigma)$$

$$\xrightarrow[r \to \infty]{} e^{ikz} \chi_{sm_s}(\sigma) + (e^{ikr}/r) k^{-1} \sum_{l,l',J,m_s'} i^{(l-l')} [4\pi(2l+1)]^{1/2}$$

$$\times \langle l, s; 0 m_s | l, s; J m_s \rangle \langle l', s; m_s - m_s', m_s' | l', s; J m_s \rangle$$

$$\times T_{l'l}^{Js} Y_{l', m_s - m_s'}(\theta, \phi) \chi_{sm_s'}(\sigma)$$

$$= e^{ikz} \chi_{sm_s}(\sigma) + (e^{ikr}/r) \sum_{m_s'} f_{m_s' m_s}(\theta, \phi) \chi_{sm_s'}(\sigma), \quad (4.17)$$

where the spin-dependent scattering amplitude is defined by

$$f_{m_s' m_s}(\theta, \phi) \equiv k^{-1} \sum_{l,l'J} i^{(l-l')} [4\pi(2l+1)]^{1/2} \langle l, s; 0 m_s | l, s; J m_s \rangle$$

$$\times \langle l', s; m_s - m_s', m_s' | l', s; J m_s \rangle T_{l'l}^{Js} Y_{l', m_s - m_s'}(\theta, \phi). \quad (4.18)$$

This scattering amplitude describes the transition from an incident plane wave moving along the z axis and having a spin projection m_s to an outgoing wave of spin projection m_s' with its momentum in the direction (θ, ϕ).

Introduction of the definition of $T_{l'l}^{Js}$, Eq. (4.2), into Eq. (4.18) leads, after some resummation of terms, to the expression

$$f_{m_s' m_s}(\theta, \phi) = \frac{-2m}{4\pi\hbar^2} \sum_{\sigma, \sigma' = -s}^{s} \int d^3r \, \chi_{sm_s'}^{\dagger}(\sigma) e^{-i\mathbf{k}' \cdot \mathbf{r}} V(\mathbf{r}, \sigma, \sigma') \psi_{m_s}^{(+)}(\mathbf{r}', \sigma')$$

(4.19)

for the scattering amplitude. This form is analogous to the results derived in Chapter 3 for the spinless case. The final state is $e^{i\mathbf{k}' \cdot \mathbf{r}} \chi_{sm_s'}(\sigma)$, and the exact wave function is $\psi_{m_s}^{(+)}(\mathbf{r}, \sigma)$, so that this has the familiar structure of a T matrix, $\langle \phi_f | V | \psi_i^{(+)} \rangle$.

TARGETS WITH NONZERO SPIN

When the target as well as the projectile have intrinsic spin, the previous formulas do not apply. However, results having very little additional complexity can be derived if the target spin and projectile spin are combined to form what is called "channel spin." For each initial channel spin s, one can define a wave function $\psi_{sm_s}^{(+)}(\mathbf{r}, \sigma_1, \sigma_2)$ which satisfies equations very similar to those developed in Section 3. The additional complexity is reflected in two properties of the channel spin description:

(1) Since the spins and magnetic quantum numbers of the individual particles are measurable before and after the experiment, while the channel spin is not, measurable quantities such as cross sections are expressed in terms of the individual spins. Appropriate sums over the allowed values of the channel spin must be performed in order to obtain quantities that are directly related to experimental observations.

(2) The channel spin in the outgoing state need not be the same as that in the incident channel. This "spin," as well as the orbital angular momentum, can change during the collision, although in some special cases, as in the scattering of identical spin-$\frac{1}{2}$ particles, selection rules may prevent a change in channel spin.

The initial plane-wave state may be represented as

$$\phi_{m_1 m_2}(\mathbf{r}, \sigma_1, \sigma_2) = e^{ikz} \chi_{s_1 m_1}(\sigma_1) \chi_{s_2 m_2}(\sigma_2). \tag{4.20}$$

The channel spin is introduced by coupling the two initial spins using an expression analogous to Eq. (2.8):

$$\chi_{s_1 m_1}(\sigma_1) \chi_{s_2 m_2}(\sigma_2) = \sum_{sm_s} \langle s_1, s_2; m_1 m_2 | s_1, s_2; sm_s \rangle \chi_{sm_s}(\sigma_1, \sigma_2). \tag{4.21}$$

This can be inverted to express the channel spin function as a sum over the individual spin functions:

$$\chi_{sm_s}(\sigma_1, \sigma_2) = \sum_{m_1 m_2} \langle s_1, s_2; m_1 m_2 | s_1, s_2; sm_s \rangle \chi_{s_1 m_1}(\sigma) \chi_{s_2 m_2}(\sigma_2). \tag{4.22}$$

A wave function for each channel spin, exactly analogous to that of Eq.

(3.3), may be defined by

$$\psi^{(+)}_{sm_s}(\mathbf{r}, \sigma_1, \sigma_2) = \phi_{sm_s}(\mathbf{r}, \sigma_1, \sigma_2) + \frac{2m}{4\pi\hbar^2} \sum_{\substack{s',m_s' \\ \sigma_1',\sigma_2' \\ \sigma_1'',\sigma_2''}} \chi_{s'm_s'}(\sigma_1, \sigma_2)$$

$$\times \chi^\dagger_{s'm_s'}(\sigma_1', \sigma_2') \int d^3r' \, \frac{\exp(ik|\mathbf{r}-\mathbf{r}'|)}{|\mathbf{r}-\mathbf{r}'|}$$

$$\times V(\mathbf{r}', \sigma_1', \sigma_2', \sigma_1'', \sigma_2'') \psi^{(+)}_{sm_s}(\mathbf{r}', \sigma_1'', \sigma_2''), \quad (4.23)$$

with the incident wave defined by

$$\phi_{sm_s}(\mathbf{r}, \sigma_1, \sigma_2) = e^{ikz} \chi_{sm_s}(\sigma_1, \sigma_2). \quad (4.24)$$

The asymptotic form of Eq. (4.23) is

$$\psi^{(+)}_{sm_s}(\mathbf{r}, \sigma_1, \sigma_2) \xrightarrow[r \to \infty]{} e^{ikz} \chi_{sm_s}(\sigma_1, \sigma_2)$$

$$+ \frac{e^{ikr}}{r} \sum_{s'm_s'} f_{s'm_s' sm_s}(\theta, \phi) \chi_{s'm_s'}(\sigma_1, \sigma_2) \quad (4.25)$$

with the scattering amplitude defined analogously to Eq. (4.19):

$$f_{s'm_s' sm_s}(\theta, \varphi) = -\frac{2m}{4\pi\hbar^2} \sum_{\sigma_1,\sigma_2,\sigma_1',\sigma_2'=-s}^{s} \int d^3r \chi^\dagger_{s'm_s'}(\sigma_1, \sigma_2)$$

$$\times e^{-i\mathbf{k}'\cdot\mathbf{r}} V(\mathbf{r}, \sigma_1, \sigma_2, \sigma_1', \sigma_2') \psi^{(+)}_{sm_s}(\mathbf{r}, \sigma_1', \sigma_2'). \quad (4.26)$$

The wave function may also be expanded in terms of angular momentum eigenfunctions, yielding an expansion of the form

$$\psi^{(+)}_{sm_s}(\mathbf{r}, \sigma_1, \sigma_2) = (kr)^{-1} \sum_{l,l's'JM} i^l [4\pi(2l+1)]^{1/2} \langle l, s; 0m_s | l, s; JM \rangle$$

$$\times u^{Js's}_{l'l}(r) \mathcal{Y}^{l's'}_{JM}(\theta, \varphi, \sigma_1, \sigma_2). \quad (4.27)$$

We could at this point derive the integral equation for the radial function $u^{Js's}_{l'l}(r)$ by the methods used in Section 3. By this time, however, the method should be clear, and we will not repeat this derivation. We merely note that the one extra feature is the possibility of different final channel spins, for a fixed incident channel spin. Radial equations such as Eq. (3.11) are then reproduced with the only change being an additional spin variable and a sum over intermediate values of the channel spin:

$$u^{Js's}_{l'l}(r) = F_l(kr) \delta_{l'l}\delta_{s's} - \frac{2m}{\hbar^2 k} \sum_{l''s''} \int_0^\infty dr' \, F_{l'}(kr_<) H^{(+)}_{l'}(kr_>)$$

$$\times V^{Js's''}_{l'l''}(r') u^{Js''s}_{l''l}(r'). \quad (4.28)$$

Since only the magnetic quantum numbers of the individual particles, and not the channel spins, are measurable, in practice scattering amplitudes describing scattering between eigenstates of these individual quantum numbers must be used. Using Eqs. (4.20) and (4.21), we see that the exact wave function for a "measurable" initial state is a superposition of channel functions:

$$\psi^{(+)}_{m_1 m_2}(\mathbf{r}, \sigma_1, \sigma_2) = \sum_{sm_s} \langle s_1, s_2; m_1 m_2 | s_1, s_2'; sm_s \rangle \psi_{sm_s}(\mathbf{r}, \sigma_1, \sigma_2). \quad (4.29)$$

Introduction of Eq. (4.22) into Eq. (4.25) leads to the asymptotic form for this wave function

$$\psi^{(+)}_{m_1 m_2}(\mathbf{r}, \sigma_1, \sigma_2) \xrightarrow[r \to \infty]{} \phi_{m_1 m_2}(\mathbf{r}, \sigma_1, \sigma_2)$$

$$+ \frac{e^{ikr}}{r} \sum_{m_1' m_2'} f_{m_1' m_2', m_1 m_2}(\theta, \phi) \chi_{s_1 m_1'}(\sigma_1) \chi_{s_2 m_2'}(\sigma_2). \quad (4.30)$$

For elastic scattering we can identify the "measurable" scattering amplitude as

$$f_{m_1' m_2', m_1 m_2}(\theta, \phi) = \sum_{\substack{sm_s \\ s'm_s'}} \langle s_1, s_2; m_1 m_2 | s_1, s_2; sm_s \rangle$$

$$\times \langle s_1, s_2; m_1' m_2' | s_1, s_2; s' m_s' \rangle f_{s'm_s', sm_s}(\theta, \phi). \quad (4.31)$$

As expected, the measured amplitude is expressed as a sum over the channel spins. Nevertheless, the channel spin form is the most convenient one for performing calculations, since the resulting integral equations have no remaining dependence on the magnetic quantum numbers. Furthermore, there are some cases where the channel spin is conserved, but, even when it is not, the analysis of transitions between different channel spin states can yield useful insights into the effect of any spin dependence of the scattering potential.

5. Cross Sections and Polarizations

It is an easy matter to derive expressions for experimentally measurable quantities when the magnetic quantum numbers in the initial beam are known precisely. However, this is seldom the case, and we must examine more carefully the common situation when these are not known.

CHARACTERIZATION OF THE BEAM

Let us first consider the example of a spin-$\frac{1}{2}$ particle scattering from a spin-zero target. We take the initial spin state to be

$$\chi(\sigma_1) = \alpha \chi_{\frac{1}{2}\frac{1}{2}}(\sigma_1) + \beta \chi_{\frac{1}{2}-\frac{1}{2}}(\sigma_1), \quad (5.1)$$

where $|\alpha|^2 + |\beta|^2 = 1$ so that χ is normalized to unity. It may be recalled that such a state is completely polarized and has its spin lying along the direction defined by Eq. (1.7). Let us for the moment consider the total outward radial current, independent of the final spin projection; that is, we assume that the detector measures total current alone and does not distinguish spin projections. Since we readily see from Eqs. (1.32)–(1.38) that

$$f_{++}(\theta, \phi) = f_{--}(\theta, \phi) \equiv g(\theta) \tag{5.2}$$

and

$$f_{-+}(\theta, \phi) e^{-i\phi} = f_{+-}(\theta, \phi) e^{i\phi} \equiv h(\theta),$$

and since we may use any convenient complete set of final spin states, we immediately find that this cross section is

$$\begin{aligned}\sigma(\theta, \phi) &= |\alpha f_{++}(\theta, \phi) + \beta f_{+-}(\theta, \phi)|^2 + |\alpha f_{-+}(\theta, \phi) + \beta f_{--}(\theta, \phi)|^2 \\ &= |\alpha g(\theta) + \beta h(\theta) e^{-i\phi}|^2 + |\alpha h(\theta) e^{i\phi} + \beta g(\theta)|^2 \\ &= |g(\theta)|^2 + |h(\theta)|^2 + 2\,\text{Re}\{g^*(\theta) h(\theta)\}(\alpha^*\beta\, e^{-i\phi} + \alpha\beta^*\, e^{i\phi}). \end{aligned} \tag{5.3}$$

The cross section is independent of ϕ only if $\alpha^*\beta$ vanishes. This is just the condition for the incident particle to have its spin aligned along the direction of the incoming beam. Since this initial state is symmetric about the z axis, the scattering must of course be independent of the azimuthal angle.

In the most general case, the incident beam will be composed of particles in states having different spin directions, with each such state given by

$$\chi^{(n)}(\sigma_1) = \alpha^{(n)} \chi_{\frac{1}{2}\frac{1}{2}}(\sigma_1) + \beta^{(n)} \chi_{\frac{1}{2}-\frac{1}{2}}(\sigma_1) \tag{5.4}$$

and $|\alpha^{(n)}|^2 + |\beta^{(n)}|^2 = 1$. Suppose we represent each particle by a wave packet that does not overlap the wave packet of any other particle. The cross section is then the weighted sum of the cross sections for scattering from each of the incident states. We immediately see that for the scattering of such a beam the cross section is

$$\begin{aligned}\sigma(\theta, \phi) &= \sum_n W_n\, \sigma^{(n)}(\theta, \phi) \\ &= |g(\theta)|^2 + |h(\theta)|^2 + 2\,\text{Re}\{g^*(\theta) h(\theta)\} \\ &\quad \times \left\{\sum_n W_n(\alpha^{(n)*}\beta^{(n)} e^{-i\phi} + \alpha^{(n)}\beta^{(n)*} e^{i\phi})\right\}, \end{aligned} \tag{5.5}$$

where the real nonnegative number W_n is a weight factor representing the fraction of incident particles in the state $\chi^{(n)}$, and the sum in Eq. (5.5) is over all these states. The sum $\Sigma_n W_n$ must be unity, of course, and the condition

5. Cross Sections and Polarizations

that the cross section be independent of ϕ is now

$$\sum_n W_n \alpha^{(n)*} \beta^{(n)} = 0. \tag{5.6}$$

The polarization of the incident beam is the average value of the spin vector $\mathbf{S}/\hbar s_1$. For a spin-$\frac{1}{2}$ particle the polarization of the beam is, from Eq. (1.7),

$$\mathbf{P} = \sum_n W_n \mathbf{P}^{(n)} = \sum_n W_n \{ \hat{\mathbf{x}}(\alpha^{(n)*}\beta^{(n)} + \beta^{(n)*}\alpha^{(n)}) - i\hat{\mathbf{y}}(\alpha^{(n)*}\beta^{(n)} - \beta^{(n)*}\alpha^{(n)}) + \hat{\mathbf{z}}(|\alpha^{(n)}|^2 - |\beta^{(n)}|^2) \}. \tag{5.7}$$

Equation (5.6) is then equivalent to the condition that $P_x = P_y = 0$; that is, if the incident beam has no component of polarization normal to the direction of propagation of the beam, the cross section is necessarily independent of azimuth.

In general, the length of the polarization vector $\mathbf{P}$ defined by Eq. (5.7) will be less than unity, corresponding to partial polarization of the beam. When the sum over n includes only a single term, as in Eqs. (5.1)–(5.3), we have 100% polarization, and the spins of all the incident particles are aligned along the same direction. Let us briefly discuss the contrary situation, when the incident beam is completely unpolarized. The conditions for $\mathbf{P} = 0$ are

$$\sum_n W_n \alpha^{(n)*} \beta^{(n)} = 0$$

and

$$\sum_n W_n (|\alpha^{(n)}|^2 - |\beta^{(n)}|^2) = 0. \tag{5.8}$$

These can be satisfied in many ways. For example, they are satisfied if there are two states and $\alpha^{(1)} = \beta^{(2)} = 0$, while $W_1 = W_2 = \frac{1}{2}$. In such a beam half the particles have their spins aligned along the positive z axis and half along the negative z axis. Alternatively, Eq. (5.8) is satisfied if $\alpha^{(1)} = \alpha^{(2)} = 1/2^{1/2}$, $\beta^{(1)} = -\beta^{(2)} = i/2^{1/2}$, and $W_1 = W_2 = \frac{1}{2}$, which corresponds to an equal mixture of the two spin orientations along the y axis. In either case, or any other case in which Eq. (5.8) is satisfied, the cross section is

$$\sigma(\theta, \phi) \equiv \sigma(\theta) = |g(\theta)|^2 + |h(\theta)|^2. \tag{5.9}$$

For spin-$\frac{1}{2}$, no experiment can distinguish between the various ways in which a given initial polarization is achieved. In order to so distinguish, we would have to measure the average value of an operator that would have different values in two cases having the same initial polarizations, but achieved in different ways. However, for spin-$\frac{1}{2}$ any operator that depends on $\mathbf{S}$ can be expressed as a linear combination of the unit operator and the Pauli matrices σ, and the average values of these operators are already determined by the beam intensity and by **P**.

In summary, then, for spin-$\frac{1}{2}$ particles specification of the polarization vector of the incident beam allows a complete prediction of the results of any scattering experiment conducted with that beam. Furthermore, in the case where the beam is completely unpolarized, it is possible to take $W_1 = W_2 = \frac{1}{2}$, and $\alpha^{(1)} = \beta^{(2)} = 0$, which we interpret as an equal mixture of each possible value of m_s. An unpolarized beam of particles also results when the spins are randomly oriented, but the description of an unpolarized beam as an equal mixture of the possible eigenstates of J_z is completely equivalent to this.

For particles of spin greater than $\frac{1}{2}$, specification of the polarization vector is not sufficient to characterize the beam. As an example, for spin-1 particles there are $(2s + 1)^2 = 9$ independent spin operators (see Section 6); for example, we may choose for these the set of operators $S_m S_{m'}^\dagger$, with $|m'| \leq s$. The average polarization vector of a beam of particles fixes just three quantities and hence does not alone identify the incident mixture of states well enough to predict the results of all possible measurements.

The particles in an incident beam of spin-1 particles are in states of the form

$$\chi^{(n)}(\sigma_1) = \sum_{m=-1}^{1} \alpha_m^{(n)} \chi_{1m}(\sigma_1). \tag{5.10}$$

So long as the wave packets for the individual particles do not overlap, there will be no interference between these different states. To characterize the beam we require the average values of the operators $\xi_{mm'} = S_m S_{m'}^\dagger$ in this beam, and these are then given by the weighted sum

$$\bar{\xi}_{mm'} = \sum_n W_n \langle \chi^{(n)} | S_m S_{m'}^\dagger | \chi^{(n)} \rangle, \tag{5.11}$$

for all choices of m and m'.

If we explicitly construct the matrix $\bar{\xi}_{mm'}$, we obtain the Hermitian matrix

$$\|\xi\| = \frac{1}{2} \sum_n W_n \begin{pmatrix} |\alpha_1^{(n)}|^2 + |\alpha_0^{(n)}|^2 & -\alpha_0^{(n)*}\alpha_{-1}^{(n)} & \alpha_1^{(n)*}\alpha_{-1}^{(n)} \\ -\alpha_{-1}^{(n)*}\alpha_0^{(n)} & |\alpha_1^{(n)}|^2 + |\alpha_{-1}^{(n)}|^2 & \alpha_1^{(n)*}\alpha_0^{(n)} \\ \alpha_{-1}^{(n)*}\alpha_1^{(n)} & \alpha_0^{(n)*}\alpha_1^{(n)} & |\alpha_0^{(n)}|^2 + |\alpha_{-1}^{(n)}|^2 \end{pmatrix}, \tag{5.12}$$

where we have taken for the spin operators

$$\begin{aligned} S_{+1} &= (1/2\hbar)(S_x + iS_y) \\ S_0 &= (1/2\hbar)S_z \\ S_{-1} &= (1/2\hbar)(S_x - iS_y). \end{aligned} \tag{5.13}$$

5. Cross Sections and Polarizations

If the spin projections in the beam are randomly oriented, this matrix is $\bar{\xi}_{mm'} = \frac{1}{3}\delta_{mm'}$. Thus if we define the "polarization tensor" to be

$$\bar{P}_{mm'} = \bar{\xi}_{mm'} - \frac{1}{3}\delta_{mm'}, \tag{5.14}$$

we achieve a definition such that $\bar{P}_{mm'} = 0$ when the spin projections are randomly oriented. When the polarization tensor vanishes, the polarization vector will likewise vanish, since one can easily see from the commutation relations for angular momentum that $\mathbf{P}$ is a linear combination of the elements of the matrix $\bar{P}_{mm'}$:

$$\begin{aligned}\mathbf{P} = \sum W_n \{ &\hat{\mathbf{x}}(\alpha_1^{(n)*}\alpha_0^{(n)} + \alpha_0^{(n)*}\alpha_{-1}^{(n)} + \alpha_0^{(n)*}\alpha_1^{(n)} + \alpha_{-1}^{(n)*}\alpha_0^{(n)}) \\ - &i\hat{\mathbf{y}}(\alpha_1^{(n)*}\alpha_0^{(n)} + \alpha_0^{(n)*}\alpha_{-1}^{(n)} - \alpha_0^{(n)*}\alpha_1^{(n)} - \alpha_{-1}^{(n)*}\alpha_0^{(n)}) \\ + &\hat{\mathbf{z}}(|\alpha_1^{(n)}|^2 - |\alpha_{-1}^{(n)}|^2)\}. \end{aligned} \tag{5.15}$$

In general, all relevant information about the incident beam is contained in the measured values of the tensor $\bar{P}_{mm'}$. For arbitrary projectile and target spins $(s_1 s_2)$, there will be a square matrix $\bar{P}_{m_1 m_2, m_1' m_2'}$ of order $(2s_1 + 1)(2s_2 + 1)$, which will in a similar way completely characterize the beam.

Henceforth, we shall refer to an unpolarized beam as one for which the polarization tensor vanishes, a condition which we have noted exists when the spin projections are randomly oriented. The polarization tensor for spin-one also vanishes if we choose $\alpha_1^{(1)} = \alpha_0^{(2)} = \alpha_{-1}^{(3)} = 1$ and $W_1 = W_2 = W_3 = \frac{1}{3}$. Since all configurations that yield the same values for $\bar{P}_{mm'}$ are equivalent, we can then represent an unpolarized beam of spin-one particles as an equal mixture of the three spin projections. This result is true in general: an unpolarized beam can always be represented as an equal mixture of the $(2s_1 + 1)(2s_2 + 1)$ possible spin states.

Differential Cross Sections

In the absence of intrinsic spin, the differential cross section is defined as the number of particles scattered per unit solid angle in a particular direction, divided by the number of particles per unit area in the incident beam. When both the incident and target particles can have intrinsic spin, the incident state may be quite complex (as we have seen), and the final state is characterized by the spin quantum numbers as well as by the outgoing momentum. There are then a number of different cross sections that may be of interest, depending upon the properties of the target used to scatter the outgoing particles. The most straightforward experiment conceptually, but the most difficult in practice, would be one in which only particles having a particular set of spin quantum numbers were accepted by the detector. We shall consider this case first, before passing to the more realistic situation in which the detector does not distinguish between particles having different spin directions.

When the system is initially in the state $\chi^{(n)}$, we may define a differential cross section $\sigma^{(n)}_{m_1'm_2'}$ for scattering into a final state in which the outgoing particles have the spin projections $(m_1'm_2')$. Using the methods of Section 4, we see that this cross section may be expressed as

$$\sigma^{(n)}_{m_1'm_2'}(\theta, \phi) = |f^{(n)}_{m_1'm_2'}(\theta, \phi)|^2, \tag{5.16}$$

where the scattering amplitude is

$$\begin{aligned} f^{(n)}_{m_1'm_2'}(\theta, \phi) &= \sum_{m_1 m_2} f_{m_1'm_2', m_1 m_2}(\theta, \phi) \langle s_1 s_2 m_1 m_2 | \chi^{(n)} \rangle \\ &= \sum_{m_1 m_2} f_{m_1'm_2', m_1 m_2}(\theta, \phi) \alpha^{(n)}_{m_1 m_2} \end{aligned} \tag{5.17}$$

and the initial spin function is written as

$$\chi^{(n)}(\sigma_1, \sigma_2) = \sum_{m_1 m_2} \alpha^{(n)}_{m_1 m_2} \chi_{s_1 m_1}(\sigma_1) \chi_{s_2 m_2}(\sigma_2). \tag{5.18}$$

For an incident beam of arbitrary complexity, the differential cross section is the weighted sum of the cross sections for all the states represented in the beam:

$$\sigma_{m_1'm_2'}(\theta, \phi) = \sum_n W_n \sigma^{(n)}_{m_1'm_2'}(\theta, \phi). \tag{5.19}$$

The simplest situation would be one in which both particles in the initial beam were polarized, so that their spin projections were fixed. There would then be only a single state in the initial beam, and the differential cross section would be

$$\sigma_{m_1'm_2', m_1 m_2}(\theta, \phi) = |f_{m_1'm_2', m_1 m_2}(\theta, \phi)|^2. \tag{5.20}$$

In a more common situation, the initial beam is completely unpolarized. As we have seen, such a beam can be represented as an equally weighted mixture of all the possible spin states, and the differential cross section in this case is

$$\sigma_{m_1'm_2'}(\theta, \phi) = \frac{1}{(2s_1 + 1)(2s_2 + 1)} \sum_{m_1 m_2} |f_{m_1'm_2', m_1 m_2}(\theta, \phi)|^2. \tag{5.21}$$

These cross sections describe the results of experiments in which the final spins are measured by the detector. In general, such a measurement would be too difficult, and one uses instead a detector that simply accepts all particles, irrespective of their spin projections. For such experiments the differential cross section is the sum of the "partial" cross sections shown above. For an unpolarized initial beam, for instance, the cross section is

$$\sigma(\theta, \phi) = \frac{1}{(2s_1 + 1)(2s_2 + 1)} \sum_{\substack{m_1'm_2' \\ m_1 m_2}} |f_{m_1',m_2',m_1 m_2}(\theta, \phi)|^2. \quad (5.22)$$

The unpolarized cross section is then obtained by summing over all possible final states and averaging over the initial states. If Eq. (4.31) is introduced and the orthogonality relation for the Clebsch–Gordan coefficient, Eq. (2.7), is used, this can be expressed in terms of the channel-spin scattering amplitude as

$$\sigma(\theta, \phi) = \frac{1}{(2s_1 + 1)(2s_2 + 1)} \sum_{\substack{s'm_s' \\ sm_s}} |f_{s'm_s', sm_s}(\theta, \phi)|^2. \quad (5.23)$$

Then it is only necessary to know the scattering amplitude for channel-spin states to obtain this cross section.

POLARIZATION

There are a number of different measurable quantities that give independent pieces of information. As Eq. (5.20) shows, there are $(2s_1 + 1)^2 \times (2s_2 + 1)^2$ different cross sections and, although some of these are related through various symmetry relations, there still remain a number of independent observable quantities. One such quantity is the polarization $\mathbf{P}(\theta, \phi)$ of the scattered beam, defined as the polarization produced in a beam of outgoing particles scattered along a ray (θ, ϕ) from an unpolarized incident beam. (Note that this "polarization" is a property of the interaction between the projectile and the target, whereas the "polarization" discussed earlier in this section depended only on the nature of the source of the beam. In practice this overlapping terminology does not cause any problem.) Other observable quantities measure the probability of transitions between different spin states.

As we have noted, direct measurements of the spin projections of particles (as in a Stern–Gerlach experiment) are sufficiently difficult that they are never performed as part of a scattering experiment. Let us then consider how the polarization of the scattered beam may be measured. For initially unpolarized spin-$\frac{1}{2}$ particles scattered from a spin-zero target, the methods developed in the next section can be used to show that the polarization of the scattered beam has components

$$P_x(\theta, \phi) = \{[2 \operatorname{Re} g^*(\theta) h(\theta)] / [\,|\,g(\theta)\,|^2 + |\,h(\theta)\,|^2\,]\} \cos \phi$$
$$P_y(\theta, \phi) = \{[2 \operatorname{Re} g^*(\theta) h(\theta)] / [\,|\,g(\theta)\,|^2 + |\,h(\theta)\,|^2\,]\} \sin \phi \quad (5.24)$$
$$P_z(\theta, \phi) = 0,$$

or

$$\mathbf{P}(\theta) = |\,P(\theta)\,|\,\hat{\xi} = \frac{2\{\operatorname{Re} g^*(\theta) h(\theta)\}}{|\,g(\theta)\,|^2 + |\,h(\theta)\,|^2}\,\hat{\xi}, \quad (5.25)$$

where $\hat{\xi}$ is a unit vector in the $(\mathbf{k} \times \mathbf{k}')$ direction, that is, perpendicular to the plane of scattering. Using Eqs. (5.5) and (5.7), we see that, in general, the cross section depends on the polarization of the incident beam, which we denote by $\mathbf{P}_{\text{inc}}$. If we combine Eqs. (5.5) and (5.7) and use Eq. (5.25), we find

$$\sigma(\theta, \phi) = |\,g(\theta)\,|^2 + |\,h(\theta)\,|^2 + 2[\operatorname{Re} g^*(\theta) h(\theta)] [(\mathbf{P}_{\text{inc}})_x \cos \phi$$
$$+ (\mathbf{P}_{\text{inc}})_y \sin \phi] = \sigma(\theta)[1 + \mathbf{P}(\theta) \cdot \mathbf{P}_{\text{inc}}], \quad (5.26)$$

where $\sigma(\theta)$ is the unpolarized cross section, Eq. (5.9). Since $\mathbf{P}(\theta)$ is normal to the scattering plane, the angular distribution is independent of the component of incident polarization lying in the scattering plane.

From Eq. (5.26) we see that we may determine the polarization $\mathbf{P}(\theta)$, provided $\mathbf{P}_{\text{inc}}$ is nonzero and we know both $\sigma(\theta)$ and $\mathbf{P}_{\text{inc}}$, by comparing scattering for two values of $\mathbf{P}_{\text{inc}} \cdot \hat{\xi}$. (Usually these are taken as "right" and "left," where "right" is defined as that value of ϕ for which $\mathbf{P}_{\text{inc}} \cdot \hat{\xi}$ is a maximum and "left" is the value of ϕ for which it is a minimum.) If we have an initially unpolarized beam, as in the usual situation, it may become polarized as a result of the scattering. (This is shown in Eq. (5.25).) If we then use the scattered beam as the incident beam for a second scattering experiment, Eq. (5.26) shows that, by examining scattering to the right and to the left in the same plane in the second scattering, we may determine the product $P_1(\theta_1) P_2(\theta_2)$, where the subscripts 1 and 2 refer to the first and second scattering (see Fig. 11.3):

$$P_1(\theta_1) P_2(\theta_2) = \frac{\sigma_L - \sigma_R}{\sigma_L + \sigma_R}. \quad (5.27)$$

This function of the cross sections is termed the "left-right asymmetry."

For higher spins there are quantities in addition to the polarization, measuring the transition rate between different spin states, which must be determined. More complex experiments must be performed, including triple scat-

6. Matrix Methods

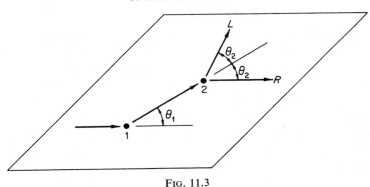

FIG. 11.3

tering experiments, and the theory for such experiments is straightforward but tedious.

6. Matrix Methods

It is apparent from the formulas developed in previous sections that the computation of observables can become quite involved when either or both incident particles have an intrinsic spin. Some simplification can often be achieved through the use of matrix methods, especially when the spins are no greater than $s = 1$.

The origin of the matrix approach can be seen quite easily from the earlier result for the unpolarized scattering cross section:

$$\sigma(\theta, \phi) = \frac{1}{(2s_1 + 1)(2s_2 + 1)} \sum_{\substack{m_1'm_2' \\ m_1 m_2}} |f_{m_1'm_2', m_1 m_2}(\theta, \phi)|^2. \quad (6.1)$$

The scattering amplitude $f_{m_1'm_2', m_1 m_2}(\theta, \phi)$ may be considered to be an element of a matrix $f(\theta, \phi)$, with this matrix having $(2s_1 + 1)(2s_2 + 1)$ rows and columns labeled by a pair of spin projection values $(m_1 m_2)$. Thus Eq. (6.1) may be written in matrix form as

$$\sigma(\theta, \phi) = \frac{1}{(2s_1 + 1)(2s_2 + 1)} \sum_{\substack{m_1'm_2' \\ m_1 m_2}} f^*_{m_1'm_2', m_1 m_2}(\theta, \phi)$$
$$\times f_{m_1'm_2', m_1 m_2}(\theta, \phi)$$
$$= \sum_{\substack{m_1'm_2' \\ m_1 m_2}} f^\dagger_{m_1 m_2, m_1'm_2'}(\theta, \phi) f_{m_1'm_2', m_1 m_2}(\theta, \phi) / \sum_{m_1 m_2} 1$$
$$= \frac{\operatorname{Tr} f^\dagger(\theta, \phi) f(\theta, \phi)}{\operatorname{Tr} 1}. \quad (6.2)$$

Here Tr denotes the trace over the $(2s_1 + 1)(2s_2 + 1)$-dimensional matrix defined by the particle spin projections, and the dagger indicates the Hermitian adjoint matrix. The matrix $f(\theta, \phi)$ is defined for each scattering angle and can be used to compute the cross section as well as all other observables. In this way a considerable simplification in notation is achieved, and calculations can be made very much simpler by using matrix algebra.

To establish a general formalism for computing observables, let us return for a moment to our earlier discussion of wave functions. Recalling Eq. (4.30), we have the asymptotic form for the wave function

$$\psi^{(+)}_{m_1 m_2}(\mathbf{r}, \sigma_1, \sigma_2) \xrightarrow[r \to \infty]{} e^{ikz} \chi_{s_1 m_1}(\sigma_1) \chi_{s_2 m_2}(\sigma_2) + \frac{e^{ikr}}{r} \sum_{m_1' m_2'}$$
$$\times f_{m_1' m_2', m_1 m_2}(\theta, \phi) \chi_{s_1 m_1'}(\sigma_1) \chi_{s_2 m_2'}(\sigma_2). \quad (6.3)$$

The two-particle spin function $\chi_{s_1 m_2}(\sigma_1) \chi_{s_2 m_2}(\sigma_2)$ can, in the matrix formulation, be reinterpreted as a unit vector having the direction in spin space associated with the set $(m_1 m_2)$. Any arbitrary vector in that space has a component $\alpha_{m_1 m_2}$ in this direction and may be written as a column vector

$$\begin{pmatrix} \alpha_{s_1 s_2} \\ \alpha_{s_1 s_2 - 1} \\ \alpha_{m_1 m_2} \\ \alpha_{-s_1 -s_2} \end{pmatrix}.$$

For simplicity, we will in some cases denote the set $(m_1 m_2)$ by the index i running from 1 to $(2s_1 + 1)(2s_2 + 1)$. The component α_i is interpreted as a probability amplitude, and $|\alpha_i|^2$ measures the probability that the system is in the state $i \equiv (m_1 m_2)$, if the vector is normalized according to $\Sigma_i |\alpha_i|^2 = 1$.

It is, of course, not necessary to use axes in spin space that are identified with eigenvalues of the two magnetic quantum numbers. It is also possible, for instance, to use axes characterized by the channel spin s and its magnetic quantum number m_s; in this case $i \equiv (sm_s)$. The choice between these two sets of axes is sometimes dictated by the particular experiment being examined. Thus, in nucleon-nucleon scattering it is frequently more convenient to use singlet and triplet states, that is, channel-spin states, to characterize the wave function, in which case the four-component spin vector is

$$\begin{pmatrix} \alpha_{11} \\ \alpha_{10} \\ \alpha_{1-1} \\ \alpha_{00} \end{pmatrix}.$$

The wave function given in Eq. (6.3) results from an initial state having well-defined values of m_1 and m_2. A more general outgoing wave function would have the asymptotic form

$$\psi^{(+)}(\mathbf{r}, \sigma_1, \sigma_2) \xrightarrow[r \to \infty]{} e^{ikz} \chi(\sigma_1, \sigma_2) + e^{ikr}/r \sum_{\substack{m_1'm_2' \\ m_1m_2}} \alpha_{m_1m_2} f_{m_1'm_2',m_1m_2}(\theta, \phi)$$

$$\times \chi_{s_1m_1'}(\sigma_1) \chi_{s_2m_2'}(\sigma_2), \qquad (6.4)$$

where the initial spin state is

$$\chi(\sigma_1, \sigma_2) = \sum_{m_1m_2} \alpha_{m_1m_2} \chi_{s_1m_1}(\sigma_1) \chi_{s_2m_2}(\sigma_2). \qquad (6.5)$$

This initial state may be represented by the spin vector

$$\chi = \begin{pmatrix} \alpha_{s_1s_2} \\ \vdots \\ \alpha_{m_1m_2} \\ \vdots \\ \alpha_{-s_1-s_2} \end{pmatrix} \qquad (6.6)$$

and Eq. (6.5) then becomes, in vector notation,

$$\psi^{(+)}(r) \xrightarrow[r \to \infty]{} e^{ikz} \chi + \frac{e^{ikr}}{r} f(\theta, \phi) \chi. \qquad (6.7)$$

Here $f(\theta, \phi)$ is the matrix introduced above whose matrix elements $f_{m_1'm_2'm_1m_2}(\theta, \phi)$ are the scattering amplitudes. Equation (6.7) implies a very simple interpretation for this matrix: it transforms the initial spin state into the final spin state; the state χ is transformed into the state $f(\theta, \phi) \chi$ by the collision process.

If desired, transformations between the wave function form of Eq. (6.4) and the spin vector form of Eq. (6.7) can be carried out using the Dirac formalism. Thus, a vector having a direction in spin space associated with the variables $(\sigma_1 \sigma_2)$ is denoted by $| \sigma_1 \sigma_2 \rangle$, and the vector χ is transformed into the wave function $\chi(\sigma_1, \sigma_2)$ by the scalar product operation: $\langle \sigma_1 \sigma_2 | \chi \rangle = \chi(\sigma_1, \sigma_2)$. In particular, the vector $| m_1 m_2 \rangle$ becomes $\langle \sigma_1 \sigma_2 | m_1 m_2 \rangle = \chi_{s_1m_2}(\sigma_1) \chi_{s_2m_2}(\sigma_2)$.

EXPECTATION VALUES OF SPIN OPERATORS

We now show how this formulation can be used to compute the results of possible experiments. Having obtained spin vectors that represent the states of the system, we use spin matrices to represent the Hermitian operators associated with the possible observables.

Since there are $(2s_1 + 1)(2s_2 + 1)$ basis states of the two-particle system, an arbitrary Hermitian matrix in this spin space contains $(2s_1 + 1)^2(2s_2 + 1)^2$ elements. It is most convenient to define a set of this many linearly independent Hermitian matrices M^μ satisfying the orthogonality relations

$$\mathrm{Tr}\, M^\mu M^\nu = (2s_1 + 1)(2s_2 + 1)\delta_{\mu\nu} = (\mathrm{Tr}\, 1)\delta_{\mu\nu}, \qquad (6.8)$$

with $1 \le \mu, \nu \le (2s_1 + 1)^2(2s_2 + 1)^2$. One of the M^μ may be taken as the unit matrix, and the others may then be chosen to have zero trace. The matrix associated with any operator may then be expressed as a linear combination of these standard matrices through the expansion

$$\mathcal{O} = \sum_\mu \left(\frac{\mathrm{Tr}\, \mathcal{O} M_\mu}{\mathrm{Tr}\, 1} \right) M^\mu. \qquad (6.9)$$

For a spin-$\tfrac{1}{2}$ projectile incident on a spinless target, $s_1 = \tfrac{1}{2}$, $s_2 = 0$, and the set M^μ may be chosen to be the unit matrix and the three Pauli matrices σ_j; if the projectile has spin one, the set M^μ may be taken as the elements of the tensor $S_m S_{m'}$, as discussed in Section 5. When both particles have spin $\tfrac{1}{2}$, the sixteen basis matrices may be taken as $1_1 1_2$, $\sigma_{j1} 1_2$, $1_1 \sigma_{j2}$, $\sigma_{j1}\sigma_{k2}$ where the first matrix operates on the projectile spin vector and the second acts on the target vector.

The average value of a particular operator M^μ in the initial state is given by

$$\langle M^\mu \rangle_i = \frac{\langle \chi | M^\mu | \chi \rangle}{\langle \chi | \chi \rangle} = \frac{\sum\limits_{jk} \alpha_k^* M_{kj}^\mu \alpha_j}{\sum\limits_{j} |\alpha_j|^2}; \qquad (6.10)$$

in the final state it is

$$\langle M^\mu \rangle_f = \frac{\langle f(\theta, \phi)\chi | M^\mu | f(\theta, \phi)\chi \rangle}{\langle f(\theta, \phi)\chi | f(\theta, \phi)\chi \rangle}$$

$$= \frac{\sum\limits_{jklm} \alpha_l^* f_{kl}^*(\theta, \phi) M_{kj}^\mu f_{jm}(\theta, \phi) \alpha_m}{\sum\limits_{jkl} \alpha_l^* f_{kl}^*(\theta, \phi) f_{kj}(\theta, \phi) \alpha_j}. \qquad (6.11)$$

These formulas are simplest to apply for low values of the spin ($s = 0, \tfrac{1}{2}, 1$), when special properties of the matrices can be used and it is not necessary explicitly to sum over these indices.

As Eq. (6.10) shows, the expectation value $\langle M^\mu \rangle_i$ depends only on the product $\alpha_k^* \alpha_j$ rather than on α_j alone. Because of this bilinear structure, it is

6. Matrix Methods

possible to describe all of the measurable properties of the initial beam in terms of a single matrix ρ_i, termed the density matrix. When the initial beam consists of particles in the state χ, the density matrix is defined by the product of the probability amplitudes:

$$(\rho_i)_{jk} = \alpha_j \alpha_k^*. \tag{6.12}$$

The expectation value of M^μ in the initial state can then be expressed in terms of the density matrix as

$$\langle M^\mu \rangle_i = \frac{\sum_{jk}(\rho_i)_{jk} M^\mu_{kj}}{\sum_j (\rho_i)_{jj}} = \frac{\text{Tr }\rho_i M^\mu}{\text{Tr }\rho_i}, \tag{6.13}$$

so that the expectation value of any operator can be determined if the density matrix is known. Conversely, if the expectation values of the complete set of matrices M^μ are known, then, from Eq. (6.9), the density matrix can be determined as

$$\rho_i = \left(\frac{\text{Tr }\rho_i}{\text{Tr }1}\right) \sum_\mu \langle M^\mu \rangle_i M^\mu. \tag{6.14}$$

The factor $(\text{Tr }\rho_i) = \sum_j |\alpha_j|^2$ gives the normalization of the state vector and thus the probability of finding a particle in the state χ. There are $(2s_1 + 1)^2 (2s_2 + 1)^2$ real measurable numbers $\langle M^\mu \rangle_i$, and the Hermitian matrix ρ_i has the same number of real constants; then there is indeed a one-to-one correspondence between the number of observables and the information contained in the density matrix.

In the final state the density matrix is

$$(\rho_f(\theta, \phi))_{jk} = (\sum_m f_{jm}(\theta, \phi) \alpha_m)(\sum_l f_{kl}(\theta, \phi) \alpha_l)^*$$
$$= \sum_{lm} f_{jm}(\theta, \phi) (\rho_i)_{ml} f_{kl}^*(\theta, \phi), \tag{6.15}$$

or

$$\rho_f(\theta, \phi) = f(\theta, \phi) \rho_i f^\dagger(\theta, \phi). \tag{6.16}$$

The scattering amplitude thus defines a transformation that carries ρ_i into $\rho_f(\theta, \phi)$. With this definition of $\rho_f(\theta, \phi)$, we see from Eq. (6.11) that, just as for the initial state, the properties of the final state are related to the final-state density matrix by

$$\langle M^\mu \rangle_f = \frac{\text{Tr }\rho_f(\theta, \phi) M^\mu}{\text{Tr }\rho_f(\theta, \phi)}. \tag{6.17}$$

One special quantity is the ratio of the traces

$$\frac{\operatorname{Tr} \rho_f(\theta, \phi)}{\operatorname{Tr} \rho_i} = \frac{\operatorname{Tr} f(\theta, \phi) \rho_i f^\dagger(\theta, \phi)}{\operatorname{Tr} \rho_i}$$

$$= \sum_i \left| \sum_j f_{ij}(\theta, \phi) \alpha_j \right|^2 = \sum_i \sigma_i(\theta, \phi), \tag{6.18}$$

if the initial state is normalized to unity. This is the differential cross section for scattering from the initial state χ, as in Eqs. (5.16) and (5.17).

The real utility of the density matrix arises when the initial beam cannot be represented by a single state vector, but must be described instead as a mixture of different states. Let us then consider a beam composed of an incoherent superposition of normalized states $\chi^{(n)}$ which need not be mutually orthogonal and need not form a complete set. We define the density matrix in this case by

$$\rho_{jk} = \sum_n W_n \alpha_j^{(n)} \alpha_k^{(n)*}, \tag{6.19}$$

where we sum over every state represented in the beam and let W_n be the relative statistical weight of each state.

With this definition the formulas obtained above hold in general. With the vectors $\chi^{(n)}$ normalized to unity, the expectation value of any operator M^μ is the weighted average

$$\langle M^\mu \rangle = \frac{\sum_n W_n \langle \chi^{(n)} | M^\mu | \chi^{(n)} \rangle}{\sum_n W_n} = \frac{\operatorname{Tr} \rho M^\mu}{\operatorname{Tr} \rho}. \tag{6.20}$$

The density matrix can in general be expressed as

$$\rho = \left(\frac{\operatorname{Tr} \rho}{\operatorname{Tr} 1}\right) \sum_\mu \langle M^\mu \rangle M^\mu, \tag{6.21}$$

and the density matrix for a final scattered state is always

$$\rho_f(\theta, \phi) = f(\theta, \phi) \rho_i f^\dagger(\theta, \phi). \tag{6.22}$$

As an example of the general form of ρ, for a spin-$\tfrac{1}{2}$ particle incident on a spinless target, Eq. (6.21) implies that the density matrix is

$$\rho = \tfrac{1}{2}(\operatorname{Tr} \rho)(1 + \mathbf{P} \cdot \boldsymbol{\sigma})$$

$$= \tfrac{1}{2}(\operatorname{Tr} \rho) \begin{pmatrix} 1 + P_z & P_x - iP_y \\ P_x + iP_y & 1 - P_z \end{pmatrix}, \tag{6.23}$$

where $\mathbf{P} = \langle \boldsymbol{\sigma} \rangle$ is the polarization of the beam described by ρ.

For an unpolarized beam it follows from the definition of the density matrix, Eq. (6.19), that ρ is diagonal and is, in fact, a multiple of the unit matrix. (As shown in Section 5, $\alpha_j^{(n)} = \delta_{jn}$ and all W_n are equal.) A corollary of this is that all the expectation values $\langle M^\mu \rangle$ vanish except for $\langle 1 \rangle$, so that in Eq. (6.21) we have

$$\rho_{\text{unpol.}} = \frac{\text{Tr}\,\rho}{\text{Tr}\,1}\,1. \tag{6.24}$$

For such a beam the differential cross section is

$$\sigma(\theta, \phi) = \frac{\text{Tr}\,\rho_f(\theta, \phi)}{\text{Tr}\,\rho_i} = \frac{\text{Tr}\,\rho_i f^\dagger(\theta, \phi) f(\theta, \phi)}{\text{Tr}\,\rho_i}$$

$$= \frac{\text{Tr}\,f^\dagger(\theta, \phi) f(\theta, \phi)}{\text{Tr}\,1}$$

$$= \frac{1}{(2s_1 + 1)(2s_2 + 1)} \sum_{\substack{m_1'm_2' \\ m_1 m_2}} |f_{m_1'm_2', m_1 m_2}(\theta, \phi)|^2. \tag{6.25}$$

Likewise, the polarization of the outgoing particle, defined as the average of the spin vector divided by $\hbar s_1$ for an initially unpolarized beam, is

$$\mathbf{P}(\theta, \phi) = \frac{\langle \mathbf{S}_1 \rangle_f}{\hbar s_1} = \frac{\text{Tr}\,\rho_f(\theta, \phi)\,\mathbf{S}_1}{\hbar s_1\,\text{Tr}\,\rho_f(\theta, \phi)} = \frac{\text{Tr}\,f(\theta, \phi) f^\dagger(\theta, \phi)\,\mathbf{S}_1}{\hbar s_1\,\text{Tr}\,f(\theta, \phi) f^\dagger(\theta, \phi)}. \tag{6.26}$$

THE POLARIZATION-ASYMMETRY RELATION

As an example of the application of matrix methods, we shall again derive the relation between the polarization and the left-right asymmetry in the double scattering of a spin-$\frac{1}{2}$ particle. First, a discussion of invariance properties will prove helpful.

Often we know on general grounds the transformation properties of a quantity, that is, whether it is a scalar, a vector, or an axial vector, without knowing all its other features. In order to learn the experimental consequences of these transformation properties, we must deal with expressions that contain only directly observable quantities such as momenta and spins. If this is done, we can sometimes develop a simple intuitive derivation of a general result which might otherwise be much more difficult to obtain.

As an illustration, consider the differential cross section. For a scalar (or spherically symmetric) potential, the cross section must be rotationally invariant; that is, no physically measurable results can depend on the over-all orientation of the experiment. Since spin is not measured in an ordinary cross section measurement, the cross section for scattering of an unpolarized beam can depend only on the incident momentum **k** and the outgoing momentum **k'**. Since the cross section is a scalar function, it can depend only on the scalars that can be formed from these vectors, namely $k^2 = k'^2$ and $\mathbf{k} \cdot \mathbf{k}' = k^2 \cos\theta$. Other scalars such as $(\mathbf{k} \cdot \mathbf{k}')^2 = k^4 \cos^2\theta$ or E are not independent of these two. Hence the differential cross section is actually a function only of k^2 (or E) and $\cos\theta$ and is independent of ϕ, the azimuthal angle. As we have seen, we may then denote the cross section by $\sigma(\theta)$.

Another example is provided by the polarization of a scattered beam. Again we observe that the vector $\mathbf{P}(\theta, \phi)$, given by Eq. (6.26), can depend only on the momenta **k** and **k'**. Furthermore, it must be an axial vector, since $\mathbf{S}_1$ is such a quantity while $f(\theta, \phi)$ transforms as a scalar for a scalar potential. (As we saw in Chapter 9, the scattering amplitude or T matrix is invariant under rotations.) The only axial vector that we can form is $\mathbf{k} \times \mathbf{k}'$, so that

(1) **P** must be composed of a scalar function multiplied by $\mathbf{k} \times \mathbf{k}'$; and
(2) **P** must lie in the direction defined by $\mathbf{k} \times \mathbf{k}'$, i.e., normal to the scattering plane and transverse to the directions of motion.

As a result we can write the polarization as

$$\mathbf{P}(\theta, \phi) = P(\theta)\, \hat{\boldsymbol{\xi}}, \tag{6.27}$$

with $\mathbf{k} \times \mathbf{k}' = k^2 \sin\theta\, \hat{\boldsymbol{\xi}}$.

For the scattering of spin-$\tfrac{1}{2}$ particles from a spinless target, the most general form of the scattering amplitude is

$$f(\theta, \phi) = g(\theta, \phi) + \mathbf{h}(\theta, \phi) \cdot \boldsymbol{\sigma}_1 \tag{6.28}$$

since, as noted earlier, unity and the Pauli matrices $\boldsymbol{\sigma}_1$ form a complete set of operators for spin $\tfrac{1}{2}$. The function $g(\theta, \phi)$ must be a scalar function and, as we saw in the case of the cross section, must therefore be independent of the azimuthal angle ϕ. The function $\mathbf{h}(\theta, \phi)$ must be an axial vector and, as in the case of the polarization, must have the form $h(\theta)\, \hat{\boldsymbol{\xi}}$. Then Eq. (6.28) can be written as

$$f(\theta, \phi) = g(\theta) + h(\theta)\, \boldsymbol{\sigma}_1 \cdot \hat{\boldsymbol{\xi}}. \tag{6.29}$$

In terms of these functions the unpolarized cross section is

$$\sigma(\theta) = \frac{\operatorname{Tr} \rho_f(\theta, \phi)}{\operatorname{Tr} \rho_i} = |g(\theta)|^2 + |h(\theta)|^2, \tag{6.30}$$

while the polarization is found to be

$$\mathbf{P}(\theta, \phi) = \frac{\text{Tr}\, \rho_f(\theta, \phi)\, \boldsymbol{\sigma}_1}{\text{Tr}\, \rho_f(\theta, \phi)} = \frac{2\, \text{Re}\{g^*(\theta)\, h(\theta)\}}{|g(\theta)|^2 + |h(\theta)|^2}\, \boldsymbol{\xi}, \qquad (6.31)$$

which confirm Eqs. (5.9) and (5.25).

Let us now compute the left-right asymmetry in the double scattering of a spin-$\frac{1}{2}$ particle (Fig. 11.3). The "incident" beam in the second scattering is described by a density matrix of the form

$$\rho_1(\theta_1, \phi_1) = \tfrac{1}{2}[\text{Tr}\, \rho_1(\theta_1, \phi_1)](1 + \mathbf{P}_1(\theta_1, \phi_1)\cdot\boldsymbol{\sigma}_1), \qquad (6.32)$$

as in Eq. (6.23). The vector $\mathbf{P}_1(\theta_1, \phi_1)$ is the polarization produced in the first scattering:

$$\mathbf{P}_1(\theta_1, \phi_1) = \frac{\text{Tr}\, f_1(\theta_1, \phi_1) f_1^\dagger(\theta_1, \phi_1)\, \boldsymbol{\sigma}_1}{\text{Tr}\, |f_1(\theta_1, \phi_1)|^2}. \qquad (6.33)$$

The differential cross section for the second scattering is

$$\begin{aligned}
\sigma_2(\theta_2, \phi_2) &= \frac{\text{Tr}\, \rho_2(\theta_2, \phi_2)}{\text{Tr}\, \rho_1(\theta_1, \phi_1)} \\
&= \tfrac{1}{2}\text{Tr}\, f_2(\theta_2, \phi_2)(1 + \mathbf{P}_1(\theta_1, \phi_1)\cdot\boldsymbol{\sigma}_1) f_2^\dagger(\theta_2, \phi_2) \\
&= \sigma_2(\theta_2)\left(1 + \mathbf{P}_1(\theta_1, \phi_1)\cdot\frac{\text{Tr}\, f_2^\dagger(\theta_2, \phi_2) f_2(\theta_2, \phi_2)\, \boldsymbol{\sigma}_1}{\text{Tr}\, |f_2(\theta_2, \phi_2)|^2}\right),
\end{aligned} \qquad (6.34)$$

where $\sigma_2(\theta_2)$ is the cross section for single scattering of an unpolarized beam by the second target. One finds using Eq. (6.29) that

$$\text{Tr}\, f_2^\dagger(\theta_2, \phi_2) f_2(\theta_2, \phi_2)\, \boldsymbol{\sigma}_1 = \text{Tr}\, f_2(\theta_2, \phi_2) f_2^\dagger(\theta_2, \phi_2)\, \boldsymbol{\sigma}_1. \qquad (6.35)$$

Since Eq. (6.29) depends upon rotation and reflection invariance, this nontrivial relation also depends upon these properties. Then, using the equivalent of Eq. (6.33) for the second scattering, Eq. (6.34) becomes simply

$$\sigma_2(\theta_2, \phi_2) = \sigma_2(\theta_2)(1 + \mathbf{P}_1(\theta_1, \phi_1)\cdot\mathbf{P}_2(\theta_2, \phi_2)). \qquad (6.36)$$

This yields the relation between the polarization and asymmetry, Eq. (5.27). If both scatterings take place in the same plane, $\mathbf{P}_1$ and $\mathbf{P}_2$ are either parallel or antiparallel, and the left-right asymmetry is

$$e_{LR} = \frac{\sigma_2(\theta_2, \pi/2) - \sigma_2(\theta_2, -\pi/2)}{\sigma_2(\theta_2, \pi/2) + \sigma_2(\theta_2, -\pi/2)} = P_1(\theta_1)\, P_2(\theta_2). \qquad (6.37)$$

If the targets are identical and the scatterings take place at the same angle θ, the asymmetry is $e_{LR} = P^2(\theta)$ and $|P(\theta)|$ can be determined by measuring e_{LR}; otherwise, $P_1(\theta_1)$ can be found if $P_2(\theta_2)$ is known.

7. Projection Operators

The connection between the matrix methods of the last section, which are most useful for empirical analysis, and the orthogonal-expansion methods of the earlier sections, which are most useful for deriving theoretical expressions, can be made by means of projection operators. We shall show how the use of such operators permits the matrix expressions to be related to the partial-wave expansions and the phase shifts.

A projection operator selects out of a state (or wave function) that component having a particular property. This property is usually related to a specific observable A, and one wishes to select out of the state that component having a particular eigenvalue a'. If the eigenstates of A are denoted by $\phi_{a'}$, an arbitrary state may be expanded as

$$\psi = \sum_{a'} C_{a'} \phi_{a'}, \tag{7.1}$$

or, in the Dirac notation,

$$|\psi\rangle = \sum_{a'} |a'\rangle \langle a'|\psi\rangle. \tag{7.2}$$

The projection operator is defined to satisfy the relation

$$P_{a'} \psi = C_{a'} \phi_{a'}, \tag{7.3}$$

or, again in the Dirac notation,

$$P_{a'} |\psi\rangle = |a'\rangle \langle a'|\psi\rangle. \tag{7.4}$$

This implies that we may write the projection operator $P_{a'}$ simply as

$$P_{a'} = |a'\rangle \langle a'|. \tag{7.5}$$

In coordinate space $P_{a'}$ can be written as a nonlocal operator

$$\langle \mathbf{r} | P_{a'} | \mathbf{r}' \rangle = \phi_{a'}(\mathbf{r}) \phi_{a'}^*(\mathbf{r}'); \tag{7.6}$$

in spin space the projection operator is simply the matrix $\phi_{a'} \phi_{a'}^\dagger$.

Several general operator relations can be used to define a projection operator. The sum of Eq. (7.4) over all eigenvalues a' must reproduce the original state vector or

$$\sum_{a'} P_{a'} = \sum_{a'} |a'\rangle \langle a'| = 1; \tag{7.7}$$

7. Projection Operators

this is the completeness or closure relation. The product of projection operators for two different eigenvalues must vanish:

$$P_{a'}P_{a''} = 0, \quad a' \neq a''; \tag{7.8}$$

this is the orthogonality condition. Finally, a projection operator acting twice must give the same result as the operator acting once, i.e.,

$$P_{a'}^2 = P_{a'}; \tag{7.9}$$

this is the normalization condition. Together, these define a projection operator.

The operator $P_{a'}$ gives the eigenvalue 1 when acting on $\phi_{a'}$ and 0 when acting on any other eigenstate of A. One can then construct this operator by recognizing that $(A - a'')$ gives 0 when acting on $\phi_{a''}$ and $(a' - a'')$ when acting on $\phi_{a'}$. Using this, we can write $P_{a'}$ as

$$P_{a'} = \prod_{a'' \neq a'} \frac{(A - a'')}{a' - a''}. \tag{7.10}$$

There is a term in this product for each state $\phi_{a''}$ other than $\phi_{a'}$, so that it gives zero when acting on any states other than $\phi_{a'}$.

We will now use this construction to derive several useful projection operators. Consider first the projection operators for the singlet ($s = 0$) and the triplet ($s = 1$) eigenstates of a system of two spin-$\frac{1}{2}$ particles. Since the operator S^2 acting on such a system has just two eigenvalues, 0 and $2\hbar^2$, the two projection operators are

$$P_{s=0} = \frac{S^2 - 2\hbar^2}{-2\hbar^2} \tag{7.11}$$

and

$$P_{s=1} = \frac{S^2}{2\hbar^2}. \tag{7.12}$$

The spin operator is $\mathbf{S} = (\hbar/2)(\sigma_1 + \sigma_2)$ and $\sigma^2 = 3$, so that

$$S^2 = (\hbar^2/4)(\sigma_1 + \sigma_2)^2 = (\hbar^2/2)(3 + \sigma_1 \cdot \sigma_2). \tag{7.13}$$

Thus we obtain

$$P_{s=0} = \frac{1 - \sigma_1 \cdot \sigma_2}{4} \tag{7.14}$$

and
$$P_{s=1} = \frac{3 + \boldsymbol{\sigma}_1 \cdot \boldsymbol{\sigma}_2}{4}. \tag{7.15}$$

We can easily verify that the properties of projection operators, Eqs. (7.7)–(7.9), are satisfied by these operators, using the identity

$$(\boldsymbol{\sigma}_1 \cdot \boldsymbol{\sigma}_2)^2 = \sigma_2{}^2 + i\boldsymbol{\sigma}_1 \cdot \boldsymbol{\sigma}_2 \times \boldsymbol{\sigma}_2 = 3 - 2\boldsymbol{\sigma}_1 \cdot \boldsymbol{\sigma}_2. \tag{7.16}$$

We shall again use the example of the scattering of a spin-$\frac{1}{2}$ particle on a spinless target to illustrate the use of projection operators in analyzing scattering amplitudes. These operators can be used to select eigenstates of quantities that are conserved in collision processes. For instance, in all cases the total angular momentum will be conserved, and one can write an operator that will project eigenstates of the total angular momentum out of states characterized by an orbital angular momentum l and a spin s.

The appropriate operator can be constructed by using the method of Eq. (7.10), giving the projection operator for a state having a total angular momentum J' as

$$P_l^{J's} = \prod_{\substack{J''=|l-s| \\ J'' \neq J'}}^{l+s} \frac{J^2 - J''(J'' + 1)\hbar^2}{J'(J' + 1)\hbar^2 - J''(J'' + 1)\hbar^2}. \tag{7.17}$$

If this operator is applied only upon wave functions $Y_{lm_l}(\theta, \phi) \chi_{sm_s}(\sigma)$ having an orbital angular momentum l and spin s, we can replace J^2 by $\{l(l + 1)\hbar^2 + s(s + 1)\hbar^2 + 2\mathbf{L} \cdot \mathbf{S}\}$. For the special case of $s = \frac{1}{2}$ this gives a very simple result, since J' can take on only two values, $J' = l \pm \frac{1}{2}$. Denoting $P_l^{J'=l \pm \frac{1}{2}, s=\frac{1}{2}}$ by $P_l^\pm$, we have

$$P_l^+ = \frac{l + 1 + \boldsymbol{\sigma} \cdot \mathbf{L}}{2l + 1} \tag{7.18}$$

and

$$P_l^- = \frac{l - \boldsymbol{\sigma} \cdot \mathbf{L}}{2l + 1}. \tag{7.19}$$

These satisfy $P_l^+ + P_l^- = 1$ as well as the other conditions required of projection operators, and one can readily verify that, when acting on $Y_{lm_l}(\theta, \phi) \chi_{\frac{1}{2}m_s}(\sigma)$, they select out the correct states. In fact, it follows from Eqs. (7.4) and (2.8) that, in general,

$$P_l^{J's} Y_{lm_l}(\theta, \phi) \chi_{sm_s}(\sigma) = \langle l, s; m_l m_s \mid l, s; JM \rangle \mathcal{Y}_{JM}^{ls}(\theta, \phi, \sigma). \tag{7.20}$$

We can see from Eqs. (4.1) and (4.9) that the asymptotic form of the radial wave function in this case is

$$u_l^{J=l\pm\frac{1}{2},s=\frac{1}{2}}(r) \xrightarrow[r\to\infty]{} F_l(kr) + \exp(i\delta_l^\pm)\sin\delta_l^\pm \exp[i(kr-\tfrac{1}{2}\pi l)]. \quad (7.21)$$

From this it follows that, if the incident wave is written as

$$\phi_{m_s}(\mathbf{r}, \sigma) = e^{ikz}\chi_{\frac{1}{2}m_s}(\sigma)$$
$$= (kr)^{-1}\sum_l i^l [4\pi(2l+1)]^{1/2} F_l(kr)(P_l^+ + P_l^-)$$
$$\times Y_{l,0}(\theta)\chi_{\frac{1}{2}m_s}(\sigma), \quad (7.22)$$

the scattered wave is asymptotically

$$\psi_{sc,m_s}(\mathbf{r}, \sigma) \xrightarrow[r\to\infty]{} (kr)^{-1} e^{ikr} \sum_{l=0}^{\infty} [4\pi(2l+1)]^{1/2}$$
$$\times \{\exp(i\delta_l^+)\sin\delta_l^+ P_l^+ + \exp(i\delta_l^-)\sin\delta_l^- P_l^-\}$$
$$\times Y_{l,0}(\theta)\chi_{\frac{1}{2}m_s}(\sigma). \quad (7.23)$$

We may then write the scattering amplitude as the spin operator

$$f(\theta,\phi) = k^{-1}\sum_{l=0}^{\infty} [4\pi(2l+1)]^{1/2}\{\exp(i\delta_l^+)\sin\delta_l^+ P_l^+$$
$$+ \exp(i\delta_l^-)\sin\delta_l^- P_l^-\}Y_{l,0}(\theta), \quad (7.24)$$

where $P_l^\pm$ are given by Eqs. (7.18) and (7.19).

We have previously shown that the scattering amplitude may be written as a spin operator in the form of Eq. (6.29). We now wish to relate the amplitudes $g(\theta)$ and $h(\theta)$ to the amplitudes given in Eq. (7.24) and hence to the phase shifts. We first observe that the following identity holds:

$$\boldsymbol{\sigma}\cdot\mathbf{L}\, Y_{l,0}(\theta) = i\frac{\partial}{\partial\theta}Y_{l,0}(\theta)\,\boldsymbol{\sigma}\cdot\hat{\boldsymbol{\xi}}. \quad (7.25)$$

Then, if we express $h(\theta)$ as

$$h(\theta) = \frac{\partial}{\partial\theta}H(\theta), \quad (7.26)$$

and expand the quantities $g(\theta)$ and $H(\theta)$ as Legendre series

$$g(\theta) = \sum_{l=0}^{\infty} g_l Y_{l,0}(\theta) \quad (7.27)$$

and

$$H(\theta) = \sum_{l=0}^{\infty} H_l Y_{l,0}(\theta), \quad (7.28)$$

the scattering amplitude is

$$f(\theta, \phi) = \sum_{l=0}^{\infty} \left(g_l + H_l \frac{\partial}{\partial \theta} \boldsymbol{\sigma} \cdot \hat{\boldsymbol{\xi}} \right) Y_{l,0}(\theta)$$

$$= \sum_{l=0}^{\infty} (g_l - iH_l \boldsymbol{\sigma} \cdot \mathbf{L}) Y_{l,0}(\theta). \tag{7.29}$$

From Eqs. (7.18) and (7.19) we have the identity

$$\boldsymbol{\sigma} \cdot \mathbf{L} = lP_l^+ - (l+1)P_l^-, \tag{7.30}$$

so that the scattering amplitude becomes

$$f(\theta, \phi) = \sum_{l=0}^{\infty} \{[g_l - ilH_l]P_l^+ + [g_l + i(l+1)H_l] P_l^-\} Y_{l,0}(\theta). \tag{7.31}$$

Comparison of this with Eq. (7.24) yields the desired relationship, viz.,

$$g_l = k^{-1} [4\pi/(2l+1)]^{1/2} \{(l+1) \exp(i\delta_l^+) \sin \delta_l^+ + l \exp(i\delta_l^-) \sin \delta_l^-\} \tag{7.32}$$

and

$$H_l = k^{-1} [(4\pi/2l+1)]^{1/2} \{\exp(i\delta_l^+) \sin \delta_l^+ - \exp(i\delta_l^-) \sin \delta_l^-\}. \tag{7.33}$$

There are many other examples where the use of projection operators can effect an economy in the arithmetical manipulation and can clarify the physical situation.

Chapter

12

Applications

In this chapter we shall apply the theoretical framework developed in previous chapters to some particular scattering situations. The discussion in each case will be carried to the point where numerical data from actual experiments could be introduced, but this will not be done in the general presentation given here.

1. The Two-Potential Formula

We have observed in earlier chapters that there are many scattering problems in which the interaction between the projectile and the target decomposes naturally into two parts: $V = V_0 + V_1$. This division is especially useful if the scattering wave function under the action of one part can be obtained exactly, while the effect of the other can be treated in some approximation.

For a complex system the unperturbed Hamiltonian H_0 is frequently taken to be the sum of a part which determines the internal characteristics of the projectile and the target and another, the kinetic energy operator, which describes the relative motion of these two particles. The usual form for the perturbing interaction is a potential V depending on some internal coordinates of the two particles and on their relative separation. The potential V_0 is chosen to depend only on the relative separation of the two particles. In that case the Hamiltonian $H_0 + V_0$ is the sum of an internal part and a part depending only on the relative separation of the projectile and target. This Hamiltonian leads to an elastic scattering problem of the type that can be solved by the methods of the earlier chapters, and the solution is a state vector describing purely elastic scattering of the two complex particles.

An important feature of the two-potential approach is that it allows us to take account of much of the elastic scattering that occurs during *any* collision process. Even if a particular reaction process occurs only weakly, there may be strong interactions between the constituents during the collision. The *elastic* scattering resulting from these interactions is important and is in large measure calculable. The two-potential formula provides a mechanism for including these elastic scattering effects. This formula also leads to the so-called "distorted-wave Born approximation" and, in other circumstances, to the method of the "final-state interaction." In each case the goal is, as far as possible, to reduce a many-body problem to a sequence of soluble problems describing two-body elastic scattering.

Derivation of the Two-Potential Formula

Our ability to solve for the scattering in a two-body potential V_0 can be exploited by introducing the outgoing- and ingoing-wave solutions $\chi_f^{(\pm)}$ which satisfy the equation

$$\chi_f^{(\pm)} = \phi_f + \frac{1}{E - H_0 \pm i\varepsilon} V_0 \chi_f^{(\pm)}, \qquad (1.1\text{a})$$

or, equivalently,

$$\chi_f^{(\pm)} = \phi_f + \frac{1}{E - H_0 - V_0 \pm i\varepsilon} V_0 \phi_f. \qquad (1.1\text{b})$$

The spatial representations of these vectors are sometimes called, for convenience, "distorted wave functions," while V_0 is the "distorting potential."

The T matrix is

$$\begin{aligned}T_{fi} &= \langle \phi_f | V | \psi_i^{(+)} \rangle \equiv \langle \phi_f | V_0 + V_1 | \psi_i^{(+)} \rangle \\ &= \langle \phi_f | V_0 | \psi_i^{(+)} \rangle + \langle \phi_f | V_1 | \psi_i^{(+)} \rangle. \end{aligned} \qquad (1.2)$$

If we use Eq. (1.1b) to replace ϕ_f with $\chi_f^{(-)}$ in the second term of this formula, we find

$$\begin{aligned}T_{fi} =\ & \langle \phi_f | V_0 | \psi_i^{(+)} \rangle + \langle \chi_f^{(-)} | V_1 | \psi_i^{(+)} \rangle \\ & - \langle \phi_f | V_0 \frac{1}{E - H_0 - V_0 \pm i\varepsilon} V_1 | \psi_i^{(+)} \rangle. \end{aligned} \qquad (1.3)$$

The first and third terms can be combined if the state vector $\psi_i^{(+)}$ is rewritten in terms of the distorted-wave functions $\chi_i^{(+)}$. To accomplish this, we replace the free-particle Green's function with the Green's function for a particle

1. The Two-Potential Formula

moving in the potential V_0 by using

$$\frac{1}{E - H_0 + i\varepsilon} = \frac{1}{E - H_0 - V_0 + i\varepsilon} - \frac{1}{E - H_0 - V_0 + i\varepsilon} V_0 \frac{1}{E - H_0 + i\varepsilon}. \quad (1.4)$$

By inserting Eq. (1.4) into the integral equation for $\psi_i^{(+)}$, we obtain the result

$$\begin{aligned}
\psi_i^{(+)} &= \phi_i + \frac{1}{E - H_0 + i\varepsilon}(V_0 + V_1)\psi_i^{(+)} \\
&= \phi_i + \frac{1}{E - H_0 - V_0 + i\varepsilon}(V_0 + V_1)\psi_i^{(+)} \\
&\quad - \frac{1}{E - H_0 - V_0 + i\varepsilon} V_0 \frac{1}{E - H_0 + i\varepsilon}(V_0 + V_1)\psi_i^{(+)} \\
&= \phi_i + \frac{1}{E - H_0 - V_0 + i\varepsilon}(V_0 + V_1)\psi_i^{(+)} \\
&\quad - \frac{1}{E - H_0 - V_0 + i\varepsilon} V_0(\psi_i^{(+)} - \phi_i) \\
&= \chi_i^{(+)} + \frac{1}{E - H_0 - V_0 + i\varepsilon} V_1 \psi_i^{(+)}. \quad (1.5)
\end{aligned}$$

Thus, if $\chi_i^{(+)}$ is treated as the incident state, scattering out of this state is produced by V_1 alone. This expression is the generalization, in state vector terms, of Eq. (1.17) of Chapter 5.

With this result, the T matrix may now be written as

$$T_{fi} = \langle \phi_f | V_0 | \chi_i^{(+)} \rangle + \langle \chi_f^{(-)} | V_1 | \psi_i^{(+)} \rangle. \quad (1.6)$$

This is the so-called "two-potential formula." It is a generalization of our previous result for two-body elastic scattering, Eq. (1.22) of Chapter 5. The first term is the scattering amplitude for the potential V_0 alone; in many cases it is either known or calculable. The second term incorporates the effect of the "residual potential" V_1 and must often be treated approximately.

Equation (1.6) may be written in a suggestive way by introducing the explicit form of Eq. (1.5),

$$\psi_i^{(+)} = \chi_i^{(+)} + \frac{1}{E - H + i\varepsilon} V_1 \chi_i^{(+)}. \quad (1.7)$$

(This can be derived, as in the simpler case treated in Section 2, Chapter 6, by expressing $1/(E - H_0 - V_0 + i\varepsilon)$ in terms of $1/(E - H + i\varepsilon)$.) Inserting this into Eq. (1.6), we have

$$T_{fi} = \langle \phi_f | V_0 | \chi_i^{(+)} \rangle$$
$$+ \langle \chi_f^{(-)} | V_1 + V_1 \frac{1}{E - H_0 - V_0 - V_1 + i\varepsilon} V_1 | \chi_i^{(+)} \rangle. \quad (1.8)$$

The operator in the second term has a structure identical with that of a transition operator for V_1 (cf. Eq. (1.17) of Chapter 7), except that the Hamiltonian is $H_0 + V_0 + V_1$ rather than simply $H_0 + V_1$. One may then interpret this term as the scattering amplitude describing scattering by the potential V_1 *in the presence of the potential V_0*.

The interaction V_1 may be sufficiently weak so that it can be treated in a first-order approximation. If we maintain our exact treatment of V_0 while treating V_1 only approximately, then Eq. (1.5) tells us to approximate $\psi_i^{(+)}$ by $\chi_i^{(+)}$, so that the two-potential formula becomes

$$T_{fi} \simeq \langle \phi_f | V_0 | \chi_i^{(+)} \rangle + \langle \chi_f^{(-)} | V_1 | \chi_i^{(+)} \rangle. \quad (1.9)$$

The first term is the exact T matrix for the potential V_0. The second term is a generalization of the Born approximation in which distorted-wave functions are used to calculate the scattering due to V_1 in the presence of V_0; it is often called the "distorted-wave Born approximation" or, more simply, the "distorted-wave" approximation. For instance, in the scattering of protons from nuclei, one might choose V_0 to be the Coulomb potential and V_1 to be the short-range nuclear force. The second term in Eq. (1.9) approximates the scattering due to the nuclear force, taking account of the Coulomb repulsion and the consequent reduction in the amplitude of the wave function at small separations.

REARRANGEMENT COLLISIONS

A result analogous to Eq. (1.6) holds for rearrangement processes also. For a rearranged system the potential decomposition is $V' = V_0' + V_1'$, and the distorted-wave function is defined by

$$\chi_{f'}^{(-)} = \phi_{f'} + \frac{1}{E - H_0' - i\varepsilon} V_0' \chi_{f'}^{(-)}$$
$$= \phi_{f'} + \frac{1}{E - H_0' - V_0' - i\varepsilon} V_0' \phi_{f'}. \quad (1.10)$$

1. The Two-Potential Formula

With these definitions the two-potential formula for rearrangement collisions is

$$\begin{aligned} T_{f'i} &= \langle \phi_{f'} | T | \phi_i \rangle \\ &= \langle \chi_{f'}^{(-)} | V_1' | \psi_i^{(+)} \rangle. \end{aligned} \quad (1.11)$$

To obtain the result given in Eq. (1.11), we proceed just as in the derivation of Eq. (1.6). The T matrix is

$$\begin{aligned} T_{f'i} &= \langle \phi_{f'} | V' | \psi_i^{(+)} \rangle \\ &= \langle \phi_{f'} | V_0' | \psi_i^{(+)} \rangle + \langle \phi_{f'} | V_1' | \psi_i^{(+)} \rangle. \end{aligned} \quad (1.12)$$

From Eq. (1.10) we write

$$\phi_{f'} = \chi_{f'}^{(-)} - \frac{1}{E - H_0' - V_0' - i\varepsilon} V_0' \phi_{f'}, \quad (1.13)$$

so that Eq. (1.12) may be re-expressed as

$$\begin{aligned} T_{f'i} = \langle \phi_{f'} | V_0' | \psi_i^{(+)} \rangle + \langle \chi_{f'}^{(-)} | V_1' | \psi_i^{(+)} \rangle \\ - \langle \phi_{f'} | V_0' \frac{1}{E - H_0' - V_0' + i\varepsilon} V_1' | \psi_i^{(+)} \rangle. \end{aligned} \quad (1.14)$$

If we now make use of the relation

$$\frac{1}{E - H_0 + i\varepsilon} = \frac{1}{E - H_0' - V_0' + i\varepsilon}$$

$$- \frac{1}{E - H_0' - V_0' + i\varepsilon} (V - V_1') \frac{1}{E - H_0 + i\varepsilon}, \quad (1.15)$$

we may express the state vector $\psi_i^{(+)}$ in the form

$$\begin{aligned} \psi_i^{(+)} &= \phi_i + \frac{1}{E - H_0 + i\varepsilon} V \psi_i^{(+)} \\ &= \phi_i + \frac{1}{E - H_0' - V_0' + i\varepsilon} (V - V_1') \phi_i \\ &\quad + \frac{1}{E - H_0' - V_0' + i\varepsilon} V_1' \psi_i^{(+)}. \end{aligned} \quad (1.16)$$

When the potential V_0' is chosen as described at the beginning of this section, so that it depends only on the relative separation of the two particles in the state $\phi_{f'}$, the first two terms on the right-hand side of Eq. (1.16) will cancel identically. This can be demonstrated by the same procedure as was used in Chapter 7 to derive the "final state" form of the scattering state vector. Equation (1.16) is just the two-potential form of that earlier result, and exactly the same arguments show that it reduces to

$$\psi_i^{(+)} = \frac{1}{E - H_0' - V_0' + i\varepsilon} V_1' \psi_i^{(+)}. \tag{1.17}$$

The physical reason for the vanishing of these terms is clear. The potential V_0' cannot alone lead to a rearrangement process. In fact, V_0' is so chosen that the scattering induced by it is simply two-body elastic scattering involving, by definition, no rearrangement of the internal constituents of the various particles concerned. Then the terms involving only this potential cannot contribute to rearrangements, and we have the simple result

$$T_{f'i} = \langle \chi_{f'}^{(-)} | V_1' | \psi_i^{(+)} \rangle. \tag{1.18}$$

An important result of these manipulations is the recognition that the potential V_0' does not contribute directly to the rearrangement process. This was not apparent in the original expression for the T matrix, but it is now clear that such a potential contributes only to elastic distortion of the wave function in the final state. We can take maximum advantage of this fact by using Eq. (1.18), which isolates the real cause of the rearrangement process, the potential V_1'.

In many applications it is convenient to use for V_0' a potential which, in coordinate space, has an imaginary component. (The rationale for this is discussed in Section 5.) Such a potential would, of course, not be Hermitian, but it is still possible to use Eq. (1.18) if an appropriate modification of the ingoing wave function is introduced. One can see, by tracing through the derivation again, that Eq. (1.18) will again be obtained if, in Eq. (1.10) defining $\chi_{f'}^{(-)}$, V_0' is replaced by its adjoint $V_0'^{\dagger}$. This adjoint is, in a sense, the time-reversed potential and is associated with the time-reversed state $\chi_{f'}^{(-)}$. (We saw in Chapter 10 that the time-reversal operation corresponds in coordinate space to complex conjugation, and an ingoing wave state is obtained by applying time reversal to an outgoing wave state.) Complex potentials can then be used in a straightforward manner.

Computations are often made simpler by using an "initial-state" form of the two-potential formula rather than the "final-state" form we have derived. Using the reciprocal relations, we obtain, instead of Eq. (1.18), the reciprocal form

$$T_{f'i} = \langle \psi_{f'}^{(-)} | V_1 | \chi_i^{(+)} \rangle, \tag{1.19}$$

with $\chi_i^{(+)}$ given by Eq. (1.1).

As in the ordinary scattering case, these results are most useful when one of the potentials in the initial or final state is weak, while the other leads to a soluble scattering problem. If this is the case, then we can, for instance, approximate $\psi_i^{(+)}$ by $\chi_i^{(+)}$ in Eq. (1.18) to obtain appropriate "distorted wave" approximations. This relation then becomes simply

$$T_{f'i} \simeq \langle \chi_{f'}^{(-)} | V_1' | \chi_i^{(+)} \rangle. \tag{1.20}$$

Similarly, Eq. (1.19) becomes

$$T_{f'i} \simeq \langle \chi_{f'}^{(-)} | V_1 | \chi_i^{(+)} \rangle. \tag{1.21}$$

In this case the interaction responsible for the rearrangement collision is the same for all final states.

One can easily show that these two results are identical, provided $\chi_i^{(+)}$ and $\chi_{f'}^{(-)}$ are chosen to be the same in both cases. There are many cases, however, when one formulation is more convenient and more accurate than the other. Furthermore, the wave functions need not be chosen this way; for instance, in Eq. (1.20), $\chi_i^{(+)}$ can be replaced by any suitable approximation to the exact scattering state, and similarly for $\chi_{f'}^{(-)}$ in Eq. (1.21).

Higher-order corrections to Eqs. (1.20) and (1.21) can be calculated in various ways, such as by including low-lying excited states of the target. These usually involve straightforward but tedious extensions of the present methods.

2. Some Examples of the Two-Potential Approach—The Distorted-Wave Approximation

Let us now consider some applications of these two-potential formulas. We can illustrate the principles involved by using the three-particle situation described at the beginning of Chapter 7. To be definite, we will discuss the case of a neutron scattering on a target consisting of a heavy core plus a proton. Of course, the same general approach is applicable to many other different physical situations.

The processes we shall examine are

$$\begin{aligned}
N + (P, C) &\to N + (P, C) & &\text{Elastic scattering} \\
&\to N + (P, C)^* & &\text{Inelastic scattering} \\
&\to P + (N, C) & &\text{Exchange scattering} \\
&\to (N, P) + C & &\text{Pickup reaction} \\
&\to P + N + C & &\text{Breakup reaction} \\
&\to N + (P, C) + \gamma & &\text{Bremsstrahlung.}
\end{aligned}$$

Since the solution of this full collision problem would require the complete solution of a three-body problem, only approximate solutions can be expected. We shall consider the case in which the neutron-proton potential V_{NP} is sufficiently weak that it need be treated only to first order, and we can use the "distorted-wave" approximation. As we shall see, this also implies that the various reaction processes are weak, so that their effects upon each other can be neglected. These couplings are at least of second order in V_{NP}. If terms of this order must be considered, some more precise method must be used.

All positions are considered to be measured from the center of the core, which is assumed to be infinitely massive. The initial-state wave function is then the product of a plane wave $e^{i\mathbf{k}\cdot\mathbf{r}_N}$ describing the motion of the neutron and a bound-state wave function $\eta_\alpha(\mathbf{r}_P)$ for the proton:

$$\phi_i(\mathbf{r}_N, \mathbf{r}_P) = e^{i\mathbf{k}\cdot\mathbf{r}_N}\, \eta_\alpha(\mathbf{r}_P). \tag{2.1}$$

This wave function satisfies the Schrödinger equation

$$\left[-\frac{\hbar^2}{2m_N}\nabla_N^2 - \frac{\hbar^2}{2m_P}\nabla_P^2 + V_{PC}(\mathbf{r}_P)\right]\phi_i(\mathbf{r}_N, \mathbf{r}_P) = E\,\phi_i(\mathbf{r}_N, \mathbf{r}_P), \tag{2.2}$$

with $V_{PC}(\mathbf{r}_P)$ the interaction between the proton and the core. Thus, if we denote the kinetic energy operator for particle j by K_j, the initial-state "free-particle" Hamiltonian is

$$H_0 = K_N + K_P + V_{PC}, \tag{2.3}$$

while in this situation the initial-state interaction is

$$V = V_{NC} + V_{NP}. \tag{2.4}$$

It follows from the form of H_0 that the distorted wave function $\chi_i^{(+)}(\mathbf{r}_N, \mathbf{r}_P)$ can be factored, and thus computed exactly in principle, if we choose the distorting potential to be

$$V_0 = V_{NC}$$

and the residual interaction to be

$$V_1 = V_{NP}. \tag{2.5}$$

As we expect, V_0 depends only on the relative coordinate $\mathbf{r}_N$ of the two initial particles, and not on the "internal" coordinate $\mathbf{r}_P$. In other words, only the neutron wave is affected by this potential. With this choice, $H_0 + V_0$ is the sum of a neutron part and a proton part, and the distorted wave function

2. The Distorted-Wave Approximation

is the product

$$\chi_i^{(+)}(\mathbf{r}_N, \mathbf{r}_P) = \chi_{\mathbf{k}}^{(+)}(\mathbf{r}_N)\, \eta_\alpha(\mathbf{r}_P). \tag{2.6}$$

The scattering function for the neutron satisfies

$$\left[-\frac{\hbar^2}{2m_N}\nabla_N^2 + V_{NC}(\mathbf{r}_N)\right]\chi_{\mathbf{k}}^{(+)}(\mathbf{r}_N) = \frac{\hbar^2 k^2}{2m_N}\chi_{\mathbf{k}}^{(+)}(\mathbf{r}_N) \tag{2.7}$$

and has the asymptotic form

$$\chi_{\mathbf{k}}^{(+)}(\mathbf{r}_N) \xrightarrow[r_N \to \infty]{} e^{i\mathbf{k}\cdot\mathbf{r}_N} + \frac{e^{ikr_N}}{r_N} f_N(\theta_N). \tag{2.8}$$

Let us see formally how this factorization comes about. The initial-state wave function is the coordinate representation of the product $\phi_{\mathbf{k}}\eta_\alpha$, where $\phi_{\mathbf{k}}$ is a momentum eigenvector in the space of all possible neutron states, and η_α is an eigenvector in the space of proton states. The product is a vector in the "product space" of all states of protons and neutrons together.

If we denote the neutron and proton parts of $H_0 + V_0$ by H_N and H_P, respectively, then the explicit form for the state vector $\chi_i^{(+)}$ is

$$\chi_i^{(+)} = \phi_{\mathbf{k}}\eta_\alpha + \frac{1}{E - H_N - H_P + i\varepsilon} V_{NC}\phi_{\mathbf{k}}\eta_\alpha. \tag{2.9}$$

Since V_{NC} does not contain the coordinates of the proton, H_P can operate "through" it onto η_α, giving the energy of the bound state, $-\varepsilon_\alpha = E - \hbar^2 k^2/2m_N$. Performing this operation, Eq. (2.9) becomes

$$\chi_i^{(+)} = \left(\phi_{\mathbf{k}} + \frac{1}{(\hbar^2 k^2/2m_N) - H_N + i\varepsilon} V_{NC}\phi_{\mathbf{k}}\right)\eta_\alpha = \chi_{\mathbf{k}}^{(+)}\eta_\alpha, \tag{2.10}$$

the state vector form of Eq. (2.6).

Elastic and Inelastic Scattering

For elastic and inelastic scattering ("ordinary" scattering), the initial- and final-state interactions are identical. The final-state free-particle wave function is

$$\phi_f(\mathbf{r}_N, \mathbf{r}_P) = \exp(i\mathbf{k}'\cdot\mathbf{r}_N)\, \eta_{\alpha'}(\mathbf{r}_P), \tag{2.11}$$

where, for elastic scattering, the quantum number α' is equal to α. Energy conservation is expressed by the relation

$$E = \frac{\hbar^2 k^2}{2m_N} - \varepsilon_\alpha = \frac{\hbar^2 k'^2}{2m_N} - \varepsilon_{\alpha'}. \tag{2.12}$$

Using the first-order approximation to the two-potential formula, Eq. (1.9), we obtain the T matrix for elastic scattering,

$$T_{fi} \simeq \int d^3r_N \exp(-i\mathbf{k}' \cdot \mathbf{r}_N) \, V_{NC}(\mathbf{r}_N) \, \chi_\mathbf{k}^{(+)}(\mathbf{r}_N)$$
$$+ \int d^3r_N \, d^3r_P \, \chi_{\mathbf{k}'}^{(-)*}(\mathbf{r}_N) \, \eta_\alpha(\mathbf{r}_P) \, V_{NP}(\mathbf{r}_N - \mathbf{r}_P) \, \chi_\mathbf{k}^{(+)}(\mathbf{r}_N) \, \eta_\alpha(\mathbf{r}_P), \quad (2.13)$$

under the assumption that the bound-state wave function is normalized to unity and is real:

$$\int d^3r_P \, \eta_\alpha^2(\mathbf{r}_P) = 1. \quad (2.14)$$

As we expected, the first term is proportional to the scattering amplitude for V_{NC}. (It is, in fact, just $-(4\pi\hbar^2/2m_N) f_N(\theta_N)$.) The second term gives the effect of the direct neutron-proton collision via V_{NP} and is a multiple integral over the positions of the proton and neutron. Only in special circumstances, such as when $V_{NP}(\mathbf{r}_N - \mathbf{r}_P)$ is a zero-range force, does it reduce to a single integral. It may be noted, though, that the proton integral gives simply the average over the bound-state wave function of the $N-P$ interaction. The first-order approximation neglects any change in the bound-state wave function during the collision with the incident neutron and leads to this simple average.

This suggests an alternative choice for the distorting potential V_0 which can simplify the calculation. Suppose we were to choose for this potential the sum of the neutron-core interaction V_{NC} and the average of V_{NP} over the proton bound-state wave function. That is, assuming that V_{NP} contains no gradients, we let

$$V_0(\mathbf{r}_N) = V_{NC}(\mathbf{r}_N) + \int d^3r_P \, \eta_\alpha^2(\mathbf{r}_P) \, V_{NP}(\mathbf{r}_N - \mathbf{r}_P)$$

and

$$V_1(\mathbf{r}_N, \mathbf{r}_P) = V_{NP}(\mathbf{r}_N - \mathbf{r}_P) - \int d^3r_P \, \eta_\alpha^2(\mathbf{r}_P) \, V_{NP}(\mathbf{r}_N - \mathbf{r}_P). \quad (2.15)$$

The distorted wave function can still be factored, since V_0 depends only on the neutron coordinates, and Eq. (1.9) now gives for the first-order approximation to elastic scattering

$$T_{fi} \simeq \int d^3r_N \exp(i\mathbf{k}' \cdot \mathbf{r}_N) \, V_0(\mathbf{r}_N) \, \chi_\mathbf{k}^{(+)}(\mathbf{r}_N). \quad (2.16)$$

There is *no* correction for the direct neutron-proton interaction, since the average of V_1 is zero. The effect of the neutron-proton force is now included, to at least first order, in the neutron scattering amplitude, and corrections arising from perturbations of the bound-state wave function occur only in second order. This result is completely analogous to the corresponding result in the Hartree approach to atomic structure, and $V_0(\mathbf{r}_N)$ in Eq. (2.15) is in fact the analog of the Hartree potential customarily used for bound-state problems.

It is important to remember that *any* potential depending solely on $\mathbf{r}_N$ could have been used for V_0. The choice of Eq. (2.5) is the simplest, and Eq. (2.15) is the choice that causes the first-order correction to elastic scattering to vanish, but any potential $V_0(\mathbf{r}_N)$ could have been used.

Turning now to inelastic scattering, any choice of V_0 that depends only on $\mathbf{r}_N$ will cause the first term in Eq. (1.9) to vanish, because of the orthogonality of the bound-state wave functions:

$$\int d^3 r_N \, d^3 r_P \, \exp(-i\mathbf{k}' \cdot \mathbf{r}_N) \, \eta_{\alpha'}(\mathbf{r}_P) \, V_0(\mathbf{r}_N) \, \chi_{\mathbf{k}}^{(+)}(\mathbf{r}_N) \, \eta_\alpha(\mathbf{r}_P) = 0 \quad \text{for} \quad \alpha' \neq \alpha. \tag{2.17}$$

The T matrix is then, for any choice of $V_0(\mathbf{r}_N)$,

$$T_{fi} \simeq \int d^3 r_N \, d^3 r_P \, \chi_{\mathbf{k}'}^{(-)*}(\mathbf{r}_N) \, \eta_{\alpha'}(\mathbf{r}_P) \, V_{NP}(\mathbf{r}_N - \mathbf{r}_P) \, \chi_{\mathbf{k}}^{(+)}(\mathbf{r}_N) \, \eta_\alpha(\mathbf{r}_P). \tag{2.18}$$

Only the neutron-proton interaction contributes directly to an inelastic process. Without this interaction, there would be no way of transferring energy from the neutron to the bound proton to bring it into an excited state. The remaining interaction of the neutron with the target, represented by $V_0(\mathbf{r}_N)$, serves only to distort the incident and outgoing neutron wave; it does not directly contribute to the excitation process.

This conclusion depends on the assumption that the core is infinitely heavy. If this is not so, then the foregoing requires some modification. A convenient set of coordinates for this general case is provided by the center-of-mass coordinate and two relative coordinates, the projectile-target separation and the internal coordinate of the target:

$$\mathbf{R} = \frac{m_N \mathbf{r}_N + m_P \mathbf{r}_P + m_C \mathbf{r}_C}{m_N + m_P + m_C}$$

$$\mathbf{r} = \mathbf{r}_N - \frac{m_P \mathbf{r}_P + m_C \mathbf{r}_C}{m_P + m_C} \tag{2.19}$$

$$\mathbf{s} = \mathbf{r}_P - \mathbf{r}_C.$$

The kinetic energy operator separates when these coordinates are used:

$$-\frac{\hbar^2}{2m_N}\nabla_N^2 - \frac{\hbar^2}{2m_P}\nabla_P^2 - \frac{\hbar^2}{2m_C}\nabla_C^2 = -\frac{\hbar^2}{2M}\nabla_R^2 - \frac{\hbar^2}{2m}\nabla_r^2 - \frac{\hbar^2}{2\mu}\nabla_s^2, \quad (2.20)$$

with

$$M = m_N + m_P + m_C$$

$$m = \frac{m_N(m_P + m_C)}{M} \quad (2.21)$$

$$\mu = \frac{m_P m_C}{m_P + m_C}.$$

Furthermore, the potential energy will not depend on **R**, the center-of-mass coordinate, because of the translation invariance of the process (see Section 1, Chapter 9). The initial-state wave function is then, in the center-of-mass system,

$$\phi_i(\mathbf{r}_N, \mathbf{r}_P, \mathbf{r}_C) = e^{i\mathbf{k}\cdot\mathbf{r}}\eta_\alpha(\mathbf{s}). \quad (2.22)$$

The considerations of Eqs. (2.9) and (2.10) imply that the distorted wave function will factor into two parts if the scattering potential V_0 depends upon **r**, the relative position of the neutron and the target, but not upon **s**, the internal coordinate of the target. If such a choice is made for V_0, the distorted wave function is

$$\chi_i^{(+)}(\mathbf{r}, \mathbf{s}) = \chi_k^{(+)}(\mathbf{r})\eta_\alpha(\mathbf{s}), \quad (2.23)$$

with the scattering function satisfying

$$\left[-\frac{\hbar^2}{2m}\nabla_r^2 + V_0(\mathbf{r})\right]\chi_k^{(+)}(\mathbf{r}) = \frac{\hbar^2 k^2}{2m}\chi_k^{(+)}(\mathbf{r}). \quad (2.24)$$

The previous discussion of elastic and inelastic scattering can now be repeated using this more general form. However, this is not necessary, since the results are identical. Suffice it to say that one can generalize from the case of an infinitely heavy core and still retain the same basic conclusions regarding the structure of the scattering amplitude.

EXCHANGE SCATTERING

Exchange scattering is the first rearrangement process we will deal with, and it is the simplest case to treat. The proton and neutron exchange roles with the neutron bound and the proton free in the final state. The over-all structure of the wave function is then identical with that which we have discussed previously.

The initial-state potential is

$$V = V_{NC} + V_{NP}; \quad (2.25a)$$

the final state-potential is

$$V' = V_{PC} + V_{NP}. \quad (2.25b)$$

The final-state free-particle Hamiltonian is

$$H_0' = K_N + K_P + V_{NC}. \quad (2.26)$$

Then any potential V_0' depending only on the coordinate of the outgoing proton will give a Hamiltonian $H_0' + V_0'$ which is the sum of a neutron part and a proton part. For such a potential the distorted-wave function is

$$\chi_{f'}^{(-)}(\mathbf{r}_N, \mathbf{r}_P) = \eta_\beta(\mathbf{r}_N) \chi_{\mathbf{k}'}^{(-)}(\mathbf{r}_P), \quad (2.27)$$

with the single-particle functions satisfying

$$\left[-\frac{\hbar^2}{2m_N}\nabla_N^2 + V_{NC}(\mathbf{r}_N)\right]\eta_\beta(\mathbf{r}_N) = -\varepsilon_\beta \eta_\beta(\mathbf{r}_N). \quad (2.28)$$

$$\left[-\frac{\hbar^2}{2m_P}\nabla_P^2 + V_0'(\mathbf{r}_P)\right]\chi_{\mathbf{k}'}^{(-)}(\mathbf{r}_P) = \frac{\hbar^2 k'^2}{2m_P} \chi_{\mathbf{k}'}^{(-)}(\mathbf{r}_P), \quad (2.29)$$

and $\hbar^2 k'^2/2m_P - \varepsilon_\beta = E$.

Using Eq. (2.6) for the initial state function, both Eq. (1.20) and Eq. (1.21) give

$$T_{f'i} \simeq \int d^3 r_N \, d^3 r_P \, \eta_\beta(\mathbf{r}_N) \chi_{\mathbf{k}'}^{(-)*}(\mathbf{r}_P) V_{NP}(\mathbf{r}_N - \mathbf{r}_P) \chi_{\mathbf{k}}^{(+)}(\mathbf{r}_N)\eta_\alpha(\mathbf{r}_P). \quad (2.30)$$

As in the case of inelastic scattering, only the direct $N-P$ interaction contributes to exchange scattering. Only a direct collision of the neutron with the proton can lead to the ejection of the proton, when the core is very heavy. Mathematically, the other contributions vanish because of the orthogonality of the wave functions given by Eqs. (2.6) and (2.27). Thus $\eta_\beta(\mathbf{r}_N)$ and $\chi_{\mathbf{k}}^{(+)}(\mathbf{r}_N)$ are eigenfunctions of the same Hamiltonian for different energies and are orthogonal.

If the core has a finite mass, then the neutron could collide with the core through V_{NC}, and the recoiling target could "shake off" the proton. As a result, in the finite-mass case there will be a nonzero contribution from V_{NC} as well as from V_{NP}, but in the infinite-mass case this does not appear.

Pickup Reaction

In this process the incident neutron captures, or picks up, the bound proton, forming a deuteron. The inverse of this process is a stripping reaction, with the proton stripped from the "incident" deuteron and captured by the core. Because of time-reversal symmetry, both of these processes are described by the same matrix element, which we will now write down in the "distorted wave" approximation.

The final-state free-particle Hamiltonian is

$$H_0' = K_N + K_P + V_{NP}, \tag{2.31}$$

and the final-state interaction is

$$V' = V_{NC} + V_{PC}. \tag{2.32}$$

Even though the neutron and proton are bound together in the final state, we assume that they continue to interact independently with the core, that is, that only two-body forces exist.

In the final state the natural coordinates are no longer the separate neutron and proton coordinates, but rather the deuteron center-of-mass coordinates and the internal coordinate of the deuteron. Denoting these by

$$\mathbf{r} = \frac{m_N \mathbf{r}_N + m_P \mathbf{r}_P}{m_N + m_P}$$

and

$$\mathbf{s} = \mathbf{r}_N - \mathbf{r}_P, \tag{2.33}$$

the kinetic energy operator becomes

$$K_N + K_P = -\frac{\hbar^2}{2m_N}\nabla_N^2 - \frac{\hbar^2}{2m_P}\nabla_P^2$$

$$= -\frac{\hbar^2}{2(m_N + m_P)}\nabla_r^2 - \frac{\hbar^2}{2\left(\frac{m_N m_P}{m_N + m_P}\right)}\nabla_s^2. \tag{2.34}$$

The final-state wave function $\phi_{f'}(\mathbf{r}_N, \mathbf{r}_P)$ is the product of a plane wave function describing the center-of-mass motion of the deuteron and the internal wave function of the deuteron:

$$\phi_{f'}(\mathbf{r}_N, \mathbf{r}_P) = \exp(i\mathbf{k}' \cdot \mathbf{r})\,\eta_D(\mathbf{s}). \tag{2.35}$$

By analogy with the previous examples, it is clear from this that the final-

2. The Distorted-Wave Approximation

state potential V_0' must be an interaction $V_{DC}(\mathbf{r})$ between the deuteron and the core which depends only on the position of the deuteron, in order that the resulting Schrödinger equation be solvable. The distorted wave function then has the form

$$\chi_{f'}^{(-)}(\mathbf{r}_N, \mathbf{r}_P) = \chi_{\mathbf{k}'}^{(-)}(\mathbf{r})\, \eta_D(\mathbf{s}), \qquad (2.36)$$

with each function separately satisfying an appropriate Schrödinger equation, as in the previous examples.

The approximate T matrix for the process is then, using Eqs. (1.20) and (1.21), respectively,

$$T_{f'i} \simeq \int d^3r_N\, d^3r_P\, \chi_{\mathbf{k}'}^{(-)*}\!\left(\frac{m_N\mathbf{r}_N + m_P\mathbf{r}_P}{m_N + m_P}\right) \eta_D(\mathbf{r}_N - \mathbf{r}_P) \bigg[V_{NC}(\mathbf{r}_N)$$

$$+ V_{PC}(\mathbf{r}_P) - V_{DC}\!\left(\frac{m_N\mathbf{r}_N + m_P\mathbf{r}_P}{m_N + m_P}\right)\bigg] \chi_{\mathbf{k}}^{(+)}(\mathbf{r}_N)\, \eta_\alpha(\mathbf{r}_P) \qquad (2.37)$$

$$\simeq \int d^3r_N\, d^3r_P\, \chi_{\mathbf{k}'}^{(-)*}\!\left(\frac{m_N\mathbf{r}_N + m_P\mathbf{r}_P}{m_N + m_P}\right) \eta_D(\mathbf{r}_P - \mathbf{r}_N)\, V_{NP}(\mathbf{r}_N - \mathbf{r}_P)$$

$$\times \chi_{\mathbf{k}}^{(+)}(\mathbf{r}_N)\, \eta_\alpha(\mathbf{r}_P). \qquad (2.38)$$

These two forms are completely equivalent, as one may see by using the Schrödinger equations for the respective wave functions. Nevertheless, the second is often far easier to use in practice. The interactions in Eq. (2.37) have ranges of the order of the size of the core, so that this form usually involves multiple integrations over large ranges. These have to be performed carefully if the result, which depends on the difference of several terms, is to be obtained with reasonable precision. The second form involves only the relatively short-range $N-P$ potential. In fact, if the range of this potential is short compared to the spacing of characteristic variations in each of the wave functions (i.e., the respective de Broglie wavelengths), Eq. (2.38) can be factored into the product

$$T_{f'i} \simeq \left[\int d^3s\, \eta_D(\mathbf{s})\, V_{NP}(\mathbf{s})\right]\!\left[\int d^3r\, \chi_{\mathbf{k}'}^{(-)*}(\mathbf{r})\, \chi_{\mathbf{k}}^{(+)}(\mathbf{r})\, \eta_\alpha(\mathbf{r})\right]. \qquad (2.39)$$

The second overlap integral, which determines the angular distribution for the pickup reaction, is relatively easy to calculate with computing machines or in various simple approximations. This approximation also has a straightforward physical interpretation; the neutron and proton must be near to each other in order to interact, and the incoming neutron then collides directly with the proton and picks it up. Together they emerge from the target in a bound state.

Corrections to Eq. (2.38) involve modifications of the outgoing deuteron by the interaction of its individual constituents with the core. The full state vector is

$$\psi_{f'}^{(-)} = \chi_{f'}^{(-)} + \frac{1}{E - H_0' - V_{DC} - i\varepsilon}(V_{NC} + V_{PC} - V_{DC})\psi_{f'}^{(-)} \quad (2.40)$$

The first term is included in Eq. (2.38); corrections to it involve the difference between the interaction of the neutron and proton separately with the core and the interaction of the composite deuteron with the core.

BREAKUP REACTION

In a breakup reaction the incoming neutron collides with the target and the neutron and proton emerge but they are not bound together. The final-state free-particle Hamiltonian is then simply the kinetic energy operator

$$H_0' = K_N + K_P, \quad (2.41)$$

and the final-state potential is the full interaction

$$V' = V_{NC} + V_{PC} + V_{NP}. \quad (2.42)$$

The free-particle wave function for the final state is then

$$\phi_{f'}(\mathbf{r}_N, \mathbf{r}_P) = \exp(i\mathbf{k}_N' \cdot \mathbf{r}_N)\exp(i\mathbf{k}_P' \cdot \mathbf{r}_P). \quad (2.43)$$

In this case several approaches are available in selecting a final distorting potential V_0'. Clearly it is not possible to include all three interactions and yet obtain a solvable problem. The interactions with the core can be included by letting $V_0' = V_{NC} + V_{PC}$. The Hamiltonian $H_0' + V_0'$ is then the sum of a neutron part and a proton part, and the distorted wave function is

$$\chi_{f'}^{(-)}(\mathbf{r}_N, \mathbf{r}_P) = \chi_{\mathbf{k}_N'}^{(-)}(\mathbf{r}_N)\chi_{\mathbf{k}_P'}^{(-)}(\mathbf{r}_P). \quad (2.44)$$

Alternatively, one could include the mutual scattering of the neutron and proton by letting $V_0' = V_{NP}$. The free-particle Hamiltonian H_0' can be written in terms of the relative and total kinetic energy of the proton-neutron pair, as in Eq. (2.34). Then the full Hamiltonian $H_0' + V_0'$ again separates and has the solution

$$\chi_{f'}^{(-)}(\mathbf{r}_N, \mathbf{r}_P) = \chi_{f'}^{(-)}(\mathbf{r}, \mathbf{s}) = \exp(i\mathbf{k}' \cdot \mathbf{r})\chi_{\mathbf{k}_{NP}'}^{(-)}(\mathbf{s}), \quad (2.45)$$

2. The Distorted-Wave Approximation

with

$$\frac{\hbar^2 k'^2}{2(m_N + m_P)} + \frac{\hbar^2 k'^2_{NP}}{2\left(\frac{m_N m_P}{m_N + m_P}\right)} = E$$

and

$$\mathbf{k}_{NP'} = \frac{m_P \mathbf{k}_N - m_N \mathbf{k}_P}{m_N + m_P}. \tag{2.46}$$

Of course, one could also add to V_0' an interaction between the outgoing particles and the core, analogous to V_{DC} in the case of a pickup reaction. This would be somewhat artificial, though, since in order to be solvable, this potential could depend only on the center-of-mass coordinate of the unbound pair of outgoing particles. Such a potential would be physically meaningless.

This second approach is a generalization of the previous treatment of the pickup reaction. There the final state was described by a wave function for the bound neutron-proton pair and a wave function describing their center-of-mass motion. Here the wave function is also the product of a wave function describing the relative motion of the neutron and proton, and a function giving the distribution of their center of mass.

If these two alternatives are introduced into Eq. (1.20), we obtain either

$$T_{f'i} \simeq \int d^3 r_N \, dr_P \, \chi_{\mathbf{k}_N}^{(-)*}(\mathbf{r}_N) \, \chi_{\mathbf{k}_P}^{(-)*}(\mathbf{r}_P) \, V_{NP}(\mathbf{r}_N - \mathbf{r}_P)$$
$$\times \chi_{\mathbf{k}}^{(+)}(\mathbf{r}_N) \, \eta_\alpha(\mathbf{r}_P) \tag{2.47}$$

or

$$T_{f'i} \simeq \int d^3 r_N \, d^3 r_P \, \exp(-i\mathbf{k}' \cdot \mathbf{r}) \, \chi_{\mathbf{k}'_{NP}}^{(-)*}(\mathbf{s})$$
$$\times [V_{NC}(\mathbf{r}_N) + V_{PC}(\mathbf{r}_P)] \, \chi_{\mathbf{k}}^{(+)}(\mathbf{r}_N) \, \eta_\alpha(\mathbf{r}_P). \tag{2.48}$$

Another form that is somewhat simpler to use than Eq. (2.48), although it is completely equivalent to it, is obtained by introducing Eq. (2.45) into the initial-state formula, Eq. (1.21):

$$T_{f'i} \simeq \int d^3 r_N \, d^3 r_P \, \exp(-i\mathbf{k}' \cdot \mathbf{r}) \, \chi_{\mathbf{k}'_{NP}}^{(-)*}(\mathbf{s}) \, V_{NP}(\mathbf{s}) \, \chi_{\mathbf{k}}^{(+)}(\mathbf{r}_N) \, \eta_\alpha(\mathbf{r}_P). \tag{2.49}$$

The approach of Eq. (2.47) is quite different from that of Eq. (2.48) and Eq. (2.49). The former includes in the final-state wave function the scattering of each particle by the core, but neglects the mutual neutron-proton interaction; the latter includes this interaction but neglects the interactions with the core. These two approximations are each appropriate in different circumstances, depending upon the momentum of each particle and the strengths of the respective interactions.

There are three separated particles in the final state, and the total energy is shared among them. One or the other particle might carry most of the kinetic energy, or the neutron-proton pair might travel off together. As we know from earlier chapters, the scattering of two particles tends to be strongest when their relative momentum is low. (The interaction "time," the force range divided by the velocity, is greater, leading to a greater chance for a collision.) Then we expect that Eq. (2.47) will be most applicable when either the neutron or the proton has a small momentum relative to the core, and it is important to include the interaction of the slow particle with the core. (Correspondingly, it is less important to include the scattering of the fast particle.) If the relative velocity of the neutron-proton pair is low, then it is more important to include the $N-P$ interaction, and Eq. (2.48) or (2.49) is more appropriate.

Clearly there is a region in the final-state spectrum where none of these approximations is valid, and all interactions must be included. However, this region is of lesser interest. In the absence of any interactions in the final state, all combinations of momenta would be equally likely. The interactions tend to distort this statistical distribution, providing a means of observing the effects of these interactions. It is in the region where one or the other interaction dominates that their most striking effects can be seen.

These effects can be computed quite simply in the special case of a final-state interaction through a short-range force. If the range of V_{NP}, for instance, is short compared to the characteristic oscillations in the radial wave functions, Eq. (2.49) is

$$T_{f'i} \simeq \left[\int d^3s\, \chi_{\mathbf{k}'}^{(-)*}{}_{NP}(\mathbf{s})\, V_{NP}(\mathbf{s})\right]\int d^3r\, \exp(-i\mathbf{k}'\cdot\mathbf{r})\, \chi_{\mathbf{k}}^{(+)}(\mathbf{r})\, \eta_\alpha(\mathbf{r}). \quad (2.50)$$

For those values of the final momenta for which k'_{NP} is small, the first integral is just the T matrix for NP scattering. That is, if R_0 is the range of V_{NP} and $k'_{NP}R_0 \ll 1$, this integral is approximately

$$\int d^3s\, \chi_{\mathbf{k}'}^{(-)*}{}_{NP}(\mathbf{s})\, V_{NP}(\mathbf{s})\, \exp(i\mathbf{k}'_{NP}\cdot\mathbf{s}) = -\frac{4\pi\hbar^2}{2\left(\dfrac{m_N m_P}{m_N + m_P}\right)}\frac{e^{i\delta_0}\sin\delta_0}{k'_{NP}}, \quad (2.51)$$

2. The Distorted-Wave Approximation

where δ_0 is the S-wave phase shift for a relative momentum k'_{NP}. In this region of low relative momentum the breakup cross section is proportional to the NP-scattering cross section,

$$\sigma_{NP} = 4\pi \sin^2 \delta_0 / k'^2_{NP},$$

and it reflects the magnitude and energy dependence of that two-body cross section. Any attraction thus enhances the breakup cross section in the region of small relative momentum as compared to the cross section in other regions of final momentum.

Bremsstrahlung

In the Bremsstrahlung process a photon is emitted in an otherwise-elastic collision. This is an example of a production process in which there are different particles in the final state than those included in the initial state. We must then introduce the electromagnetic interaction that produces the photon.

This interaction is derived in the familiar way by requiring that the Hamiltonian be gauge invariant. The result in nonrelativistic theory for the electromagnetic interaction with a set of particles of charge $Q_i e$ is

$$V_A = -\sum_{i=1} \frac{Q_i}{m_i} \frac{e}{c} \mathbf{A}(\mathbf{r}) \cdot \mathbf{p}_i, \qquad (2.52)$$

where $\mathbf{p}_i$ is the momentum operator for the ith particle and Q_i is its charge in units of the electron charge e. We will treat the electromagnetic potential $\mathbf{A}(\mathbf{r})$ as an operator acting on a quantized photon field. If the state $|0\rangle$ has no photons, while the state $|\mathbf{q}, \hat{\varepsilon}\rangle$ has one photon of momentum $\hbar\mathbf{q}$ and polarization $\hat{\varepsilon}$, the nonvanishing matrix elements of $\mathbf{A}(\mathbf{r})$ are:

$$\langle \mathbf{q}, \hat{\varepsilon} | \mathbf{A}(\mathbf{r}) | 0 \rangle = \langle 0 | \mathbf{A}(\mathbf{r}) | \mathbf{q}, \hat{\varepsilon} \rangle^* = \left(\frac{2\pi\hbar c}{q}\right)^{1/2} \hat{\varepsilon} \, e^{-i\mathbf{q}\cdot\mathbf{r}}. \qquad (2.53)$$

The final-state Hamiltonian is the sum of the Hamiltonians describing the target, the free scattered "neutron," and the free electromagnetic field. The final-state interaction is

$$V' = V_{NC} + V_{NP} + V_A. \qquad (2.54)$$

Choosing $V_0(\mathbf{r}_N)$ to have any convenient form (e.g., Eq. (2.5) or (2.15)), the final-state distorted wave function is $\chi_{\mathbf{k}}^{(-)}(\mathbf{r}_N)$ and Eqs. (1.20) and (2.53) give

$$T_{f'i} \simeq -i\frac{e\hbar}{c}\frac{Q_N}{m_N}\left(\frac{2\pi\hbar c}{q}\right)^{1/2}\int d^3r_N\,\chi_{\mathbf{k}'}^{(-)*}(\mathbf{r}_N)$$
$$\times \exp(-i\mathbf{q}\cdot\mathbf{r}_N)\,(\hat{\mathbf{\varepsilon}}\cdot\nabla_N)\,\chi_{\mathbf{k}}^{(+)}(\mathbf{r}_N). \qquad (2.55)$$

Here Q_N is the charge of the "neutron" in units of the electron charge. (The "neutron" could in any particular experiment have a charge, of course.) The neutron potential terms vanish in using Eq. (1.20) simply because the one-photon state is orthogonal to the zero-photon state. In Eq. (2.55) the neutron wave is distorted by the interaction with the target both before and after emission of the photon. This is an important example historically, since study of it led to the first realization that scattering of the charged particle after emission of the photon should be described by an *ingoing* wave function.

A related process is the radiative capture process, in which the "neutron" is captured and a photon is emitted. The final state then consists of a bound state of the neutron, proton, and core plus an outgoing photon. This system is described by the full Hamiltonian H including all three potentials and the Hamiltonian of the free electromagnetic field. The final-state interaction V' is simply V_A, the interaction of the charged particles with the electromagnetic field. The resulting first-order T matrix is

$$T_{f'i} \simeq -\frac{ie\hbar}{c}\left(\frac{2\pi\hbar c}{q}\right)^{1/2}\int d^3r_N\,d^3r_P\,\eta_f(\mathbf{r}_N,\mathbf{r}_P)\left[\frac{Q_N}{m_N}e^{i\mathbf{q}\cdot\mathbf{r}_N}(\hat{\mathbf{\varepsilon}}\cdot\nabla_N)\right.$$
$$\left.+\frac{Q_P}{m_P}e^{-i\mathbf{q}\cdot\mathbf{r}_P}(\hat{\mathbf{\varepsilon}}\cdot\nabla_P)\right]\chi_{\mathbf{k}}^{(+)}(\mathbf{r}_N)\,\eta_\alpha(\mathbf{r}_P). \qquad (2.56)$$

The final bound state is here denoted by $\eta_f(\mathbf{r}_N,\mathbf{r}_P)$, and Q_N and Q_P are the charges of the "neutron" and "proton" in units of the electron charge.

There is one important difference between the otherwise-similar "neutron" and "proton" terms. It is customary to use for the bound-state wave function a shell-model approximation in which the $N-P$ interaction is neglected and the wave function is a product:

$$\eta_f(\mathbf{r}_N,\mathbf{r}_P) = \eta_\beta(\mathbf{r}_N)\,\eta_\alpha(\mathbf{r}_P). \qquad (2.57)$$

In this approximation the proton term gives no contribution since, without an interaction between the neutron and the proton, there is no way to transfer the kinetic energy of the incident neutron to the proton, and thence to the photon. (Formally, the initial and final neutron wave functions are mutually orthogonal.) However, if the "neutron" is charged, it can transfer energy directly to the electromagnetic field and be captured into a bound state. If it is neutral (and its magnetic interaction is neglected), radiative capture can occur only after a portion of its initial kinetic energy is transformed to a proton; this occurs only in a higher-order approximation than is represented by Eq. (2.57), or through motion of the center-of-mass of the target, which we have not included here.

3. The Impulse Approximation

In all the examples of the previous section we assumed that V_{NP} was weak and could be treated to first order. If, instead, this potential is strong, the scattering that it causes must be treated more exactly and some other approach must be used. The "impulse approximation" provides a method that sometimes can be applied in this case. As in the examples treated previously, it replaces a many-body problem with a set of two-body problems, so that all quantities that enter the theory can in principle be computed exactly.

To illustrate the principles involved, we will examine the impulse approximation for the particular case of inelastic scattering in the three-body situation discussed in the previous section. The method can be easily generalized to other situations. In using the notation of Section 2, the reader should remember that the terms "neutron," "proton," and "core" are used only as a handy way of identifying the particles, which can actually be any particles, whether nucleons, electrons, mesons, atoms, or molecules. The only requirement is that the particles interact with each other exclusively through two-body interactions. This allows us to factor the wave function into a product of two-body wave functions, and thus to obtain a calculable description of any process.

FORMAL DEVELOPMENT

The exact expression for the T matrix is given by Eq. (1.8). For any inelastic or rearrangement process V_0 can be chosen so that the first term vanishes, and the T matrix can be expressed as

$$T_{fi} = \langle \chi_f^{(-)} | t_1 | \chi_i^{(+)} \rangle, \tag{3.1}$$

where the operator t_1 is given by

$$t_1 = V_1 + V_1 \frac{1}{E - H_0 - V_0 - V_1 + i\varepsilon} V_1$$

$$= V_1 + V_1 \frac{1}{E - H_0 - V_0 + i\varepsilon} t_1. \tag{3.2}$$

Using the notation of Section 2 and choosing $V_0 = V_{NC}$, this operator becomes

$$t_{NP} = V_{NP} + V_{NP} \frac{1}{E - K_N - K_P - V_{PC} - V_{NC} + i\varepsilon} t_{NP}. \qquad (3.3)$$

As pointed out in Section 1, this describes the scattering of the neutron by the bound proton in the presence of the interactions with the core. This three-body scattering problem cannot be solved exactly, but a soluble description can be obtained if the interactions with the core can be neglected or included in some particularly simple way.

If the core interactions V_{NC} and V_{PC} are neglected, the resulting operator is

$$t_{NP} \approx V_{NP} + V_{NP} \frac{1}{E - K_N - K_P + i\varepsilon} t_{NP}. \qquad (3.4)$$

This is just the T matrix describing the collision of a neutron and proton having a total energy E equal to the initial energy of the system (Eq. (2.12)). For ordinary two-body potentials V_{NP} this can be obtained exactly; since both $\chi_i^{(+)}$ and $\chi_f^{(-)}$ can be found exactly, this would provide a calculable approximation. However, to obtain this result we had to neglect completely the effects of the core interactions, and these interactions are rarely sufficiently weak to allow this (e.g., V_{PC} is strong enough to bind the proton). A related difficulty follows from the fact that the total energy E, rather than representing the kinetic energy of the neutron and proton, actually includes as well the potential energy due to the core interactions. To be consistent, we should remove this potential energy to obtain the true kinetic energy of the particles.

This can be done if the complete set of free-particle states ϕ_g, satisfying

$$(E_g - K_N - K_P)\phi_g = 0, \qquad (3.5)$$

is introduced into the T matrix:

$$T_{fi} = \sum_{g,h} \langle \chi_f^{(-)} | \phi_h \rangle \langle \phi_h | t_{NP} | \phi_g \rangle \langle \phi_g | \chi_i^{(+)} \rangle. \qquad (3.6)$$

If the free-particle states are chosen to be momentum eigenstates (plane waves), the amplitudes $\langle \phi_g | \chi_i^{(+)} \rangle$ and $\langle \chi_f^{(-)} | \phi_h \rangle$ are the Fourier transforms of the initial- and final-state distorted-wave functions. For instance, using Eq. (2.6), we have for the initial-state amplitude

$$\langle \phi_g | \chi_i^{(+)} \rangle = \int d^3r_N \, e^{-i\mathbf{k}_N \cdot \mathbf{r}_N} \chi_k^{(+)}(\mathbf{r}_N) \int d^3r_P \, e^{-i\mathbf{k}_P \cdot \mathbf{r}_P} \eta_\alpha(\mathbf{r}_P), \qquad (3.7)$$

3. The Impulse Approximation

where the momenta in the state ϕ_g are denoted by $\mathbf{k}_N$ and $\mathbf{k}_P$:

$$\phi_g(\mathbf{r}_N, \mathbf{r}_P) = \exp[i(\mathbf{k}_N \cdot \mathbf{r}_N + \mathbf{k}_P \cdot \mathbf{r}_P)].$$

We now introduce the *impulse approximation* by replacing the matrix element $\langle \phi_h | t_{NP} | \phi_g \rangle$ with the free-particle matrix element $\langle \phi_h | t_{NP}^{(f)} | \phi_g \rangle$, where the free-particle transition operator $t_{NP}^{(f)}$ is defined by an expression similar to Eq. (3.4) except that E is replaced by E_g:

$$t_{NP}^{(f)} = V_{NP} + V_{NP} \frac{1}{E_g - K_N - K_P + i\varepsilon} t_{NP}^{(f)}. \tag{3.8}$$

We then have the impulse approximation to the T matrix

$$T_{fi}^{(IA)} = \sum_{g,h} \langle \chi_f^{(-)} | \phi_h \rangle \langle \phi_h | t_{NP}^{(f)} | \phi_g \rangle \langle \phi_g | \chi_i^{(+)} \rangle. \tag{3.9}$$

This expression gives the T matrix for a many-body process as a superposition of T matrices for a two-body scattering process. These matrices in turn are weighted by the overlap of the initial state $\chi_i^{(+)}$ with the free-particle state ϕ_g and of the final state $\chi_f^{(-)}$ with the state ϕ_h. Thus the effect of the core interaction is to establish the proper initial and final states; *while* the particles are interacting with each other, the presence of the core is ignored. The particles interact with each other as free particles having a kinetic energy determined by the incident energy and by the core potentials; together these determine the kinetic energy of the neutron and proton in the region of the core. As we see, the energy in the free-particle T matrix is now, correctly, the total kinetic energy of the particles; we will also see shortly that Eq. (3.9) does not require that the core interactions be weak. Thus the difficulties mentioned above are avoided by this approach.

If E_h were equal to E_g, the matrix element $\langle \phi_h | t_{NP}^{(f)} | \phi_g \rangle$ would be simply the amplitude for the elastic scattering of two particles having an energy

$$E_g = \frac{\hbar^2}{2m_N} k_N^2 + \frac{\hbar^2}{2m_P} k_P^2.$$

It could be obtained mathematically by the methods of Chapters 3-5, or directly from elastic scattering experiments. However, there is no requirement that $E_g = E_h$ in the double summation of Eq. (3.9), so that the matrix element $\langle \phi_h | t_{NP}^{(f)} | \phi_g \rangle$ is needed both on and off the energy shell. This means that the necessary matrix elements of $t_{NP}^{(f)}$ cannot be obtained directly from the experimental scattering amplitude, which involves only on-the-energy-shell matrix elements. To evaluate Eq. (3.9) in general, we must then know the potential so that we can solve Eq. (3.8). In some cases, however, the off-the-energy-shell problem can be avoided; these are also the situations in which the impulse approximation is most nearly correct.

The impulse approximation can be written more concisely if we introduce the formal device of using the label E_K to denote the energy of the *free-particle* state to its right. In the general case we then define

$$t_1^{(f)} = V_1 + V_1 \frac{1}{E_K - K + i\varepsilon} t_1^{(f)}$$

$$= V_1 + V_1 \frac{1}{E_K - K - V_1 + i\varepsilon} V_1 \qquad (3.10)$$

and perform the sums over the free-particle states to obtain

$$T_{fi}^{(IA)} = \langle \chi_f^{(-)} | t_1^{(f)} | \chi_i^{(+)} \rangle, \qquad (3.11)$$

or, more simply,

$$t_1 \simeq t_1^{(f)}. \qquad (3.12)$$

This is the impulse approximation. In practical applications it is used by reintroducing the plane-wave expansions and returning to Eq. (3.9).

Detailed Formulas

Let us now examine the detailed form that the matrix element takes in this approximation. If the states ϕ_g are momentum eigenstates, we have, according to Sections 1 and 3 of Chapter 9,

$$\langle \phi_h | t_{NP}^{(f)} | \phi_g \rangle = \langle \mathbf{k}_N', \mathbf{k}_P' | t_{NP}^{(f)} | \mathbf{k}_N, \mathbf{k}_P \rangle$$
$$= (2\pi)^3 \, \delta(\mathbf{K}_{NP}' - \mathbf{K}_{NP}) \, \langle \mathbf{k}_{NP}' | t_{NP}^{(f)} | \mathbf{k}_{NP} \rangle. \qquad (3.13)$$

The total and relative momenta are defined by

$$\mathbf{K}_{NP} = \mathbf{k}_N + \mathbf{k}_P$$

and

$$\mathbf{k}_{NP} = \frac{m_P \mathbf{k}_N - m_N \mathbf{k}_P}{M}, \qquad (3.14)$$

with $M = m_N + m_P$. Denoting the Fourier transform of $\eta_\alpha(\mathbf{r}_P)$ by $\tilde{\eta}_\alpha(\mathbf{k}_P)$ and similarly for $\chi_k^{(+)}(\mathbf{r}_N)$, the scalar product becomes

$$\langle \phi_g | \chi_i^{(+)} \rangle = \langle \mathbf{k}_N, \mathbf{k}_P | \chi_i^{(+)} \rangle = \tilde{\chi}_k^{(+)}(\mathbf{k}_N) \, \tilde{\eta}_\alpha(\mathbf{k}_P) \qquad (3.15)$$

(cf. Eq. (3.7)), and the T matrix, Eq. (3.9), becomes

$$T_{fi}^{(IA)} = \int \frac{d^3K_{NP}}{(2\pi)^3} \frac{d^3k_{NP}}{(2\pi)^3} \frac{d^3k'_{NP}}{(2\pi)^3} \tilde{\chi}_{k'}^{(-)*}\left(\frac{m_N}{M}\mathbf{K}_{NP} + \mathbf{k}'_{NP}\right)$$

$$\times \tilde{\eta}_{\alpha'}^*\left(\frac{m_P}{M}\mathbf{K}_{NP} - \mathbf{k}'_{NP}\right) <\mathbf{k}'_{NP}|\, t_{NP}^{(f)}\,|\mathbf{k}_{NP}> \tilde{\chi}_{k}^{(+)}\left(\frac{m_N}{M}\mathbf{K}_{NP} + \mathbf{k}_{NP}\right)$$

$$\times \tilde{\eta}_\alpha\left(\frac{m_P}{M}\mathbf{K}_{NP} - \mathbf{k}_{NP}\right). \tag{3.16}$$

This expression seems fairly complex, but every quantity in it can be calculated exactly.

It is sometimes simpler to use the coordinate representation, since the spatial wave functions are often simpler or better known than the momentum functions. Using the identity

$$\langle \mathbf{k}'_{NP} | t_{NP}^{(f)} | \mathbf{k}_{NP} \rangle = \int d^3r \, \exp(-i\mathbf{k}'_{NP}\cdot\mathbf{r})\, V_{NP}(\mathbf{r})\, \psi_{\mathbf{k}_{NP}}^{(+)}(\mathbf{r}), \tag{3.17}$$

where $\psi_{\mathbf{k}_{NP}}^{(+)}(\mathbf{r})$ describes the relative motion of the neutron and proton, Eq. (3.16) becomes

$$T_{fi}^{(IA)} = \int \frac{d^3k_{NP}}{(2\pi)^3} d^3R\, d^3r\, d^3r'\, \chi_{k'}^{(-)*}\left(\mathbf{R} + \frac{m_P}{M}\mathbf{r}'\right)$$

$$\times \eta_{\alpha'}\left(\mathbf{R} - \frac{m_N}{M}\mathbf{r}'\right) V_{NP}(\mathbf{r}')\, \psi_{\mathbf{k}_{NP}}^{(+)}(\mathbf{r}') \exp(-i\mathbf{k}_{NP}\cdot\mathbf{r})$$

$$\times \chi_k^{(+)}\left(\mathbf{R} + \frac{m_P}{M}\mathbf{r}\right)\eta_\alpha\left(\mathbf{R} - \frac{m_N}{M}\mathbf{r}\right). \tag{3.18}$$

If the NP scattering is treated in Born approximation, $\psi_{\mathbf{k}_{NP}}^{(+)}(\mathbf{r}')$ is replaced by $\exp(i\mathbf{k}_{NP}\cdot\mathbf{r}')$, and Eq. (3.18) becomes identical with Eq. (2.18).

A number of approximations can be used to simplify these results. If the energy of the incident neutron is sufficiently high, it may be possible to neglect the effect of its interaction with the core and replace $\chi_k^{(+)}(\mathbf{r}_N)$ by its Born approximation $\exp(i\mathbf{k}\cdot\mathbf{r}_N)$. Making this approximation in the initial and final states and changing variables, Eq. (3.16) becomes

$$T_{fi}^{(IA)} = \int \frac{d^3k_P\, d^3k_P'}{(2\pi)^3}\, \delta(\mathbf{k}_P + \mathbf{k} - \mathbf{k}_P' - \mathbf{k}')\, \tilde{\eta}_\alpha^*(\mathbf{k}_P')$$

$$\times \left\langle \frac{m_P\mathbf{k}' - m_N\mathbf{k}_P'}{M} \left| t_{NP}^{(f)} \right| \frac{m_P\mathbf{k} - m_N\mathbf{k}_P}{M} \right\rangle \tilde{\eta}_\alpha(\mathbf{k}_P). \tag{3.19}$$

One can see in this expression how the T matrix for scattering of the neutron and proton is weighted by the probability amplitudes for finding the bound proton with a particular momentum. That is, although the neutron's interaction with the core may be neglected, the proton's interaction may still have an effect.

Since $\eta_\alpha(\mathbf{r}_P)$ is the wave function for a bound state, $\tilde{\eta}_\alpha(\mathbf{k}_P)$ will be large only for wave numbers less than $1/R$, where R is the size of the target. For instance, in the case of the hydrogen atom, with a_0 the Bohr radius,

$$\tilde{\eta}_\alpha(\mathbf{k}_P) = \frac{(64\pi a_0^3)^{1/2}}{(1 + (k_P a_0)^2)^2}$$

and $k_P \lesssim 1/a_0$. Hence the integral in Eq. (3.19) will span a finite range of momentum such that $|\mathbf{k}_P|, |\mathbf{k}_P'| \lesssim 1/R$.

If the momentum of the incident neutron is large compared to this momentum (specifically, if $k \gg (m_N/m_P)R^{-1}$, one can consider the bound proton initially to be at rest. Then setting $\mathbf{k}_P = 0$, the neutron-proton T matrix is

$$\left\langle \frac{m_P \mathbf{k}' - m_N \mathbf{k}_P'}{M} \middle| t_{NP}^{(f)} \middle| \frac{m_P \mathbf{k} - m_N \mathbf{k}_P}{M} \right\rangle \simeq \left\langle \mathbf{k}' - \frac{m_N}{M} \mathbf{k} \middle| t_{NP}^{(f)} \middle| \frac{m_P}{M} \mathbf{k} \right\rangle. \quad (3.20)$$

With the bound proton at rest, $(m_P/M)\hbar \mathbf{k}$ is the relative momentum in the neutron-proton center-of-mass system, and the momentum transferred to the neutron is $\hbar(\mathbf{k}' - \mathbf{k})$. Since this T matrix no longer depends on the momentum of the bound particle, it can be removed from the integral and the result expressed as a product:

$$T_{fi}^{(IA)} \simeq \left\langle \mathbf{k}' - \frac{m_N}{M} \mathbf{k} \middle| t_{NP}^{(f)} \middle| \frac{m_P}{M} \mathbf{k} \right\rangle \int \frac{d^3 k_P}{(2\pi)^3} \tilde{\eta}_\alpha^*(\mathbf{k}_P + \mathbf{k} - \mathbf{k}') \tilde{\eta}_\alpha(\mathbf{k}_P)$$

$$= \left\langle \mathbf{k}' - \frac{m_N}{M} \mathbf{k} \middle| t_{NP}^{(f)} \middle| \frac{m_P}{M} \mathbf{k} \right\rangle F_{\alpha'\alpha}(\mathbf{k} - \mathbf{k}'). \quad (3.21)$$

The first term here has a simple interpretation. If the incident neutron energy is much greater than the binding energy of the proton, energy conservation requires $E_{k'} \simeq E_k$, and the matrix element of $t_{NP}^{(f)}$ is on the energy shell. The inelastic scattering cross section is proportional to $|T_{fi}^{(IA)}|^2$ and thus, from Eq. (3.21), to $|\langle \mathbf{k}' - (m_N/M)\mathbf{k} | t_{NP}^{(f)} | (m_P/M)\mathbf{k}\rangle|^2$; the latter in turn is proportional to the neutron-proton scattering cross section in the laboratory, where the proton is at rest. Thus free particle neutron-proton scattering data can be used to determine this matrix element. We conclude that at sufficiently high energies the impulse approximation yields a direct relation between scattering on a complex target composed of many particles and scattering on a single free particle.

The remaining part of Eq. (3.21) is usually called the target "form factor"; it is characteristic of the shape and other properties of the target. Introducing the inverse Fourier transforms, we can write it as

$$F_{\alpha'\alpha}(\mathbf{k} - \mathbf{k}') = \int \frac{d^3k_P}{(2\pi)^3} \tilde{\eta}_{\alpha'}^*(\mathbf{k}_P + \mathbf{k} - \mathbf{k}') \tilde{\eta}_\alpha(\mathbf{k}_P)$$

$$= \int d^3r \, \eta_{\alpha'}(\mathbf{r}) \, \eta_\alpha(\mathbf{r}) \exp[i(\mathbf{k} - \mathbf{k}') \cdot \mathbf{r}]. \quad (3.22)$$

In an operator notation this can be written concisely as

$$F_{\alpha'\alpha}(\mathbf{k} - \mathbf{k}') = \langle \eta_{\alpha'} | \exp[i(\mathbf{k} - \mathbf{k}') \cdot \mathbf{r}] | \eta_\alpha \rangle. \quad (3.23)$$

(Here $\mathbf{r}$ is the position operator.) The form factor generally has a diffraction-like shape which limits the scattering to the forward direction with $|\mathbf{k} - \mathbf{k}'| \lesssim 1/R$. As an example, for the $1s \to 2s$ transition in hydrogen, the form factor is

$$\frac{8^{1/2} (\mathbf{k} - \mathbf{k}')^2 a_0^2}{[9/4 + (\mathbf{k} - \mathbf{k}')^2 a_0^2]^3}.$$

As we see, all the available information about the target is now contained in the form factor, since the remainder of Eq. (3.21) is determined from free-particle scattering.

With this result for the T matrix, the inelastic scattering cross section is

$$\frac{d\sigma_{\alpha'\alpha}^{(IA)}}{d\Omega} = \left(\frac{2m_N}{4\pi\hbar^2}\right)^2 |T_{fi}^{(IA)}|^2 \frac{k'}{k}$$

$$\simeq \left(\frac{d\sigma_{NP}}{d\Omega}\right)_{\text{lab}} |F_{\alpha'\alpha}(\mathbf{k} - \mathbf{k}')|^2, \quad (3.24)$$

with the neutron-proton scattering cross section in the laboratory system denoted by $(d\sigma_{NP}/d\Omega)_{\text{lab}}$. The many-body cross section is equal to the two-body cross section modulated by the target form factor, an effect that can be attributed to the internal momentum of the bound particles. This effect was observed very early in studies of the Compton scattering of photons by atomic electrons, where it was found that the Compton "line"—corresponding to two-body scattering—was smeared out by the internal momentum of the bound electrons. It has been observed more recently in neutron scattering on deuterons and in the photoejection of nucleons from nuclei.

Validity of the Impulse Approximation

Let us now examine the validity of the impulse approximation, Eq. (3.12). That is, when can we replace the transition operator in the presence of the core interactions by the free-particle transition operator $t_1^{(f)}$?

A comparison of Eqs. (3.2) and (3.10) shows that these operators differ in the definitions of the Green's function and of the energy E. The intermediate states in t_1 are eigenstates of $H_0 + V_0$, which includes the interaction with the core, and the energy E is the initial energy of the system. The intermediate states entering into $t_1^{(f)}$ are free-particle states, and the energy is that of the free-particle states inserted into Eq. (3.9). We must be able to neglect the effect of the core interaction $V_{NC} + V_{PC} \equiv V_C$ in intermediate states and use this free-particle energy, if the impulse approximation is to be valid.

The term "impulse approximation" arises out of this condition in the following way. Energy is not conserved in intermediate states, and the spread in intermediate energies ΔE is determined by the strength and shape of the two-body potential. Intermediate states having an energy much greater than the core potential are essentially free-particle states, and V_C may be neglected in these states. If the interaction is pictured as taking place over a time interval τ, this "interaction period" or "collision time" is $\tau \sim \hbar/\Delta E$, and we expect the impulse approximation to be valid when $V_C \ll \Delta E$ or $\tau \ll \hbar/V_C$. Since $\hbar/V_C$ is a characteristic oscillation time resulting from the core interaction, this condition means that this motion may be neglected during the scattering; this is an adiabatic condition and suggests an impulsive collision. In the sense that the collision time must be short and the scattering process abrupt and impulsive, the term "impulse approximation" is an appropriate one. However, we shall see that this condition is unnecessarily restrictive.

The explicit formula for t_1 is

$$t_1 = V_1 + V_1 \frac{1}{E - K - V_C - V_1 + i\varepsilon} V_1, \qquad (3.25)$$

with V_C defined in general by $V_C = H_0 + V_0 - K$. The Green's function that enters here may be expressed in terms of the free-particle Green's function using the formal relation

$$\frac{1}{E - K - V_C - V_1 + i\varepsilon} = \frac{1}{E_K - K - V_1 + i\varepsilon} - \frac{1}{E - K - V_C - V_1 + i\varepsilon}$$

$$\times (E - E_K - V_C) \frac{1}{E_K - K - V_1 + i\varepsilon}. \qquad (3.26)$$

3. The Impulse Approximation

We have here used the symbol E_K, as in Eq. (3.10), to signify that the energy is that of the free-particle state to the right. Using this result and the definition of $t_1^{(f)}$, Eq. (3.10), we find

$$t_1 - t_1^{(f)} = -V_1 \frac{1}{E - K - V_C - V_1 + i\varepsilon}(E - E_K - V_C)$$
$$\times \frac{1}{E_K - K - V_1 + i\varepsilon} V_1. \qquad (3.27)$$

This is the correction to the impulse approximation.

A simple example will show why we expect this correction to be small. Suppose V_C is constant inside a large target. Then, within the target, $\chi_i^{(+)}(\mathbf{r}_N, \mathbf{r}_P)$ should be a free-particle wave function with an energy displaced from the incident energy by the amount V_C. This energy shift is automatically included in the impulse approximation by weighting the free-particle matrix $\langle \phi_h | t_1^{(f)} | \phi_g \rangle$ with the scalar products for the initial and final states. If E is the total energy of the initial state, the scalar product $\langle \phi_g | \chi_i^{(+)} \rangle$ will be nonzero only if E_g is equal to this energy, $E + |V_C|$. Referring to Eq. (3.27), we see that the correction to the impulse approximation is zero in this case.

For a constant potential, then, there is no correction to the impulse approximation. We may expect that, in general, this approximation will be valid if V_C is slowly varying over the range of the interaction V_1. It is thus not necessary that V_C be small, only that it be smooth or that the force $(= -\nabla V_C)$ be small. This conclusion can also be stated in momentum space terms: It follows from Eq. (3.27) that, roughly speaking, the impulse approximation will be valid if the expectation value of V_C, $\langle \phi_g | V_C | \phi_g \rangle$, does not vary appreciably over the range of energies ΔE.

These conclusions can be seen in another way using a slightly different expression for $t_1 - t_1^{(f)}$. Since $(E_g - K)\phi_g = 0$ and $(E - K - V_C)\chi_i^{(+)} = 0$, we have the identity

$$(E - E_g)\langle \phi_g | \chi_i^{(+)} \rangle = \langle \phi_g | V_C | \chi_i^{(+)} \rangle. \qquad (3.28)$$

Then, acting on the distorted wave state $\chi_i^{(+)}$, we have

$$(E - E_K)\chi_i^{(+)} = \sum_g (E - E_g) | \phi_g \rangle \langle \phi_g | \chi_i^{(+)} \rangle$$
$$= \sum_g |\phi_g \rangle \langle \phi_g | V_C | \chi_i^{(+)} \rangle = V_C \chi_i^{(+)}. \qquad (3.29)$$

Further, we recall that the free-particle wave matrix is

$$\Omega_K^{(+)} = 1 + \frac{1}{E_K - K - V_1 + i\varepsilon} V_1. \qquad (3.30)$$

Recognizing that in practice both t_1 and $t_1^{(f)}$ act on $\chi_i^{(+)}$ (Eqs. (3.1) and (3.11)), we can use these results in Eq. (3.27) to obtain

$$t_1 - t_1^{(f)} = V_1 \frac{1}{E - K - V_C - V_1 + i\varepsilon} [V_C, \Omega_K^{(+)}]. \qquad (3.31)$$

The correction is proportional to the commutator of V_C with the wave matrix. Thus, as we saw before, a constant potential will give no correction. Conversely, there will be important corrections to the impulse approximation whenever V_C is a sensitive function of position, is strongly nonlocal (giving energy-dependent matrix elements), or is dependent on internal variables such as the spin that can undergo a change in an intermediate state. (Of course, this assumes V_1 is large enough that second-order contributions are significant.)

In general terms, we see that the criterion for validity of the impulse approximation is that $[V_C, \Omega_K^{(+)}]$ be much smaller than ΔE, the spread in intermediate energies. This is obviously a far weaker condition than our earlier intuitive guess, which required $V_C \ll \Delta E$. If the variation of V_C in space, energy, or spin is represented symbolically by ΔV_C, one can show from Eq. (3.31) that the fractional error in the impulse approximation is $\sim (\Delta V_C/E)(f(0)/\lambda)$, if the range of V_1 is small compared to the de Broglie wavelength λ, and is $(\Delta V_C/E)(f(0)/R_1)$ if R_1, the range of V_1, is large compared to λ ($f(0)$ is the forward scattering amplitude). Clearly, if ΔV_C is small or if $f(0)$ is small (implying that the Born approximation holds), the impulse approximation will be valid.

4. Scattering by a Many-Body System

In the previous sections of this chapter we presented theoretical prescriptions for computing scattering amplitudes when the target can be described as a two-body system (in the preceding the two bodies were called the "core" and the "proton"). In this and succeeding sections we shall apply formal scattering theory to the scattering of a particle by a general many-body system. We shall, in particular, show the relation of this scattering to scattering by the separate constituents of the target.

The formal theory discussed in previous chapters permits a concise and formally exact solution to this many-body problem. Such *formal* solutions do not bring one very much closer to *actual* solutions. Nevertheless, they provide a mathematical formulation that closely parallels the physical picture one has of scattering in a many-body system. This in turn leads to concise derivations of various approximations to the exact equations. Some of the approximations were used before the development of the formal theory, but this theory gives exact formulas from which criteria for the validity of these approximations, as well as corrections to them, may be obtained.

All approximations to the exact equations seek to reduce the many-body problem to a series of two-body problems. While exact solutions to many-body problems are usually unattainable, progress can be made if their salient features can be included in a description that involves the scattering of only two bodies at a time. The oldest example of this technique is the use of an index of refraction to represent the effect of a many-particle medium upon the passage of light.

The multiple-scattering equations that we shall derive are particularly adapted to another type of two-body reduction, in which the collision is pictured as a sequence of two-body scatterings that take place in the presence of the remainder of the target. In Section 3 we discussed the scattering of two particles in the neighborhood of another body. While this is a three-body problem, the two-body characteristics will often dominate, as at high incident energies or when the target particle is weakly bound. The physical situation can then still be described in terms of two-body collisions.

We shall express the multiple-scattering equations in terms of two-body scattering amplitudes appropriate to the interior of the target. From these equations we can obtain the index of refraction and general optical models. Several such useful approximations are described in Section 5. In Section 6 we discuss the problem of determining the scattering amplitude when the projectile can scatter many times inside the target. In this situation there is the possibility of forming a long-lived compound state, and this will appear in the scattering process as a resonance.

THE MULTIPLE-SCATTERING EQUATIONS

The development here is time independent, although we will use a time-dependent language in describing the "sequence" of collisions. It is of course always possible to construct a wave packet that will actually undergo this series of collisions, but the physical scattering process can be understood without introducing this added complexity.

We consider then the scattering of a projectile by a complex target composed of many particles which may each interact with the projectile. If we suppose that the target has a finite size, then the projectile will initially be a free particle and will again be free when it has emerged from the target and is detected. Then the formal scattering theory of Chapter 7 may be applied directly to this problem.

The initial state as described by the Hamiltonian is

$$H_0 = K_0 + H_T, \tag{4.1}$$

where K_0 is the kinetic energy operator for the projectile and H_T is the Hamiltonian of the target, including whatever interactions bind its constituents together. To distinguish many-body operators from two-body operators, we shall denote the latter by lower-case symbols. The projectile-target interaction is then written as a sum of two-body interactions:

$$V = \sum_{n=1}^{N} v_n, \tag{4.2}$$

where v_n is the interaction between the projectile and particle n of the target, and the index n runs over all the constituents of the target. This interaction may be a very general operator; for instance, scattering from a lattice is most conveniently described in terms of phonon excitation processes, so that v_n would be an operator that creates or annihilates phonons. In meson- or photon-production processes, v_n is an operator that can create or destroy these Bose particles. In general, we suppose that v_n is known, perhaps from two-particle scattering experiments, and that it is a true two-body operator, depending only on the coordinates of the nth particle and the projectile. Three- or more-body interactions can also be treated, if one wishes, by methods similar to those used below.

The exact state vector can be written, using the results of Chapter 7, as

$$\psi_i = \phi_i + \frac{1}{E - H_0 + i\varepsilon} \sum_{n=1}^{N} v_n \psi_i, \tag{4.3}$$

and, as an example, the transition amplitude for elastic or inelastic scattering is

$$T_{fi} = \langle \phi_f | \sum_{n=1}^{N} v_n | \psi_i \rangle. \tag{4.4}$$

Unless there is a possibility of confusion, we shall not explicitly indicate the outgoing-wave boundary condition on the state vectors.

One straightforward way to solve these equations is of course to expand them in powers of $V = \sum_{n=1}^{N} v_n$. However, this procedure, which generates

4. Scattering by a Many-Body System

the Born series, is useful only if the potential is sufficiently weak, and it lacks a simple physical interpretation. We can obtain a much more general result, which separates two-body effects from multiple-scattering effects, if we rearrange the Born series so that all terms involving the scattering of the projectile by a particular target particle are grouped together. In this way we can completely describe the scattering by that single target particle and will generate a series showing a succession of scatterings by *different* target particles.

We want to gather together all terms of the form

$$v_n \frac{1}{E - H_0 + i\varepsilon} v_n \frac{1}{E - H_0 + i\varepsilon} v_n \cdots = v_n \left(\frac{1}{E - H_0} v_n \right)^j.$$

The reader will recognize the sum of all such terms, for all j, as a transition operator, which we denote by t_n:

$$t_n = v_n + v_n \frac{1}{E - H_0 + i\varepsilon} t_n. \tag{4.5}$$

As in the previous section, the nature of the medium in which the particles interact determines the energy levels and intermediate-state wave functions via the Green's function. In many applications t_n can be replaced by the free-particle scattering amplitude (this is the impulse approximation), but for this formal development we will retain the exact form.

The desired rearrangement of the power series is facilitated by the introduction of a set of state vectors ψ_n that describe the waves that are incident on each target particle n. These states play a role analogous to that played by the incident state ϕ_i in the two-body problem. In that earlier case we related the exact scattering state to the incident state by

$$V\psi_i = T\phi_i, \tag{4.6}$$

where T is the transition operator. In the present case we introduce t_n, the two-body transition operator inside the target, rather than T. Correspondingly, we must replace ϕ_i in Eq. (4.6) by the appropriate "incident state" within the target. Viewing this in the coordinate representation, we may guess that the wave that is incident on particle n is made up of the incident wave ϕ_i plus the wavelets produced by all the *other* particles in the target. The key feature here is that this effective incident wave does not include the wavelet produced by particle n itself.

The use of an effective incident wave that omits the self-action of particle n is familiar in electrostatics, where the "effective field" acting on a molecule is evaluated in order to relate the dielectric constant of the medium to the polarizability of the individual molecules. In that case one finds that the effective field differs from the field outside the medium because of the induced polarization of the medium. The correspondence with our problem is immediate. The effective field corresponds to the effective incident wave, while the external field is comparable with the original incident wave ϕ_i. The polarizability of a molecule is analogous to the two-body scattering amplitude, and the dielectric constant, to the full many-body transition amplitude. In fact, of course, electrostatics is merely the zero energy limit of our multiple-scattering problem for the special case in which the projectile is a photon.

Let us now use Eq. (4.5) for the transition amplitude to demonstrate that ψ_n is indeed the incident wave we have described. By analogy with Eq. (4.6), we want to find a state vector ψ_n such that the relation

$$v_n \psi_i = t_n \psi_n \tag{4.7}$$

is satisfied. With such a definition, the exact scattering state is

$$\psi_i = \phi_i + \frac{1}{E - H_0 + i\varepsilon} \sum_{n=1}^{N} t_n \psi_n. \tag{4.8}$$

Thus the scattered wave is a superposition of wavelets emitted by each target particle, with the amplitude of each wavelet determined by the effective incident wave ψ_n and by t_n, the effective two-body interaction inside the target.

Suppose we introduce a wave matrix describing the two-body interaction inside the target. We denote this by ω_n and have

$$t_n = v_n \omega_n, \tag{4.9}$$

with

$$\omega_n = 1 + \frac{1}{E - H_0 + i\varepsilon} t_n. \tag{4.10}$$

Then Eqs. (4.7) and (4.10) together imply the relation

$$\psi_i = \omega_n \psi_n = \psi_n + \frac{1}{E - H_0 + i\varepsilon} t_n \psi_n. \tag{4.11}$$

With Eq. (4.8), this gives the result

$$\psi_n = \phi_i + \frac{1}{E - H_0 + i\varepsilon} \sum_{m \neq n}^{N} t_m \psi_m. \tag{4.12}$$

This has the form that we guessed above, in which the effective incident wave ψ_n is formed from the actual incident wave ϕ_i and the wavelets scattered by all target particles other than particle n.

4. Scattering by a Many-Body System

It should be noted that ψ_n does include some effects of particle n, since the scattered wavelet from this particle contributes to the waves that are incident upon other particles, and these in turn contribute to the wave incident on n. This reactive effect, which is of at least second order in the transition operator t_n, occurs also in electrostatics where it yields the polarization induced in the surrounding medium by each molecule.

By introducing the transition operator t_n and the effective incident wave ψ_n, we have replaced the two fundamental equations for this scattering problem, Eqs. (4.3) and (4.4), by the four equations

$$\psi_i = \phi_i + \frac{1}{E - H_0 + i\varepsilon} \sum_{n=1}^{N} t_n \psi_n$$

$$\psi_n = \phi_i + \frac{1}{E - H_0 + i\varepsilon} \sum_{m \neq n}^{N} t_m \psi_n$$

$$t_n = v_n + v_n \frac{1}{E - H_0 + i\varepsilon} t_n$$

and

$$T_{fi} = \langle \phi_f | \sum_{n=1}^{N} t_n | \psi_n \rangle. \tag{4.13}$$

This additional complexity is justified by the fact that these equations provide a description of the scattering process in terms of a multiple-scattering sequence. If we expand Eq. (4.12) in powers of the transition operator and insert this into Eq. (4.13), we find

$$T_{fi} = \langle \phi_f | \sum_{n=1}^{N} t_n + \sum_{n,m \neq n}^{N} t_n \frac{1}{E - H_0 + i\varepsilon} t_m$$

$$+ \sum_{\substack{n \\ m \neq n \\ l \neq m}}^{N} t_n \frac{1}{E - H_0 + i\varepsilon} t_m \frac{1}{E - H_0 + i\varepsilon} t_l + \cdots | \phi_i \rangle. \tag{4.14}$$

Each term in this series is a multiple-scattering sequence in which the projectile scatters successively from *different* particles in the medium. In the first term the projectile enters the target, scatters from particle n, and emerges. In the double-scattering term the projectile scatters from m, propagates to particle n where it scatters again, and then emerges.

If we had instead expanded Eq. (4.4) in powers of v_n, we would have found in the double-scattering term expressions of the form

$$\langle \phi_f | v_n \frac{1}{E - H_0 + i\varepsilon} v_n | \phi_i \rangle,$$

which represent successive interactions with the *same* particle. All of these multiple interactions with the same particle have been gathered together into t_n, so that we obtain instead Eq. (4.14), which has only true sequential multiple scattering. As in the case of two-particle scattering, the transition operator t_n includes all the multiple interactions of the projectile with particle n and describes exactly their complete interaction.

In using this result we can, in favorable circumstances, approximate the two-body transition operator inside the target by the transition operator in free space and thus relate scattering from complex targets to free two-particle scattering. To achieve this we would introduce the impulse approximation discussed in the previous section. Since only in very special cases can t_n be calculated exactly, the situations that are generally of practical interest are those in which the impulse approximation holds.

SINGLE SCATTERING

If the target is sufficiently small that only one scattering is likely to occur within it, T_{fi} may be approximated by the first term of Eq. (4.14). This neglect of multiple scattering is commonly called the "transparency assumption"; it is a valid approximation if the thickness of the target is small compared to the mean free path of the projectile (an approximate expression for this will be given in the next section). With this assumption T_{fi} becomes

$$T_{fi}^{(1)} = \sum_{n=1}^{N} \langle \phi_f | t_n | \phi_i \rangle. \tag{4.15}$$

This can be evaluated by the methods of Section 3 if the impulse approximation is invoked to determine the transition matrix for scattering of the projectile by particle n.

We shall discuss a few common situations to illustrate the implications of Eq. (4.15). To simplify the formulas we will assume that the target particles are very much heavier than the projectile, e.g., electrons scattering on atoms in a crystal or mesons on nucleons in a nucleus. In the case of elastic scattering the impulse approximation, Eqs. (3.21) and (3.22), then gives

$$T_{fi}^{(1)} = \sum_{n=1}^{N} \langle \mathbf{k}' | t_n^{(f)} | \mathbf{k} \rangle \int d^3r \, \rho_n(\mathbf{r}) \exp[i(\mathbf{k} - \mathbf{k}') \cdot \mathbf{r}], \tag{4.16}$$

where $\rho_n(\mathbf{r})$, the spatial density distribution of particle n, is defined by

$$\rho_n(\mathbf{r}) = \int d^3r_1\, d^3r_2 \ldots d^3r_{n-1}\, d^3r_{n+1} \ldots d^3r_n$$
$$\times |\eta_\alpha(\mathbf{r}_1, \mathbf{r}_2, \ldots, \mathbf{r}_{n-1}, \mathbf{r}, \mathbf{r}_{n+1}, \ldots, \mathbf{r}_N)|^2, \quad (4.17)$$

provided the target wave function $\eta_\alpha(\mathbf{r}_1, \mathbf{r}_2, \ldots, \mathbf{r}_N)$ is normalized to unity. If the target wave function can be represented by a product of single-particle wave functions, $\rho_n(\mathbf{r})$ is just the square of the single-particle wave function for particle n. If the target is not in a pure state, e.g., if it is a crystal at a finite temperature, $T_{fi}^{(1)}$ will be a sum of terms such as Eq. (4.16), with each term weighted by the probability of finding the target in a particular state.

If all the constituents of the target are identical, we can introduce the form factor for the target

$$F_\alpha(\mathbf{k} - \mathbf{k}') = N \int d^3r\, \rho_1(\mathbf{r}) \exp[i(\mathbf{k} - \mathbf{k}')\cdot\mathbf{r}] = N\langle \eta_\alpha | \exp[i(\mathbf{k} - \mathbf{k}')\cdot\mathbf{r}] | \eta_\alpha \rangle \quad (4.18)$$

and write the T matrix as

$$T_{fi}^{(1)} = \langle \mathbf{k}' | t^{(f)} | \mathbf{k} \rangle F_\alpha(\mathbf{k} - \mathbf{k}'). \quad (4.19)$$

The elastic scattering cross section is then

$$\frac{d\sigma^{el}}{d\Omega} = \left(\frac{d\sigma_2}{d\Omega}\right)_{lab} |F_\alpha(\mathbf{k} - \mathbf{k}')|^2 \quad (4.20)$$

with $(d\sigma_2/d\Omega)_{lab}$ the two-body elastic scattering cross section in the laboratory system. This result depends upon the validity of the impulse approximation and is thus most reliable at high energies.

Consider the special case that the target particles are each fixed at a particular site, as in a crystal lattice. The many-particle density distribution is then

$$|\eta_\alpha(\mathbf{r}_1, \mathbf{r}_2, \ldots, \mathbf{r}_N)|^2 = \prod_{n=1}^{N} \delta(\mathbf{r}_n - \mathbf{s}_n), \quad (4.21)$$

where $\mathbf{s}_n$ is the location of the nth site. In a lattice this position is usually specified by the coordinates along a set of base vectors

$$\mathbf{s}_n = d_1\mathbf{a}_1 + d_2\mathbf{a}_2 + d_3\mathbf{a}_3, \quad (4.22)$$

with (d_1, d_2, d_3) a set of integers. Equation (4.19) is in this case

$$T_{fi}^{(1)} = \langle \mathbf{k}' | t^{(f)} | \mathbf{k} \rangle \sum_{d_1,d_2,d_3} \exp[i(\mathbf{k} - \mathbf{k}')\cdot(d_1\mathbf{a}_1 + d_2\mathbf{a}_2 + d_3\mathbf{a}_3)]. \quad (4.23)$$

The sum in this result gives the coherent superposition of waves scattered at each lattice site. For a large lattice the sum will be appreciable only at scattering angles for which the Laue conditions are satisfied:

$$(\mathbf{k} - \mathbf{k}') \cdot \mathbf{a}_1 = 2\pi h$$
$$(\mathbf{k} - \mathbf{k}') \cdot \mathbf{a}_2 = 2\pi k \quad (4.24)$$
$$(\mathbf{k} - \mathbf{k}') \cdot \mathbf{a}_3 = 2\pi l,$$

with h, k, and l integers. These conditions ensure that each source contributes constructively to the sum. The sum over lattice sites produces a sharp diffraction pattern as a result of the interference of waves from individual scatterers separated by macroscopic distances. The factor $\langle \mathbf{k}' | t^{(f)} | \mathbf{k} \rangle$, sometimes called the "atomic scattering factor," determines the relative height of successive diffraction maxima. We may also think of $\langle \mathbf{k}' | t^{(f)} | \mathbf{k} \rangle$ as giving a diffraction pattern as a result of interferences within a single scatterer. This diffraction pattern is a very broad one since, in general, the characteristic angles of a diffraction pattern are of order λ/R, with R the size of the diffracting medium.

Analogous results can be derived for inelastic scattering. One does not encounter the Fourier transform of the spatial density distribution but instead the inelastic form factor

$$F_{\alpha'\alpha}(\mathbf{k} - \mathbf{k}') = \langle \eta_{\alpha'} | \sum_{n=1}^{N} \exp[i(\mathbf{k} - \mathbf{k}') \cdot \mathbf{r}_n] | \eta_\alpha \rangle \quad (4.25)$$

(cf. Eq. (3.22)). The single-scattering approximation to the T matrix is then

$$T_{fi}^{(1)} = \langle \mathbf{k}' | t^{(f)} | \mathbf{k} \rangle F_{\alpha'\alpha}(\mathbf{k} - \mathbf{k}') \quad (4.26)$$

if all the constituents of the target are identical. Similar formulas can be derived for such processes as exchange scattering and knockout collisions.

Very often, in performing experimental measurements, one makes no attempt to determine the state in which the target is left after the scattering. Not only is no measurement performed directly on the residual target, but the energy resolution of the detector is often so crude that one cannot, from measuring the energy of the scattered particle, determine the excitation of the residual target. This is the situation, for instance, when the energy of the incident projectile is much greater than the level spacing in the target. By detecting the projectile with a low-resolution detector, one is in effect summing over those states of the residual target that can be excited in the collision.

Using Eq. (4.26), the differential cross section for this situation (often called the "collision cross section") is

$$\frac{d\sigma^{coll}}{d\Omega} = \frac{2\pi}{\hbar v} |\langle \mathbf{k}' | t^{(f)} | \mathbf{k} \rangle|^2 \sum_{\alpha'} |F_{\alpha'\alpha}(\mathbf{k} - \mathbf{k}')|^2$$

$$\times \rho_{\alpha'}(E_{k'}) \big|_{E_{k'} = E_k + E_\alpha - E_{\alpha'}}. \qquad (4.27)$$

The sum extends over all final states permitted by energy conservation, i.e., $E_{\alpha'} \leq E_k + E_\alpha$. In general, $F_{\alpha'\alpha}(\mathbf{k} - \mathbf{k}')$ is expected to decrease rapidly as the excitation energy of the final state increases, because of the reduced overlap between the initial and final bound-state wave functions. If the cross section in this way becomes negligible at excitation energies that are small compared with the initial energy (that is, $|E_\alpha - E_{\alpha'}| << E_k$), we can make two approximations in the sum in Eq. (4.27): we can neglect the variation in the final energy $E_{k'}$, and we can extend the sum to include *all* final states of the target. (Those which are added will make no contribution to the cross section, by assumption.) If we then sum over all final states and use Eq. (4.25), together with the completeness or closure relation for the final states, we find the "closure approximation" to the collision cross section:

$$\frac{d\sigma^{coll}}{d\Omega} = \left(\frac{d\sigma_2}{d\Omega}\right)_{lab} \sum_{n,n'=1}^{N} \langle \eta_\alpha | \exp[i(\mathbf{k} - \mathbf{k}')\cdot(\mathbf{r}_n - \mathbf{r}_{n'})] | \eta_\alpha \rangle. \qquad (4.28)$$

In the forward direction ($\mathbf{k}' = \mathbf{k}$), the matrix element here is unity, and the double sum gives N^2. Thus in that case the cross section is a *coherent* superposition of contributions from each of the particles in the target. Away from the forward direction the character of the cross section changes. For $|\mathbf{k} - \mathbf{k}'| \geq N^{1/2}/R$, with R the radius of the target, the contributions from terms with $n \neq n'$ tend to cancel each other, and the double sum is just N (from the terms with $n = n'$). In that case the cross section is an *incoherent* superposition of contributions from each particle in the target:

$$\frac{d\sigma^{coll}}{d\Omega} = N\left(\frac{d\sigma_2}{d\Omega}\right)_{lab} \qquad (4.29)$$

This result provides a simple expression for the collision cross section on targets of arbitrary complexity, and it has wide applicability. In fact, since elementary particles are found only in complex targets, this formula provides the basis for all measurements of the scattering properties of these particles (see also Section 5 of Chapter 1).

Double Scattering

From Eq. (4.14) the double-scattering contribution to the T matrix is

$$T^{(2)}_{fi} = \langle \phi_f | \sum_{n,m \neq n} t_n \frac{1}{E - H_0 + i\varepsilon} t_m | \phi_i \rangle. \qquad (4.30)$$

The Green's function appearing here is a many-particle Green's function describing the propagation of all the particles between collisions. Even if the two-body T matrices are treated in the free-particle or impulse approximation, some additional approximation must be introduced to permit evaluation of this Green's function.

Since our objective here is only to illustrate the general structure of this formula, we will assume that the energy of the incident projectile is large compared to the energy-level spacing of the target, so that a "closure approximation" can be made. The eigenstates of $H_0 = K_0 + H_T$ are $\phi_g = \phi_{\mathbf{k}''}\eta_{\alpha''}$, so that the Green's function is

$$\frac{1}{E - H_0 + i\varepsilon} = \int \frac{d^3k''}{(2\pi)^3} \sum_{\alpha''} \frac{\phi_{\mathbf{k}''}\eta_{\alpha''}\eta^{\dagger}_{\alpha''}\phi^{\dagger}_{\mathbf{k}''}}{E_k + E_\alpha - E_{k''} - E_{\alpha''} + i\varepsilon}. \qquad (4.31)$$

If only states of small excitation contribute to the matrix element in Eq. (4.30), we can neglect $E_\alpha - E_{\alpha''}$ and use the completeness of the set of target states $\eta_{\alpha''}$ to obtain

$$\frac{1}{E - H_0 + i\varepsilon} \simeq \int \frac{d^3k''}{(2\pi)^3} \frac{\phi_{\mathbf{k}''}\phi^{\dagger}_{\mathbf{k}''}}{E_k - E_{k''} + i\varepsilon} = \frac{1}{E_k - K_0 + i\varepsilon}. \qquad (4.32)$$

Thus this approximation replaces the many-particle Green's function by the free-particle Green's function for the projectile.

Introducing the impulse approximation and using the same approach as in the discussion of single scattering, we find for the T matrix

$$T^{(2)}_{fi} = \sum_{n,m \neq n} \int \frac{d^3k''}{(2\pi)^3} \frac{\langle \mathbf{k}' | t_n^{(f)} | \mathbf{k}'' \rangle \langle \mathbf{k}'' | t_m^{(f)} | \mathbf{k} \rangle}{E_k - E_{k''} + i\varepsilon}$$
$$\times \langle \eta_{\alpha'} | \exp[i(\mathbf{k}'' - \mathbf{k}')\cdot \mathbf{r}_n] \exp[i(\mathbf{k} - \mathbf{k}'')\cdot \mathbf{r}_m] | \eta_\alpha \rangle. \qquad (4.33)$$

This matrix element describes the scattering of the projectile from the initial state into a state $\phi_{\mathbf{k}''}$ and thence into the state $\phi_{\mathbf{k}'}$. The form factor gives the probability amplitude for target particles m and n, respectively, to transfer the necessary momentum to the projectile. Two-body and many-body effects have indeed been separated, but the result is not so simple as to allow clear conclusions to be drawn without using specific assumptions concerning the two-body forces.

A Soluble Model

The exact multiple-scattering equations cannot be solved for the case of a general many-body system. It is usually necessary to expand the equations into a multiple-scattering sequence and to use the first few terms, as we have done in the preceding discussion. However, if special models are adopted or if particularly simple reactions are studied, one may sometimes use a more exact approach. In the next section we will investigate the simplifications that arise from considering only the elastically scattered wave. Here we discuss one model that can be solved exactly.

Consider the scattering of a particle by an assembly of infinitely heavy fixed scatterers. The positions of these scatterers are assumed to be known, so that only the wave function of the projectile is needed. As a result, we may consider the state vector ψ_i to describe only the state of the projectile, with the positions of the scatterers taken as fixed parameters.

Even if the positions of the scatterers are fixed, it is still not possible in general to solve the scattering equations exactly. For instance, a common condition under which solutions can be obtained is that the wave equation may be separated into a system of one-dimensional differential equations as, for instance, in the case of a central force. When the scattering potentials are located at different centers, as they are here, this separation of variables is impossible.

To obtain a solution, we must impose a further restriction on our model. We will assume that the range of the force v_n is short compared to the de Broglie wavelength of the projectile and to the distance between the scatterers. This assumption means that the model is not applicable to very high energy processes, where the de Broglie wavelength is short, but it does provide a simple model for the scattering of light, of slow neutrons, of medium-energy mesons, etc. With this assumption the T matrix will have a short range which, for convenience, we take to be zero. We shall not attempt here to relate the T matrix to the potential. However, the infinite mass of the scatterers assures us that the binding of the target particles has no effect on the scattering, since the scatterers would not recoil even if their positions were not fixed. Then the force that determines their positions will not affect their scattering properties, and free-particle scattering data may be used to determine t_n. Formally, we may observe that infinitely heavy particles cannot absorb any recoil energy, so that the Green's function $1/(E - H_0 + i\varepsilon)$ is equivalent to the free-particle Green's function for the projectile, $1/(E - K_0 + i\varepsilon)$.

For fixed centers Eqs. (4.8) and (4.12) become in the spatial representation

$$\psi_i(\mathbf{r}) = e^{i\mathbf{k}\cdot\mathbf{r}} - \frac{2m}{4\pi\hbar^2} \sum_{n=1}^{N} \int d^3r'\, d^3r''\, \frac{\exp(ik|\mathbf{r}-\mathbf{r}'|)}{|\mathbf{r}-\mathbf{r}'|}$$
$$\times \langle \mathbf{r}' | t_n | \mathbf{r}'' \rangle \psi_n(\mathbf{r}'') \tag{4.34}$$

and

$$\psi_n(\mathbf{r}) = e^{i\mathbf{k}\cdot\mathbf{r}} - \frac{2m}{4\pi\hbar^2} \sum_{m\neq n} \int d^3r'\, d^3r''\, \frac{\exp(ik|\mathbf{r}-\mathbf{r}'|)}{|\mathbf{r}-\mathbf{r}'|}$$
$$\times \langle \mathbf{r}' | t_m | \mathbf{r}'' \rangle \psi_m(\mathbf{r}''). \tag{4.35}$$

We have here introduced the spatial representation of the free-particle Green's function. If the T matrix is assumed to have a zero range, its spatial representation may be written as

$$\langle \mathbf{r}' | t_n | \mathbf{r}'' \rangle = -\frac{4\pi\hbar^2}{2m} a_n \delta(\mathbf{r}' - \mathbf{s}_n) \delta(\mathbf{r}'' - \mathbf{s}_n), \tag{4.36}$$

where a_n is the scattering length. With this assumption Eqs. (4.34) and (4.35) become

$$\psi_i(\mathbf{r}) = e^{i\mathbf{k}\cdot\mathbf{r}} + \sum_{n=1}^{N} a_n \frac{\exp(ik|\mathbf{r}-\mathbf{s}_n|)}{|\mathbf{r}-\mathbf{s}_n|} \psi_n(\mathbf{s}_n) \tag{4.37}$$

and

$$\psi_n(\mathbf{r}) = e^{i\mathbf{k}\cdot\mathbf{r}} + \sum_{m\neq n}^{N} a_m \frac{\exp(ik|\mathbf{r}-\mathbf{s}_m|)}{|\mathbf{r}-\mathbf{s}_m|} \psi_m(\mathbf{s}_m). \tag{4.38}$$

Equation (4.38) yields a set of coupled equations for the wave functions $\psi_n(\mathbf{r})$, which describe the effective incident wave on center n. Only the value of this wave function at $\mathbf{s}_n$ is needed to obtain its value everywhere, because of the zero-range assumption. Thus we have

$$\psi_n(\mathbf{s}_n) = e^{i\mathbf{k}\cdot\mathbf{s}_n} + \sum_{m\neq n}^{N} a_m \frac{\exp(ik|\mathbf{s}_n-\mathbf{s}_m|)}{|\mathbf{s}_n-\mathbf{s}_m|} \psi_m(\mathbf{s}_m). \tag{4.39}$$

This is a set of N algebraic equations for the numbers $\psi_n(\mathbf{s}_n)$, and it can be solved by standard algebraic methods. Once these numbers are known, $\psi_i(\mathbf{r})$ can be determined and the scattering amplitude found from the asymptotic form of Eq. (4.37):

$$f(\theta) = \sum_{n=1}^{N} a_n \exp(-i\mathbf{k}'\cdot\mathbf{s}_n) \psi_n(\mathbf{s}_n). \tag{4.40}$$

For two scattering centers we can easily obtain the solution. Equation

(4.39) gives the two equations

$$\psi_1(\mathbf{s}_1) = e^{i\mathbf{k}\cdot\mathbf{s}_1} + a_2 \frac{e^{ikR}}{R} \psi_2(\mathbf{s}_2)$$

and

$$\psi_2(\mathbf{s}_2) = e^{i\mathbf{k}\cdot\mathbf{s}_2} + a_1 \frac{e^{ikR}}{R} \psi_1(\mathbf{s}_1), \qquad (4.41)$$

where $R = |\mathbf{s}_1 - \mathbf{s}_2|$ is the distance between the two centers. The solution of these equations is the effective incident wave

$$\psi_1(\mathbf{s}_1) = \frac{1 + a_2 \dfrac{e^{ikR}}{R} e^{-i\mathbf{k}\cdot\mathbf{R}}}{1 - \dfrac{a_1 a_2}{R\, R} e^{2ikR}} e^{i\mathbf{k}\cdot\mathbf{s}_1}, \qquad (4.42)$$

with a similar equation for $\psi_2(\mathbf{s}_2)$. The first term in the numerator arises from the incident wave. The second term is a consequence of the wave emitted by center 2 and impinging upon center 1. The correction term in the denominator represents the multiple scattering of the wave back and forth between the two centers. If it is expanded into a Taylor expansion, the multiple-scattering sequence is reproduced.

This result can be inserted into Eq. (4.37) to obtain the wave function and into Eq. (4.40) to obtain the scattering amplitude. The latter is

$$f(\theta) = a_1 \left[\frac{1 + a_2 \dfrac{e^{ikR}}{R} e^{-i\mathbf{k}\cdot\mathbf{R}}}{1 - \dfrac{a_1 a_2}{R\, R} e^{2ikR}} \right] \exp[i(\mathbf{k}-\mathbf{k}')\cdot\mathbf{s}_1] + 1 \leftrightarrow 2. \qquad (4.43)$$

It is clear from this result that the multiple-scattering corrections, given by the ratio in brackets, will be important if and only if the scattering lengths a_n are of the order of the interparticle spacing R.

5. The Optical Potential and the Elastically Scattered Wave

In this section we consider the elastically scattered wave. By "elastic" scattering we mean processes in which none of the kinetic energy of the projectile is transferred into excitations of either the projectile or the target. The elementary two-body scattering process could be intrinsically elastic, but the recoil of the target particle might lead to a transfer of energy to the target. In the present context "elastic" scattering specifically excludes such effects, so that the internal states of the projectile and the target remain unchanged.

There are several physical reasons why elastic scattering is especially important. First, in an elastic scattering process the momentum transferred in the collisions with the target particles must be small, in order that no recoil energy be transferred to the target. This implies that long-range collisions can contribute strongly to the scattering, and we may then expect that elastic scattering will constitute a large fraction of the observed forward scattering cross section.

Second, because there is no change in the internal state of either the target or the projectile, the elastically scattered wavelets from different target particles will contribute coherently. In elastic scattering processes the phase relations existing in the incident wave are retained between the waves emerging from each scattering center. Because of this coherence, the total elastic scattering amplitude is of order N (the number of target constituents) times the two-particle scattering amplitude, and the elastic scattering cross section is proportional to N^2. On the other hand, inelastic scattering processes yield incoherent contributions from individual scattering centers, so that the differential cross section for these processes is only proportional to N.

The Coherent Wave

To obtain the elastically scattered wave, we must select out of the total state vector ψ_i that part which contains the target in its initial state η_α. For a system at a finite temperature, the initial state cannot be specified; in this case one would perform this operation for each initial state, and then average the resulting elastic cross sections over the ensemble of states. The separation of the coherent or elastic wave is very simply performed in the coordinate representation; we merely compute the overlap of the complete wave function $\psi_i(\mathbf{r}, \mathbf{r}_1, \mathbf{r}_2, \ldots, \mathbf{r}_N)$ with initial target wave function $\eta_\alpha(\mathbf{r}_1, \mathbf{r}_2, \ldots, \mathbf{r}_N)$. As before, $\mathbf{r}$ is the relative coordinate of the projectile and the target center of mass, and $(\mathbf{r}_1, \mathbf{r}_2, \ldots, \mathbf{r}_N)$ are the coordinates of the target particles relative to the same center of mass. For convenience we will denote this set of target coordinates by ξ, so that the target wave function is written $\eta_\alpha(\xi)$. We can then obtain the "coherent" part of the wave function, denoted by $\hat{\psi}_i(\mathbf{r})$, via

$$\hat{\psi}_i(\mathbf{r}) = \int d^3\xi \, \eta_\alpha^*(\xi) \, \psi_i(\mathbf{r}, \xi). \tag{5.1}$$

5. Elastically Scattered Wave

Since we are using a time-independent treatment, this operation selects that part of the wave function in which the target remains in its initial state *at all times*. This will be called the "coherent wave."

In order to deal more easily with the coherent wave, we shall express this and succeeding relations in the more compact language of state vectors. We define a state vector $\hat{\psi}_i$ from which the coherent wave function can be obtained by the standard rule

$$\hat{\psi}_i(\mathbf{r}) = \langle \mathbf{r} | \hat{\psi}_i \rangle. \tag{5.2}$$

This vector can be obtained conveniently if we define two separate abstract vector spaces which describe the target states and the projectile states, respectively. The coherent state vector $\hat{\psi}_i$ is a state vector in the projectile space, while η_α is a vector in the target space. The vector ψ_i is a vector in the "product space" composed of these two separate spaces; it has nonzero projections in each of them. This is analogous to the description of a particle with intrinsic spin, having a Schrödinger wave function in ordinary coordinate space, a spin wave function in spin space, and a total wave function with components in both.

In this language $\psi_i(\mathbf{r}, \mathbf{r}_1, \mathbf{r}_2, ..., \mathbf{r}_N)$ is the component of ψ_i along the "direction" in projectile space defined by the position eigenvector $|\mathbf{r}\rangle$ and the direction in target space defined by the position eigenvector $|\mathbf{r}_1, \mathbf{r}_2, ..., \mathbf{r}_N\rangle \equiv |\xi\rangle$. According to Eq. (5.1), $\hat{\psi}_i(\mathbf{r})$ is the component of ψ_i along $|\mathbf{r}\rangle$ in projectile space and along the energy eigenvector η_α in target space. This illustrates how different representations can be used in these two distinct spaces. Just as the full wave function can be written as

$$\psi_i(\mathbf{r}, \xi) = \langle \mathbf{r}, \xi | \psi_i \rangle, \tag{5.3}$$

so $\hat{\psi}_i(\mathbf{r})$ can be written as the function

$$\hat{\psi}_i(\mathbf{r}) = \langle \mathbf{r}, \eta_\alpha | \psi_i \rangle \tag{5.4}$$

in a mixed representation.

The two subspaces are completely separate, so that one may take the scalar product in one space, say, the target space, while leaving the other space untouched. This is the abstract vector space analog of the process of integrating over the target coordinates while leaving the position of the projectile fixed, as in Eq. (5.1). Adopting this approach and denoting the state vector for the target by $|\eta_\alpha\rangle$, we have for the coherent wave

$$\hat{\psi}_i = (\eta_\alpha | \psi_i \rangle. \tag{5.5}$$

This is the abstract representation of Eq. (5.1). The use of the rounded bracket is intended to emphasize that the right-hand side of Eq. (5.5) is not a number, but is instead a vector in the projectile subspace. As an example, we may write for the initial state of a system such as we have been discussing, consisting of an elementary particle in a plane-wave state and a heavy target in a state η_α, in the form

$$\phi_i = |\mathbf{k}, \eta_\alpha\rangle = \phi_\mathbf{k} \eta_\alpha. \tag{5.6}$$

The coherent part of this is

$$\hat{\phi}_i = (\eta_\alpha | \phi_i) = \phi_\mathbf{k}, \tag{5.7}$$

the state vector for a projectile of momentum $\hbar \mathbf{k}$.

We define the coherent part of an operator as the average of the operator over the initial state of the target. Then the coherent part of an operator $\mathcal{O}$ is, in the coordinate representation,

$$\langle \mathbf{r}' | \hat{\mathcal{O}} | \mathbf{r} \rangle = \int d^3\xi \, d^3\xi' \, \eta_\alpha^*(\xi') \langle \mathbf{r}', \xi' | \mathcal{O} | \mathbf{r}, \xi \rangle \eta_\alpha(\xi). \tag{5.8}$$

The coherent part $\hat{\mathcal{O}}$ is an operator in the projectile space which we write as

$$\hat{\mathcal{O}} = (\eta_\alpha | \mathcal{O} | \eta_\alpha), \tag{5.9}$$

where, as before, the matrix element is with respect to target variables only. As examples, the coherent part of H_0, the initial-state Hamiltonian, is $\hat{H}_0 = K_0 - \varepsilon_\alpha$, and the coherent part of the initial-state Green's function is

$$(\eta_\alpha | \frac{1}{E - H_0 + i\varepsilon} | \eta_\alpha) = \frac{1}{E_k - K_0 + i\varepsilon}, \tag{5.10}$$

the free-particle Green's function for the projectile.

Approximate Form of the Optical Potential

The Schrödinger equation for the full state vector may be written as

$$(E - H_0)\psi_i = \sum_{n=1}^{N} v_n \psi_i = \sum_{n=1}^{N} t_n \psi_n, \tag{5.11}$$

through the use of Eq. (4.7). Taking the coherent part of this equation, we have the Schrödinger equation for the projectile state vector

$$(E_k - K_0)\hat{\psi}_i = \sum_{n=1}^{N} (\eta_\alpha | t_n | \psi_n). \tag{5.12}$$

5. Elastically Scattered Wave

We will show later in this section that this can be written, in general, as the Schrödinger equation for a particle moving in a single-particle potential. In order to explore some general properties of the coherent wave, we will here derive an approximation that leads to such a result.

If we introduce the complete set of target states into one term on the right-hand side of Eq. (5.12), we obtain

$$(\eta_\alpha | t_n | \psi_n) = \sum_{\alpha'} (\eta_\alpha | t_n | \eta_{\alpha'})(\eta_{\alpha'} | \psi_n). \tag{5.13}$$

Let us assume that we can neglect all terms in the sum except the one for which $\alpha' = \alpha$, that is,

$$(\eta_\alpha | t_n | \psi_n) \simeq (\eta_\alpha | t_n | \eta_\alpha)(\eta_\alpha | \psi_n) = \hat{t}_n \hat{\psi}_n. \tag{5.14}$$

The effective incident wave ψ_n is thus replaced by its coherent part $\hat{\psi}_n$. This, in turn, is an average of the effective incident wave over the target wave function and differs from $\hat{\psi}_i$, the average of the full wave, by only the contribution of a single wavelet (cf. Eq. (4.11)). The second approximation we make is then to use $\hat{\psi}_n \simeq \hat{\psi}_i$, so that the same coherent wave is incident on each particle.

With these two approximations, Eq. (5.12) becomes

$$(E_k - K_0)\hat{\psi}_i = \left(\sum_{n=1}^{N} \hat{t}_n \right) \hat{\psi}_i. \tag{5.15}$$

This is a single-particle Schrödinger equation in which the potential energy of the projectile is

$$V_0^{(1)} = \sum_{n=1}^{N} \hat{t}_n. \tag{5.16}$$

This is the first approximation to the "optical potential" which, as we shall see, describes the elastic scattering of the projectile.

This approximate result suggests that the coherent wave can be considered to move in a potential that is determined by the coherent part of the interaction between the projectile and each particle in the target. As indicated by Eq. (5.8), these interactions involve an average over the target wave function of the transition operator for two-body scattering inside the target.

The spatial dependence of the optical potential $V_0^{(1)}$ is governed by the shapes of the two-body interaction and of the target particle density distribution. Thus, according to Eq. (5.8), the coordinate representation of the potential is

$$\langle \mathbf{r}' | V_0^{(1)} | \mathbf{r} \rangle = \sum_{n=1}^{N} \int d^3\xi \, d^3\xi' \, \eta_\alpha^*(\xi') \langle \mathbf{r}', \xi' | t_n | \mathbf{r}, \xi \rangle \eta_\alpha(\xi). \quad (5.17)$$

This is a nonlocal potential, but, if t_n is a local operator, the integrand will be proportional to $|\eta_\alpha(\xi)|^2$, the target density distribution. With $\langle \mathbf{r}', \xi' | t_n | \mathbf{r}, \xi \rangle = t_n(\mathbf{r}, \xi) \delta(\mathbf{r} - \mathbf{r}') \delta(\xi - \xi')$, a local approximation, the potential becomes

$$\langle \mathbf{r}' | V_0^{(1)} | \mathbf{r} \rangle = V_0^{(1)}(\mathbf{r}) \delta(\mathbf{r} - \mathbf{r}'), \quad (5.18)$$

with

$$V_0^{(1)}(\mathbf{r}) = \sum_{n=1}^{N} \int d^3\xi \, |\eta_\alpha(\xi)|^2 \, t_n(\mathbf{r}, \xi). \quad (5.19)$$

This can be further simplified if the forces between the target particles are not strong, since then $t_n(\mathbf{r}, \xi)$ will depend only on the positions of the projectile and particle n, i.e., $t_n(\mathbf{r}, \xi) = t_n(\mathbf{r}, \mathbf{r}_n)$. Then the optical potential becomes

$$V_0^{(1)}(\mathbf{r}) = \sum_{n=1}^{N} \int d^3r_n \, \rho_n(\mathbf{r}_n) \, t_n(\mathbf{r}, \mathbf{r}_n), \quad (5.20)$$

with the single-particle density distribution $\rho_n(\mathbf{r}_n)$ defined by Eq. (4.17).

The optical potential is then the average of the T matrix over the target density distribution. It is evident that for a short-range two-body potential, for which $|\mathbf{r} - \mathbf{r}_n|$ is small, the optical potential $V_0^{(1)}(\mathbf{r})$ will have a shape very similar to that of the density distribution and will be roughly proportional to the density in the neighborhood of $\mathbf{r}$. In fact, for a zero-range transition operator $t_n(\mathbf{r}, \mathbf{r}_n) \propto \delta(\mathbf{r} - \mathbf{r}_n)$, the spatial dependence of $V_0^{(1)}(\mathbf{r})$ is identical with that of the density distribution, while for a finite-range interaction $V_0^{(1)}(\mathbf{r})$ will usually extend somewhat beyond the outer limits of the target.

The integral equation corresponding to Eq. (5.15) is a one-particle equation such as we considered in Chapters 5 and 6:

$$\hat{\psi}_i = \phi_\mathbf{k} + \frac{1}{E_k - K_0 + i\varepsilon} \left(\sum_{n=1}^{N} \hat{t}_n \right) \hat{\psi}_i. \quad (5.21)$$

If this integral equation for $\hat{\psi}_i$ is expanded in a Born series and compared with the corresponding multiple-scattering expansion of ψ_i, the effect of the approximations can be seen. In each term in the expansion the projectile interacts with all the target particles; the restriction that it must interact successively with different particles has disappeared. This results from the replacement of $\hat{\psi}_n$ with $\hat{\psi}_i$, a substitution that therefore is expected to introduce an error that is of order $1/N$, because the scattering from only one particle is in error in each term. More importantly, the target is left in its initial state after *every* collision. This means that the potential of Eq. (5.16) neglects the effects of "hard" collisions, which will tend to leave the target in excited states. Thus, while Eq. (5.21) includes the effects of elastic scattering to all orders in the interaction, it does not include such "multiple inelastic scattering."

The Complex Momentum-Dependent Potential

Before developing a more exact formulation, let us examine some of the properties of the optical potential in this approximation. This expression for the optical potential allows us to relate elastic scattering from the full target to the scattering of the projectile by the individual particles that make up the target. In fact, in the impulse approximation, Eq. (5.16) provides a direct relation between many-body scattering and the two-body scattering amplitude.

The transition operator gives the scattering amplitude in the momentum representation. If we use this representation for $V_0^{(1)}$, assume that all the target particles are identical, and use the impulse approximation, as in Eqs. (4.16)–(4.20), we find

$$\langle \mathbf{k}' | V_0^{(1)} | \mathbf{k} \rangle = \langle \mathbf{k}' | t^{(f)} | \mathbf{k} \rangle F_\alpha(\mathbf{k} - \mathbf{k}'), \tag{5.22}$$

where the form factor $F_\alpha(\mathbf{k} - \mathbf{k}')$ is defined in Eq. (4.18). This is identical with the elastic scattering amplitude in the single-scattering approximation, Eq. (4.19), and would give that result if scattering by $V_0^{(1)}$ were treated in Born approximation. However, the use of the optical potential to compute $\hat{\psi}_i$ permits a much more exact calculation of the elastic scattering including in a simple way all multiple elastic scattering.

As we pointed out in Section 4, the separation of $\langle \mathbf{k}' | V_0^{(1)} | \mathbf{k}' \rangle$ into a product of two distinct factors, namely, the two-particle transition amplitude and the target form factor, separates two distinct physical effects. For short-range forces the angular distribution is largely controlled by the target form factor, which gives a diffraction pattern characteristic of the shape of the target, while the two-body amplitude affects primarily the magnitude of the elastic scattering cross section. If the characteristic radius of the target is R, momentum transfers of the order of $|\mathbf{k} - \mathbf{k}'| \lesssim 1/R$ will be permitted. If the target is sufficiently large ($kR \gg 1$), only scattering in the forward direction will be allowed. Thus, for a large target of uniform particle density ρ, the form factor is

$$F_\alpha(\mathbf{k} - \mathbf{k}') = N \int d^3 r_1 \, \rho_1(\mathbf{r}_1) \exp[i(\mathbf{k} - \mathbf{k}') \cdot \mathbf{r}_1] \simeq (2\pi)^3 \rho \, \delta(\mathbf{k} - \mathbf{k}'), \quad (5.23)$$

and only forward scattering will occur. Using the known relation between the scattering amplitude and the T matrix, we then have

$$\langle \mathbf{k}' | V_0^{(1)} | \mathbf{k} \rangle \simeq (2\pi)^3 \langle \mathbf{k} | t^{(f)} | \mathbf{k} \rangle \rho \, \delta(\mathbf{k} - \mathbf{k}')$$

$$= -(2\pi)^3 \frac{4\pi \hbar^2}{2m} \rho f_k(0) \, \delta(\mathbf{k} - \mathbf{k}'). \quad (5.24)$$

The strength of the optical potential is in this approximation determined by the forward scattering amplitude for two-particle scattering.

Let us now consider the Schrödinger equation for the coherent wave in the momentum representation:

$$\frac{\hbar^2}{2m}(k^2 - k'^2) \langle \mathbf{k}' | \hat{\psi}_i \rangle = \int \frac{d^2 k''}{(2\pi)^3} \langle \mathbf{k}' | V_0^{(1)} | \mathbf{k}'' \rangle \langle \mathbf{k}'' | \hat{\psi}_i \rangle. \quad (5.25)$$

Introducing Eq. (5.24), which uses the impulse approximation and the relations $m \ll m_n$ and $kR \gg 1$, we see that this has a solution only when

$$k'^2 = k^2 + 4\pi \rho f_{k'}(0). \quad (5.26)$$

This is an eigenvalue equation for the wave number k'. Since $\langle \mathbf{k}' | \hat{\psi}_i \rangle$ is zero unless this equation is satisfied, $\hat{\psi}_i(\mathbf{r})$ will be a superposition of momentum eigenfunctions $e^{i\mathbf{k}' \cdot \mathbf{r}}$, with k' given by Eq. (5.26). Thus the particle propagates inside the target as a free particle but with a modified wave number or kinetic energy.

This result has a number of interesting features. The wave number inside the medium is derived from the two-body scattering amplitude rather than, as one might have expected, from the two-body potential. This means that k' may be inferred directly from two-body scattering data without a prior computation of the two-body potential, provided the two-body scattering amplitude can be obtained from the data. In doing this, though, the scattering amplitude is evaluated at the energy that the projectile has within the target, not at its initial external energy.

Since the scattering amplitude is complex (with a negative imaginary part), k' and the optical potential are also complex, and the coherent wave will be attenuated as it passes through the target. This attenuation results from inelastic scattering within the target, as well as any absorption of the projectile that may be possible. For large uniform targets, in the absence of any true absorption we do not expect such attenuation, since the incident projectile must eventually traverse the target and emerge in the forward direction without a significant loss in momentum. However, in this approximation the imaginary part of the potential is proportional to $\operatorname{Im} f_{k'}(0) = k'\sigma_2/4\pi$, where σ_2 is the two-body total cross section, so that Eq. (5.26) predicts attenuation even if there is no true absorption. Higher-order corrections are necessary to ensure that, in a large uniform medium, the only attenuation is that arising from true absorption of the projectile.

In practice, data from elastic scattering on many-body targets are fitted using a phenomenological single-particle potential of the form $V_0(\mathbf{r}) = V_{\text{real}}(\mathbf{r}) - iV_{\text{imag}}(\mathbf{r})$. The parameters so obtained are compared with those given by a relation such as Eq. (5.24), which makes use of two-body scattering data. These comparisons have been moderately successful in a number of cases involving nucleons and mesons scattering on nuclei.

Equations (5.24) and (5.26) hold only if the first approximation to $V_0^{(1)}$, Eq. (5.22), is valid. Before deriving a more exact expression for V_0, we should observe that the *form* of these results holds in general for scattering in a large uniform medium. In such a medium the potential energy is unchanged under a spatial translation or a rotation. Except for possible spin dependences, the nonlocal potential $\langle \mathbf{r}' | V_0 | \mathbf{r} \rangle$ can then depend only on the scalar quantity $(\mathbf{r} - \mathbf{r}')^2$ and must have the form

$$\langle \mathbf{r}' | V_0 | \mathbf{r} \rangle = V_0((\mathbf{r} - \mathbf{r}')^2). \tag{5.27}$$

A local potential in a uniform medium must be constant independent of position, but a nonlocal potential can have the dependence indicated by Eq. (5.27).

Assuming a single-particle Schrödinger equation of the form

$$(E_k - K_0)\hat{\psi}_i = V_0\hat{\psi}_i \tag{5.28}$$

and introducing the momentum representation, as in Eqs. (5.25) and (5.26), we obtain the eigenvalue equation

$$\frac{\hbar^2 k'^2}{2m} + \tilde{V}_0(k'^2) = \frac{\hbar^2 k^2}{2m}, \tag{5.29}$$

where

$$\langle \mathbf{k}' | V_0 | \mathbf{k} \rangle = (2\pi)^3 \, \tilde{V}_0(k'^2) \, \delta(\mathbf{k} - \mathbf{k}'),$$

or

$$\tilde{V}_0(k'^2) = \int d^3 r \, e^{i\mathbf{k}' \cdot \mathbf{r}} \, V_0(r^2), \tag{5.30}$$

using Eq. (5.27). As Eq. (5.29) shows, $\tilde{V}_0(k'^2)$ is simply the depth of the potential for a particle of momentum $\hbar k'$.

The effect of the nonlocal nature of V_0 is then to make the depth of V_0 dependent on the incident momentum; conversely, any experimentally observed momentum dependence of the optical potential implies that it is a nonlocal operator. In the approximation of Eq. (5.26), this dependence follows from the energy dependence of the free two-particle scattering cross section.

A common way of treating the momentum dependence of the optical potential is to include only the first two terms in the Taylor expansion of the potential:

$$\tilde{V}_0(k'^2) = a + bk'^2.$$

The eigenvalue equation then becomes

$$\frac{\hbar^2}{2m^*} k'^2 + a = \frac{\hbar^2}{2m} k^2, \tag{5.31}$$

where m^*, the "effective mass," is defined by $1/m^* = (1/m) + 2b/\hbar^2$. The effective-mass approximation is especially convenient in calculations because it gives an energy-momentum relation having the same structure as for a free particle, so that free-particle wave functions and Green's functions can be used even when the particle is moving under the influence of the potential.

For a large medium the coherent wave $\hat{\psi}_i(\mathbf{r})$ is simply a superposition of plane waves $e^{i\mathbf{k}'\cdot\mathbf{r}}$ with k' satisfying Eq. (5.29). For the plane-wave solution e^{ikz}, Eq. (5.29) gives an equivalent index of refraction for the medium:

$$n = \frac{k'}{k} = \left(1 - \frac{2m}{\hbar^2 k^2} \tilde{V}_0(k'^2)\right)^{1/2}. \tag{5.32}$$

The intensity of the coherent wave decreases according to $|\hat{\psi}_i(\mathbf{r})|^2 = e^{-z/\lambda}$, where the mean free path λ is given by

$$\lambda = \frac{1}{2 \operatorname{Im} k'} \simeq -\frac{\hbar v_{k'}}{2} \frac{1}{\operatorname{Im} \tilde{V}_0(k'^2)}, \tag{5.33}$$

assuming that the attenuation is small. In the first approximation of Eq. (5.24), we find that

$$\operatorname{Im} V_0^{(1)}(k'^2) = -\frac{\hbar v_{k'}}{2} \rho \sigma_2, \tag{5.34}$$

5. Elastically Scattered Wave

so that we have for the mean free path

$$\lambda^{(1)} = \frac{1}{\rho \sigma_2}, \tag{5.35}$$

which is the classical result.

Exact Form of the Optical Potential

We now derive a single-particle potential that gives an exact description of the elastically scattered wave. This can be used to test the single-scattering approximation given above and to find corrections to it.

If we use the "two-potential formula," Eq. (1.6), the T matrix for elastic scattering is

$$T_{fi} = \langle \phi_f | \sum_{n=1}^{N} v_n | \psi_i^{(+)} \rangle = \langle \phi_f | V_0 | \chi_i^{(+)} \rangle$$
$$+ \langle \chi_f^{(-)} | \left(\sum_{n=1}^{N} v_n - V_0 \right) | \psi_i^{(+)} \rangle. \tag{5.36}$$

We want to choose V_0 so that the first term, the amplitude for scattering by the potential V_0, completely and correctly describes the elastic scattering, and the second term vanishes. The optical potential will then govern the motion of the coherent, or elastically scattered, wave.

The exact state vector that describes all aspects of the scattering process is

$$\psi_i = \phi_i + \frac{1}{E - H_0 + i\varepsilon} \sum_{n=1}^{N} v_n \psi_i. \tag{5.37}$$

Using Eqs. (5.5), (5.7), and (5.10), the coherent part of this state vector is

$$\hat{\psi}_i = \phi_k + \frac{1}{E_k - K_0 + i\varepsilon} \langle \eta_\alpha | \sum_{n=1}^{N} v_n | \psi_i \rangle, \tag{5.38}$$

where we recall that $\langle \eta_\alpha | \sum_{n=1}^{N} v_n | \psi_i \rangle$ is a vector in the projectile space. The optical potential V_0 is a single-particle potential acting on the projectile coordinates and not upon the internal degrees of freedom of the target. We then want to choose V_0 so that the coherent wave satisfies the usual equation of motion in a potential:

$$\hat{\psi}_i = \phi_k + \frac{1}{E_k - K_0 + i\varepsilon} V_0 \hat{\psi}_i. \tag{5.39}$$

If we compare Eqs. (5.38) and (5.39), we see that the optical potential satisfies the condition

$$V_0 \hat{\psi}_i = \langle \eta_\alpha | \sum_{n=1}^{N} v_n | \psi_i \rangle. \tag{5.40}$$

This condition guarantees that the second term in Eq. (5.36) vanishes for elastic scattering, that is, when the target is left in its initial state.

To obtain an explicit expression for the optical potential, it is most convenient to express ψ_i in terms of states describing motion in the optical potential, that is, in terms of eigenstates of $H_0 + V_0$. This problem was discussed in Section 1, and from Eq. (1.5) we have the result

$$\psi_i = \chi_i + \frac{1}{E - H_0 - V_0 + i\varepsilon} \left(\sum_{n=1}^{N} v_n - V_0 \right) \psi_i, \tag{5.41}$$

where χ_i satisfies the integral equation

$$\chi_i = \phi_i + \frac{1}{E - H_0 + i\varepsilon} V_0 \chi_i. \tag{5.42}$$

This state vector describes the passage of the incident wave through the optical potential. Since V_0 acts only on the projectile space, the coherent part of Eq. (5.42) is identical with Eq. (5.39), and the state vector χ_i is just the product

$$\chi_i = \eta_\alpha \hat{\psi}_i. \tag{5.43}$$

We now define a wave operator Ω that transforms χ_i into ψ_i:

$$\psi_i = \Omega \chi_i, \tag{5.44}$$

with

$$\Omega = 1 + \frac{1}{E - H_0 - V_0 + i\varepsilon} \left(\sum_{n=1}^{N} v_n - V_0 \right) \Omega. \tag{5.45}$$

Since χ_i describes the coherent or elastically scattered wave, the wave matrix Ω introduces the inelastic components into the state vector ψ_i. Introducing Eq. (5.44) into Eq. (5.40) and using Eq. (5.43), we obtain for the optical potential the expression

$$V_0 = \langle \eta_\alpha | \sum_{n=1}^{N} v_n \Omega | \eta_\alpha \rangle. \tag{5.46}$$

While this result is not unique (we could add to it any operator W satisfying the condition $W \chi_i = 0$), it is the simplest expression satisfying Eq. (5.40).

It may be noted that V_0 is intrinsically energy dependent, and therefore nonlocal, since the wave matrix is a function of the energy.

Equation (5.46) is not very satisfactory for performing calculations, since an expansion of it in powers of v_n may not converge sufficiently rapidly to be of practical use. A more useful expression can be obtained if we introduce a multiple-scattering formulation for this problem. There are many ways of doing this, with each approach corresponding to a different reordering of the perturbation series. Since generally only the first term in the reordered series is used in actual calculations, it is desirable to include the major contributions to elastic scattering in this first term.

One approach is simply to use the effective two-body interaction of the last section, Eq. (4.5), as the first term; this gives the approximate form of V_0 discussed earlier in this section. The exact form of V_0 obtained in this way is

$$V_0 = (\eta_\alpha | \sum_{n=1}^{N} t_n \Omega_n | \eta_\alpha), \qquad (5.47)$$

with

$$\Omega_n = 1 + \frac{1}{E - H_0 + i\varepsilon} \left(\sum_{m \neq n} t_m \Omega_m - V_0 \Omega \right). \qquad (5.48)$$

The first approximation to Eq. (5.47) is just Eq. (5.16). However, an examination of the higher-order corrections shows that these include much of the elastic scattering that takes place during the course of the scattering process and can be quite large in some cases.

An alternative approach would include as much of this scattering as possible in the first term. In such an approach it is convenient to consider the optical potential to be composed of contributions $\mathscr{V}_n$ from scattering on each target particle:

$$V_0 = \sum_{n=1}^{N} \mathscr{V}_n. \qquad (5.49)$$

Each element $\mathscr{V}_n$ is a single-particle operator acting on the projectile state vector. We will use these in defining the two-body transition operator that describes interactions inside the target.

In Section 4 we defined a transition operator t_n in which the projectile and particle n interacted in the presence of the interactions of the target particles with each other, via H_T. Here we want to include as well the elastic scattering of the projectile by the target. When the projectile is interacting directly with particle n, this scattering will arise from interactions with the remaining particles in the target. A single-particle potential that describes this residual scattering is

$$V_{0(n)} = V_0 - \mathscr{V}_n = \sum_{m \neq n}^{N} \mathscr{V}_m. \tag{5.50}$$

The difference between this potential and V_0 is proportional to $1/N$ for a large uniform medium and can usually be neglected. However, in some cases, as when N is small, when the forces have a short range, or when the target particles are widely separated (as in a crystal), this difference can be important.

Let us then define a two-body transition operator by

$$t_n = v_n + v_n \frac{1}{E - H_0 - V_{0(n)} + i\varepsilon} t_n. \tag{5.51}$$

(To avoid proliferating notation, we use the same symbol as previously, although this definition differs from that of Eq. (4.5).) This operator describes the interaction of the projectile and particle n while *both* are interacting with the remaining particles in the target, although only the average interaction of the projectile is included, via the single-particle potential $V_{0(n)}$.

We now follow the derivation of the multiple-scattering equations given in Eqs. (4.7)–(4.12). An effective incident wave ψ_n is introduced via the relation

$$v_n \psi_i = t_n \psi_n. \tag{5.52}$$

Defining the wave operator Ω_n through

$$\psi_n = \Omega_n \chi_i, \tag{5.53}$$

and using Eq. (5.51), we find it to be related to the wave operator introduced in Eq. (5.45) by

$$\Omega = \omega_n \Omega_n, \tag{5.54}$$

with

$$\omega_n = 1 + \frac{1}{E - H_0 - V_{0(n)} + i\varepsilon} t_n. \tag{5.55}$$

Using these relations, we find that the effective incident wave operator can be expressed as

$$\Omega_n = \left[1 - \frac{1}{E - H_0 - V_{0(n)} + i\varepsilon} \mathscr{V}_n \right] \\ \times \left[1 + \frac{1}{E - H_0 - V_0 + i\varepsilon} \sum_{m \neq n}^{N} (t_m - \mathscr{V}_m \omega_m) \Omega_m \right]. \tag{5.56}$$

5. Elastically Scattered Wave

As in the previous approach, the optical potential is now

$$V_0 = (\eta_\alpha | \sum_{n=1}^{N} t_n \Omega_n | \eta_\alpha). \tag{5.57}$$

This is an exact expression for the optical potential, and it can be expanded in a multiple-scattering series using Eq. (5.56) for Ω_n.

APPROXIMATIONS

The expression in the first bracket in Eq. (5.56) ensures that the effective incident wave on particle n is moving in an optical potential that does not include particle n itself. If the contribution of particle n to the optical potential is small (of order $1/N$) compared to the contributions of the remaining particles, this correction need not be included. In the first approximation Ω_n is then set equal to unity and

$$V_0^{(1)} = \sum_{n=1}^{N} (\eta_\alpha | t_n | \eta_\alpha) = \sum_{n=1}^{N} \hat{t}_n, \tag{5.58}$$

or $\mathscr{V}_n^{(1)} = t_n$. This is similar to the result found earlier, in Eq. (5.16), except that now the intermediate states in t_n include the elastic interaction of the projectile with the remainder of the target. This permits the energy shift resulting from elastic scattering to be included and, as we shall see, reduces the magnitude of the higher-order corrections.

If the $1/N$ correction is neglected, the second term in the multiple-scattering expansion of the optical potential is

$$V_0^{(2)} = \sum_{n=1}^{N} (\eta_\alpha | t_n \frac{1}{E - H_0 - V_0^{(1)} + i\varepsilon} \sum_{m \neq n} (t_m - \hat{t}_m) | \eta_\alpha), \tag{5.59}$$

where the first approximation to V_0 has been inserted where it is needed. This double-scattering correction, and in fact all higher-order corrections, involve only "hard" collisions via the residual interaction $(t_m - \hat{t}_m)$. If the collision is a "soft" or elastic collision, this difference will vanish, so that elastic scattering in the intermediate states has been included in the first approximation. In addition, only *excited* states of the target are involved in the correction terms. The full contribution of scattering when the target is in its ground state has already been included in the coherent wave.

The restriction to excited intermediate states is a consequence of the fact that $(\eta_\alpha | t_m - \hat{t}_m | \eta_\alpha) = 0$, which implies that only states $\eta_{\alpha'}$ other than η_α can contribute to the sum over intermediate states. Defining the projection operator for the initial target state by $P_\alpha = |\eta_\alpha)(\eta_\alpha|$ and using the fact that $\mathscr{V}_n^{(1)}$ acts only on projectile coordinates, so that $(\eta_{\alpha'}|\mathscr{V}_n^{(1)}|\eta_\alpha) = \mathscr{V}_n^{(1)}\delta_{\alpha\alpha'}$, we have for the double-scattering correction

$$V_0^{(2)} = \sum_{n,m \neq n}^{N} (\eta_\alpha | t_n \frac{1 - P_\alpha}{E - H_0 - V_0^{(1)} + i\varepsilon} t_m | \eta_\alpha). \tag{5.60}$$

The target is excited by the collision with particle m and must then be de-excited by the collision with a different particle n. This double-scattering correction will then require forces acting between the target particles; in the absence of such forces, the second collision could not remove the excitation generated in the first collision with a different particle. Higher-order terms in the multiple-scattering expansion of V_0 all involve target excitation and interparticle correlations, so that the series is often rapidly convergent.

We may also write a multiple-scattering expansion for the T matrix, as we did in Section 4 when no optical potential was introduced. Using Eq. (5.36) we find

$$T_{fi} = \langle \phi_f | V_0 | \chi_i^{(+)} \rangle + \langle \chi_f^{(-)} | \sum_{n=1}^{N} (t_n - \mathscr{V}_n \omega_n) | \psi_n^{(+)} \rangle. \tag{5.61}$$

If the optical potential satisfies Eq. (5.57), the second term is identically zero for elastic scattering, so that the optical potential gives an exact description of elastic scattering. For inelastic or reaction processes the first or optical-potential term vanishes. Suppose for these processes we require only the first-order in t_n and thus approximate Ω_n by the first bracket in Eq. (5.56). Then

$$\psi_n^{(+)} = \Omega_n \chi_i^{(+)} \simeq \left[1 - \frac{1}{E - H_0 - V_{0(n)} + i\varepsilon} \mathscr{V}_n \right] \chi_i^{(+)}$$

$$= \phi_i + \frac{1}{E - H_0 - V_{0(n)} + i\varepsilon} V_{0(n)} \phi_i \equiv \chi_{i(n)}^{(+)}, \tag{5.62}$$

which is the outgoing-wave solution for projectile motion in the potential $V_{0(n)}$. Using the definition of ω_n, Eq. (5.55), and the relation

$$\langle \chi_f^{(-)} | \left(1 - \mathscr{V}_n \frac{1}{E - H_0 - V_{0(n)} + i\varepsilon} \right) = \langle \chi_{f(n)}^{(-)} |, \tag{5.63}$$

analogous to Eq. (5.62), Eq. (5.61) then gives for inelastic scattering

$$T_{fi}^{(1)} = \sum_{n=1}^{N} \langle \chi_{f(n)}^{(-)} | t_n | \chi_{i(n)}^{(+)} \rangle. \tag{5.64}$$

This is the distorted-wave approximation discussed in Section 2, with the distortion of the incident and outgoing waves now provided by the optical potential. It includes all elastic scattering of the projectile during the process, but neglects multiple inelastic collisions involving excited intermediate states. The methods of Section 2 may be directly taken over in evaluating it. In most cases the $1/N$ correction can be dropped and $V_{0(n)}$ replaced by V_0.

6. Resonances

An important phenomenon in scattering from complex systems is the occurrence of resonances in the scattering cross section. In the neighborhood of a resonance, the cross sections for all allowed reactions and for elastic scattering show pronounced peaks as the incident energy is varied. All these peaks have nearly identical shapes, and the angular distributions for these processes generally have simple shapes characteristic of the angular momentum associated with the resonance.

These observations can be interpreted as arising from long-lived virtual states of the compound system that are formed when the projectile and the target coalesce. The enhanced cross section is a result of the greater probability of interacting when such long-lived states can be formed, and the widths of the peaks are associated with the finite lifetimes of these states, via the uncertainty relation $\Delta E \, \Delta t \sim \hbar$. If scattering resonances are in fact related to long-lived states, a description using a sequence of multiple scatterings, as in Section 4, is obviously impractical. Only if the lifetime of the state is comparable with the time between collisions of the projectile will such a description be successful. For states having longer lifetimes, the projectile will collide many times with particles in the target, so that the multiple-scattering series must be summed to take account of the higher-order collisions.

This summation is usually accomplished by using a description that emphasizes the existence of the virtual levels rather than the multiple collisions. A boundary condition at a finite radius is introduced to obtain a discrete set of eigenstates of the compound system which in some cases approximate the resonant states. The resonance energies and widths are considered to be experimentally determined parameters and are not calculated from first principles. The purpose of these models is then to provide a convenient empirical description by means of a small number of energy-independent parameters, This technique has been described in many publications, and we shall not repeat these derivations. Instead, we shall limit ourselves to a brief discussion of the relation between the results obtained in this book and the existence of scattering resonances.

THE BREIT–WIGNER FORMULA

The virtual states of many-particle systems are in general very complex, and only very short-lived states may have sufficiently simple structures that explicit calculations of their properties can be made. We shall then not go into the details of these states or of particular target systems, but shall simply show the form of the T matrix in the neighborhood of a resonance.

Since a unique angular momentum is associated with each virtual state, we will always be discussing the T matrix elements for a particular angular-momentum state. The initial and final states may contain different kinds of particles in various states of internal excitation. For a given value of J, each possible configuration is for convenience called a "channel," by an analogy with waveguide problems. The initial state is the "entrance channel," while the final states are "exit channels."

The free-particle state in a given channel will be denoted by $\phi_{E\alpha}$, where E is the total energy and α is the channel index, indicating the angular momentum, parity, and type of particles in the state. The scattering state vector corresponding to the initial state $\phi_{E\alpha}$ is

$$\psi_{E\alpha}^{(+)} = \phi_{E\alpha} + \frac{1}{E - H + i\varepsilon} V_\alpha \phi_{E\alpha}, \tag{6.1}$$

where V_α is the interaction between the separated particles in channel α.

We saw in Chapter 7 that in a rearrangement collision the transition operator for a transition from channel α to channel β has the form

$$T_{\beta\alpha}(E) = \langle \phi_{E\beta} | V_\beta | \psi_{E\alpha}^{(+)} \rangle. \tag{6.2}$$

Using Eq. (6.1), the T matrix element may then be written as

$$T_{\beta\alpha}(E) = \langle \phi_{E\beta} | V_\beta | \phi_{E\alpha} \rangle + \langle \phi_{E\beta} | V_\beta \frac{1}{E - H + i\varepsilon} V_\alpha | \phi_{E\alpha} \rangle. \tag{6.3}$$

(The first term here could also have been written as the matrix element of V_α, since $\langle \phi_{E\beta} | V_\beta | \phi_{E\alpha} \rangle = \langle \phi_{E\beta} | V_\alpha | \phi_{E\alpha} \rangle$.) If we introduce the complete

6. *Resonances* 381

set of outgoing-wave states $\psi_{E'\gamma}^{(+)}$, for all E' and γ, this becomes

$$T_{\beta\alpha}(E) = \langle \phi_{E\alpha} | V_\beta | \psi_{E\alpha}^{(+)} \rangle$$
$$= \langle \phi_{E\beta} | V_\alpha | \phi_{E\alpha} \rangle$$
$$+ \sum_\gamma \int dE' \, \rho_\gamma(E') \frac{\langle \phi_{E\beta} | V_\beta | \psi_{E'\gamma}^{(+)} \rangle \langle \psi_{E'\gamma}^{(+)} | V_\alpha | \phi_{E\alpha} \rangle}{E - E' + i\varepsilon}. \quad (6.4)$$

The second term involves a product of two quantities that are very nearly elements of the T matrix itself; they are elements of the T matrix evaluated off the energy shell, for $E' \neq E$.

If we examine the quantity $\langle \phi_{E\beta} | V_\beta | \psi_{E'\gamma}^{(+)} \rangle$ as a function of E and E', it is evident that the dependence on the energy E will be very simple and regular, while the most striking energy dependence near a resonance will be the dependence on the variable E'. To obtain an approximate representation of the amplitude $\langle \phi_{E\beta} | V_\beta | \psi_{E'\gamma}^{(+)} \rangle$, we shall make the following assumptions, consistent with this:

(1) In a resonance region, the resonant portion of $\langle \phi_{E\beta} | V_\beta | \psi_{E'\gamma}^{(+)} \rangle$ has the form $A_{\beta\gamma}/(E' - E_r + i\Gamma/2)$, where the resonance energy E_r, the width Γ, and the amplitude $A_{\beta\gamma}$ are in general slowly varying functions of E and E'. We shall demonstrate that this form gives an approximate solution to Eq. (6.3) in the neighborhood of the resonance.

(2) Transitions between different channels are at least of second order in the potentials, so that the Born term in Eq. (6.4) contributes only to elastic scattering: $\langle \phi_{E\beta} | V_\alpha | \phi_{E\alpha} \rangle = \mathscr{V}_\alpha \delta_{\beta\alpha}$. Because of this first-order contribution, the elastic scattering amplitude will have a slowly varying, nonresonant term. Similar nonresonant contributions to reaction processes may also be included without essential modification of our results.

We then write for the transition amplitude

$$\langle \phi_{E\beta} | V_\beta | \psi_{E'\gamma}^{(+)} \rangle = \mathscr{T}_\gamma(E, E') \delta_{\beta\gamma} + \frac{A_{\beta\gamma}}{E' - E_r + i\Gamma/2}. \quad (6.5)$$

Inserting this into the integral of Eq. (6.4) and neglecting all but the resonant energy variation in the resonance region, we find

$$\mathscr{T}_\alpha(E, E) \delta_{\beta\alpha} + \frac{A_{\beta\alpha}}{E - E_r + i\Gamma/2}$$
$$= \mathscr{V}_\alpha \delta_{\beta\alpha} + \int dE' \, \rho_\alpha(E') \frac{|\mathscr{T}_\alpha(E, E')|^2}{E - E' + i\varepsilon} \cdot \delta_{\beta\alpha}$$
$$- 2\pi i \frac{[\rho_\alpha A_{\beta\alpha} \mathscr{T}_\alpha^*(E, E) - (1/i\Gamma) \sum_\gamma \rho_\gamma A_{\beta\gamma} A_{\alpha\gamma}^*]}{E - E_r + i\Gamma/2}. \quad (6.6)$$

If we equate the nonresonant parts, we have

$$\mathscr{T}_\alpha(E, E) = \mathscr{V}_\alpha + \int dE'\, \rho_\alpha(E') \frac{|\mathscr{T}_\alpha(E', E)|^2}{E - E' + i\varepsilon}. \qquad (6.7)$$

This is the equation that would be satisfied by the elastic scattering amplitude in channel α if no reactions were allowed. The imaginary part of Eq. (6.7) is

$$\operatorname{Im} \mathscr{T}_\alpha(E, E) = -\pi \rho_\alpha(E)\, |\mathscr{T}_\alpha(E, E)|^2, \qquad (6.8)$$

which implies that the nonresonant elastic scattering amplitude may be written as

$$\mathscr{T}_\alpha(E, E) = -\frac{e^{i\delta_\alpha(E)} \sin \delta_\alpha(E)}{\pi\, \rho_\alpha(E)}, \qquad (6.9)$$

with δ_α a real phase shift. This result is a direct consequence of Assumption (2), which implies that there are no nonresonant contributions in exit channels other than the elastic scattering channel. If there were such contributions, δ_α would be complex, since there would then be nonresonant absorption from the entrance channel.

If we equate the resonant terms on each side of Eq. (6.6), we see that our choice for the form of the resonant amplitude has been a judicious one, for we have obtained, through the integration over intermediate states, a function having the same energy dependence as the original amplitude. This result holds, though, only if we neglect the energy dependences of $\mathscr{T}_\gamma$, $A_{\alpha\gamma}$, Γ, and ρ_γ. More detailed theories must be used to show when this approximation is valid. In general, the result will hold if the resonance is sufficiently narrow and is far from other resonances. Under these conditions Eq. (6.6) yields the identity

$$A_{\beta\alpha} = -2\pi i\left[\rho_\alpha A_{\beta\alpha}\, \mathscr{T}_\alpha^*(E, E) - \frac{1}{i\Gamma} \sum_\gamma \rho_\gamma A_{\beta\gamma} A_{\alpha\gamma}^*\right]. \qquad (6.10)$$

This equation has a factorable solution

$$A_{\beta\alpha} = \frac{1}{2\pi} \left(\frac{\Gamma_\beta}{\rho_\beta}\right)^{1/2} \left(\frac{\Gamma_\alpha}{\rho_\alpha}\right)^{1/2} \exp[i(\delta_\beta + \delta_\alpha)], \qquad (6.11)$$

where the real numbers Γ_α satisfy the relation

$$\sum_\alpha \Gamma_\alpha = \Gamma. \qquad (6.12)$$

This factorable form is obtained also in more detailed theories. It implies that capture from the entrance channel and decay into an exit channel are independent processes, so that the ratios of the amplitudes for scattering into different exit channels are independent of the entrance channel. This will be the case if the resonant state is truly an isolated virtual level; there is then no "memory" of the initial state.

This result for $A_{\beta\alpha}$ is symmetric, as required by time-reversal invariance (cf. Chapter 10). The phase shift associated with a given channel is just the elastic scattering phase shift for that channel; we saw this earlier for the special case of two channels in Eq. (6.50) of Chapter 10. The constants Γ_β, called the "partial widths," are proportional to the probability of capture from, or decay to, channel β, as we shall see below. Thus Eq. (6.12) is an expression of the conservation of probability; it says that the total decay probability of the virtual level is the sum of the probabilities of decay into the individual channels.

We have found, then, that in the neighborhood of a resonance a solution for the transition amplitude is

$$T_{\beta\alpha}(E) = \langle \phi_{E\beta} | V_\beta | \psi_{E\alpha}^{(+)} \rangle = -\frac{e^{i\delta_\alpha} \sin \delta_\alpha}{\pi \rho_\alpha} \delta_{\alpha\beta}$$

$$+ \frac{1}{2\pi} \frac{(\Gamma_\beta/\rho_\beta)^{1/2}(\Gamma_\alpha/\rho_\alpha)^{1/2}}{E - E_r + i\Gamma/2} \exp[i(\delta_\beta + \delta_\alpha)]. \quad (6.13)$$

The cross sections in the angular momentum state containing the resonance may be obtained from this by using the results of Chapter 7. The total cross sections are

$$\sigma_{\alpha\alpha}^J = \frac{2J+1}{(2s_1+1)(2s_2+1)} \pi \lambdabar_\alpha^2 \left| e^{-2i\delta_\alpha} - 1 + \frac{i\Gamma_\alpha}{E - E_r + i\Gamma/2} \right|^2 \quad (6.14)$$

and

$$\sigma_{\beta\alpha}^J = \frac{2J+1}{(2s_1+1)(2s_2+1)} \pi \lambdabar_\alpha^2 \frac{\Gamma_\beta \Gamma_\alpha}{(E - E_r)^2 + (\Gamma/2)^2}, \quad (6.15)$$

where s_1 and s_2 are the spins of the two particles in entrance channel α, and $\lambdabar_\alpha$ is the center-of-mass wave number in that channel ($\lambdabar_\alpha = 1/k_\alpha$). These are the well-known Breit–Wigner formulas.

RELATION BETWEEN RESONANCE PARAMETERS AND INTERACTIONS

These formulas for the resonance cross sections involve the resonance energy E_r and a phase shift δ_α and partial width Γ_α for each channel. For many-body systems the calculation of these parameters from first principles is generally impractical, and they are simply computed from the observed resonance cross sections. However, the elastic scattering phase shifts δ_α may be calculated from many-body scattering theory using, for instance, an optical potential. The partial widths and resonance energies may also be calculated in some simple cases where it is possible to derive a knowledge of the wave functions in the region of interaction. Such information is usually difficult to obtain, but estimates of the widths may be made using approximate forms for the wave function. The following shows one means of relating the resonance parameters to the interactions causing the collision process; it also shows the relation between the resonance amplitude and the nonresonant terms.

The resonance amplitudes are most easily separated from the remainder of full scattering amplitude by using the two-potential formula developed in Section 1. For each outgoing channel β we introduce an optical potential $V_{0\beta}$ describing the coherent motion of the particles in that channel. (If there are more than two particles in the channel, then, as noted in Section 2, only a portion of the motion can be included in this way.) According to Section 1, the scattering amplitude for ordinary, nonrearrangement scattering is

$$T_{\alpha'\alpha}(E) = \langle \phi_{E\alpha'} | V_{0\alpha} | \chi_{E\alpha}^{(+)} \rangle + \langle \chi_{E\alpha'}^{(-)} | T_{1\alpha} | \chi_{E\alpha}^{(+)} \rangle, \tag{6.16}$$

where the transition operator $T_{1\alpha}$ is

$$T_{1\alpha} = V_{1\alpha} + V_{1\alpha} \frac{1}{E - H_{0\alpha} - V_{0\alpha} + i\varepsilon} T_{1\alpha}, \tag{6.17}$$

$V_{1\alpha} = V_\alpha - V_{0\alpha}$, and $H_{0\alpha}$ is the free-particle Hamiltonian for the channel α.

As we saw in Section 2, the first term in Eq. (6.16) contributes only to elastic scattering. Furthermore, if it describes scattering by an optical potential $V_{0\alpha}$ that is smoothly varying with energy, then it depends only on the over-all size and shape of the target and does not contain any of the detailed resonance structure that is of primary concern in this section. (It may contain the gross structure or "shape" resonances that are characteristic of optical potential scattering.) Thus, for our purposes only the second term in Eq. (6.16) is of interest. This term describes the scattering of the projectile by the residual potential $V_{1\alpha}$ in the presence of the interactions with the target, described through the Hamiltonian $H_{0\alpha} + V_{0\alpha}$.

We may view the Hamiltonian $H_{0\alpha} + V_{0\alpha}$ as establishing a set of intermediate states through which the system passes during the interaction process. It is convenient to consider separately the several classes of such intermediate

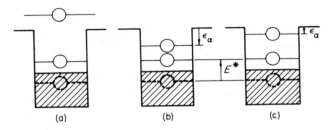

FIG. 12.1

states; these are illustrated schematically in Fig. 12.1. The first set of such intermediate states is the set of continuum states. One continuum is formed from the positive energy states of the incident projectile combined with the ground state of the target. (As usual, the zero of total energy is taken as the energy of a configuration in which the target is in its ground state and the projectile is at rest at infinity.) There will be other continua as well, for example, those formed from positive energy states of the projectile and each excited state of the target. One of these continua is shown in Fig. 12.1(a), which shows the excitation of one particle in the target.

The second set of states consists of the negative-energy bound states formed by binding the projectile within the potential $V_{0\alpha}$ with sufficient energy to exceed the excitation energy that is transferred to the target. Thus, if E^* is the excitation energy of the target ($E^* \geq 0$) and ε_α is the binding energy of the projectile ($\varepsilon_\alpha \geq 0$), then the total energy (relative to the zero of energy defined above) is $E^* - \varepsilon_\alpha \leq 0$. This case is illustrated in Fig. 12.1(b).

There is in this complex situation a third set of states which is seldom discussed, namely, positive-energy, discrete bound states. These are formed when the projectile is captured into a bound state, with the target in one of its excited states such that the target excitation energy is greater than the projectile binding energy, that is, $E^* - \varepsilon_\alpha > 0$. This is shown in Fig. 12.1(c). The cross section will be enhanced at incident energies near the energy at which each of these bound states can be formed. While these eigenstates of $H_{0\alpha} + V_{0\alpha}$ are themselves discrete states, through the coupling provided by the residual interaction $V_{1\alpha}$ they are mixed with the continuum states, yielding a broadened, but still enhanced, spectrum. The continuum states contain finite contributions at large distances from the target so that, in physical terms, the bound states will then be able to decay. They will therefore give rise to the kind of unstable states which we expect to lead to resonant enhancement of the cross section.

We shall assume that all the resonances of interest can be identified as arising from such positive-energy bound states of the projectile, admixed with continuum states. While this seems a reasonable assumption, the systems involved are generally so complex that its validity can be tested only for simple cases involving a small number of particles.

Let us then define a projection operator P_r that selects the positive-energy bound states of the Hamiltonian $H_{0\alpha} + V_{0\alpha}$, which we associate with the resonances. The operator $P_{nr} = 1 - P_r$ selects the continuum and negative-energy states that do not contribute directly to the formation of resonant states. We now define a transition operator $\hat{T}_{1\alpha}$ that contains only these non-resonant intermediate states:

$$\hat{T}_{1\alpha} = V_{1\alpha} + V_{1\alpha} \frac{P_{nr}}{E - H_{0\alpha} - V_{0\alpha} + i\varepsilon} \hat{T}_{1\alpha}. \qquad (6.18)$$

This operator is expected to vary slowly with changes in the total energy, and, in practice, it can usually be approximated by an energy-independent "effective interaction" between the two particles in channel α.

We now want to eliminate the residual interaction $V_{1\alpha}$ from $T_{1\alpha}$, in favor of the effective interaction $T_{1\alpha}$. To accomplish this, we note that a T matrix satisfying an equation of the familiar form

$$T = V + V \frac{1}{E - H_0 + i\varepsilon} T \qquad (6.19)$$

has the explicit solution

$$T = \left(1 - V \frac{1}{E - H_0 + i\varepsilon}\right)^{-1} V. \qquad (6.20)$$

This in turn implies the identity

$$\left(1 - V \frac{1}{E - H_0 + i\varepsilon}\right)^{-1} = 1 + \left(1 - V \frac{1}{E - H_0 + i\varepsilon}\right)^{-1} V \frac{1}{E - H_0 + i\varepsilon}$$

$$= 1 + T \frac{1}{E - H_0 + i\varepsilon}. \qquad (6.21)$$

Using this, and Eq. (6.18), Eq. (6.17) can be expressed as

$$T_{1\alpha} = \left(1 - V_{1\alpha} \frac{1}{E - H_{0\alpha} - V_{0\alpha} + i\varepsilon}\right)^{-1} V_{1\alpha}$$

$$= \left(1 - V_{1\alpha} \frac{1}{E - H_{0\alpha} - V_{0\alpha} + i\varepsilon}\right)^{-1} \left(\hat{T}_{1\alpha} - V_{1\alpha} \frac{P_{nr}}{E - H_{0\alpha} - V_{0\alpha} + i\varepsilon} \hat{T}_{1\alpha}\right)$$

$$= \hat{T}_{1\alpha} + T_{1\alpha} \frac{1}{E - H_{0\alpha} - V_{0\alpha} + i\varepsilon} \hat{T}_{1\alpha} - T_{1\alpha} \frac{P_{nr}}{E - H_{0\alpha} - V_{0\alpha} + i\varepsilon} \hat{T}_{1\alpha}$$

$$= \hat{T}_{1\alpha} + T_{1\alpha} \frac{1}{E - H_{0\alpha} - V_{0\alpha} + i\varepsilon} P_r \hat{T}_{1\alpha}. \tag{6.22}$$

This is an integral equation for $T_{1\alpha}$ in which the first-order term is the effective interaction $\hat{T}_{1\alpha}$ and the only possible intermediate states are the positive-energy bound states, associated with resonant cross sections.

We can find an explicit solution to this equation by using the same methods as in Chapter 6. We introduce the operator identity

$$\frac{1}{E - H_{0\alpha} - V_{0\alpha} + i\varepsilon} = \frac{1}{E - H_{0\alpha} - V_{0\alpha} - P_r \hat{T}_{1\alpha} + i\varepsilon}$$

$$- \frac{1}{E - H_{0\alpha} - V_{0\alpha} + i\varepsilon} P_r \hat{T}_{1\alpha} \frac{1}{E - H_{0\alpha} - V_{0\alpha} - P_r \hat{T}_{1\alpha} + i\varepsilon}$$
(6.23)

into Eq. (6.22) and obtain our final result

$$T_{1\alpha} = \hat{T}_{1\alpha} + T_{1\alpha} \frac{1}{E - H_{0\alpha} - V_{0\alpha} - P_r \hat{T}_{1\alpha} + i\varepsilon} P_r \hat{T}_{1\alpha}$$

$$- (T_{1\alpha} - \hat{T}_{1\alpha}) \frac{1}{E - H_{0\alpha} - V_{0\alpha} - P_r \hat{T}_{1\alpha} + i\varepsilon} P_r \hat{T}_{1\alpha}$$

$$= \hat{T}_{1\alpha} + \hat{T}_{1\alpha} \frac{1}{E - H_{0\alpha} - V_{0\alpha} - P_r \hat{T}_{1\alpha} + i\varepsilon} P_r \hat{T}_{1\alpha}. \tag{6.24}$$

This has the form we are seeking, since the potential has been eliminated and the first term contains only nonresonant contributions, while the second contains only resonant terms.

A similar result can be obtained for rearrangement collisions. If $T_{1\alpha}$ is written as $\hat{T}_{1\alpha} = V_{1\alpha} \hat{\Omega}_{1\alpha}$, with

$$\hat{\Omega}_{1\alpha} = 1 + \frac{P_{nr}}{E - H_{0\alpha} - V_{0\alpha} + i\varepsilon} V_{1\alpha} \hat{\Omega}_{1\alpha}, \tag{6.25}$$

then Eq. (6.24) can be written as

$$T_{1\alpha} = V_{1\alpha}\left(\hat{\Omega}_{1\alpha} + \hat{\Omega}_{1\alpha}\frac{1}{E - H_{0\alpha} - V_{0\alpha} - P_r\hat{T}_{1\alpha} + i\varepsilon}P_r\hat{T}_{1\alpha}\right)$$
$$\equiv V_{1\alpha}\Omega_{1\alpha}. \tag{6.26}$$

For rearrangement collisions the T matrix is then

$$T_{\beta\alpha}(E) = \langle \chi_{E\beta}^{(-)} | V_{1\beta}\Omega_{1\alpha} | \chi_{E\alpha}^{(+)} \rangle. \tag{6.27}$$

The resonant structure of these results can be exhibited by introducing the eigenfunctions of the energy denominator. These are linear combinations of the positive-energy bound-state wave functions and satisfy the equation

$$(E_j - H_{0\alpha} - V_{0\alpha} - P_r\hat{T}_{1\alpha})\eta_{j\alpha} = 0. \tag{6.28}$$

Both the effective interaction and the optical potential are complex, so that the energy eigenvalues are also complex. This reflects the fact that these eigenstates are in fact not stationary states but are unstable and will decay into one or more outgoing channels. Often the effective interaction and the imaginary part of the optical potential are relatively small compared to $H_{0\alpha}$ and to the real part of the optical potential. In this case they can be neglected in computing bound-state wave functions, but the energy eigenvalues will be complex, reflecting the coupling to the continuum.

If the energy is written as $E_j = E_{rj} - (i/2)\Gamma_j$, the resonant part of the T matrix takes the general form

$$\langle \chi_{E\beta}^{(-)} | V_{1\beta}\hat{\Omega}_{1\alpha}\frac{1}{E - H_{0\alpha} - V_{0\alpha} - P_r\hat{T}_{1\alpha} + i\varepsilon}P_r\hat{T}_{1\alpha} | \chi_{E\alpha}^{(+)} \rangle$$
$$= \sum_j \frac{\langle \chi_{E\beta}^{(-)} | V_{1\beta}\hat{\Omega}_{1\alpha} | \eta_{j\alpha}\rangle \langle \eta_{j\alpha} | \hat{T}_{1\alpha} | \chi_{E\alpha}^{(+)}\rangle}{E - E_{rj} + \tfrac{1}{2}i\Gamma_j}. \tag{6.29}$$

Comparing this with the Breit–Wigner formula, Eq. (6.13), we see that the partial widths are given by

$$\Gamma_{\beta\alpha} = 2\pi \,|\, \langle \chi_{E\beta}^{(-)} | V_{1\beta}\hat{\Omega}_{1\alpha} | \eta_{j\alpha}\rangle \,|^2 \, \rho_\beta(E), \tag{6.30}$$

and the resonance energy and full width are given by the eigenvalue relation

$$E_{rj} - \frac{i}{2}\Gamma_j = \langle \eta_{j\alpha} | H_{0\alpha} + V_{0\alpha} + \hat{T}_{1\alpha} | \eta_{j\alpha}\rangle. \tag{6.31}$$

In general, the states $\eta_{j\alpha}$, and thus the partial widths $\Gamma_{\beta\alpha}$, will depend on the entrance channel α. However, when the resonances do not overlap (so that there is no mixing of bound states), the partial widths will not depend upon which entrance channel leads to the formation of the bound state, and they may be written as

$$\Gamma_\beta = 2\pi \left| \langle \chi_{E\beta}^{(-)} | \hat{T}_{1\beta} | \eta_j \rangle \right|^2 \rho_\beta(E). \tag{6.32}$$

The partial width gives the rate of decay of the bound state into each outgoing channel, and this result has a form that is familiar for such a decay rate from time-dependent perturbation theory.

When the excitations of the target are relatively simple, as in the single-particle excitations depicted in Fig. 12.1 these results provide a means of computing the resonance parameters directly from the interactions between the particles.

Subject Index

Abstract vector space, *see* Vector space
Addition of angular momentum, *see* Angular momentum
Adiabatic switching, 218
Adjoint matrix, 171
Adjoint operator, 171
Adjoint space, 165
Adjoint vector, 165
Analiticity, *see* Radial Green's function
Angular momentum, 278-320
  addition of, 286-299
  barrier, 28
  conservation of, 261-263
  eigenfunctions of, 285-287
  orbital, 27
    eigenstates of, 241, 245
  projection operator, *see* Projection operators
  raising and lowering operators, *see* Raising and lowering operators
  representation, 241
Antilinear operator, 268
Antiunitary operator, 269
Antisymmetrization of state vector, 206-208
Asymmetry left-right, 306, 313-316
Attenuation, 370-371

Beam monitor, 9
Bessel functions
  integral representation of, 123
  spherical, *see* Spherical Bessel functions
Born approximation, 144-149, 158, 183, 291, 324
  distorted wave, 147-149, 324, 327-341, 379

  to phase shift, 147
  validity of, 145-146
Born expansion, 143-149, 158, 177, 178, 237, 353
  convergence of, 144, 153
Boundary condition, 105, 201
  incoming-wave, *see* Incoming wave
  outgoing-wave, *see* Outgoing wave
  periodic, 108
  regularity at origin, 29
  relation to experimental situation, 3
  for scattering, 13
Bragg scattering, 12
BRA, 165
BRA-KET Symbolism, 164 ff.
Breakup reaction, 187 ff., 336-339
Breit-Wigner formula, 380-383, 388
Bremsstrahlung, 339-341

Causality condition, 136-137
Center of mass, 2
Center-of-mass motion, 260-261, 331
Center-of-mass coordinate, 250
Center-of-mass system, 260
Channel, 380
  entrance, 380
  exit, 380
Channel spin, 297-299, 308
Charge exchange operator, 243
Charge exchange scattering, 243 ff.
Clebsch-Gordan coefficients, 287-288
Closure, 85, 109, 115
Closure approximation, 359
Coherent part, 366 ff.
Coherent wave, 364-366

SUBJECT INDEX

Collimator, 4
Collision cross section. *see* Cross section, differential
Commutator, 254
Commuting observables, 166
Comparison function, *see* Free particle comparison function
Competing processes, 244
Complete set, 166
Completeness, 85, 168, 232, 234, 245
Complex conjugate operator, 270
Complex potential, *see* Potential
Compound elastic scattering, 388
Compton scattering, 347
Conjugate variable, 254
Conservation laws, 249 ff.
Conservative potential, 2
Continuous spectrum, 109
Continuous variables, 166
Contour integration, 104
Contour of integration, 101, 110
Coordinate representation, 168, 345
Coulomb force, 51
Coulomb function, 63-72
  in parabolic coordinates, 66
  radial, 65
Coulomb interference, 71
Coulomb phase, 72
Coulomb potential, *see* Scattering
Coulomb scattering amplitude, *see* Scattering amplitude, Coulomb
Classical limit, *see* JWKB Approximation
Classical scattering formula, 63
Classical turning points, *see* Turning points
Cross section, 6, 187 ff., 225-226. 299-307
  differential, 8, 21, 195, 303-305, 312-313
  experimental definition of, 21
  for identical particles, 208-210
  for general scattering problem, 193
  for rearrangement collisions, 202
  total, 8
  unpolarized, 305, 315
Current, 91, 124
  conservation of, 9, 10, 22, 25, 33, 293-295
Current density, 6, 8

Damping equation, 238, 240-244, 247
Delta function, 166
  in angular momentum representation, 242
  source, 128
Density matrix, 311-313
Density of states, 195-196, 202
Detailed balance theorem, 276
Detection probability, 20
Detector, 4, 194
Determinantal method, 156
Differential cross section, *see* Cross section
Diffraction, 61, 358
Dirac delta function, *see* Delta function
Dirac notation, *see* BRA-KET Symbolism
Dirac operator formalism, *see* Operator formalism
Discrete spectrum, 84, 109
Distorted wave approximation, *see* Born approximation
Distorted wave Born approximation, *see* Born approximation
Distorting potential, *see* Potential
Double scattering, 306
  contribution to T matrix, 359-360

Effective field, 353
Effective incident wave, 353, 355, 363, 367
Effective incident wave operator, 376
Effective interaction, 386
Effective mass, 372
Effective range, 47
  generalized, 156
Effective range expansion, *see* Phase shift
Eigenfunction expansion, 100, *see also* Radial Green's function, other individual topics
Eigenfunctions, 84 ff.
  normalized, 85
Eigen-phase shift, 246
Eigenvalue equation, 166
Eigenvector expansion, 167-169
Elastic scattering, 2, 26, 187ff., 329-332, 357, 363-379, 381
Electromagnetic interaction, 339
Energy conservation, 186, 255

SUBJECT INDEX

Energy shell, 160, 343, *see also* K Matrix, T Matrix
Entrance channel, *see* Channel
Exchange scattering, 187 ff., 332-336
Exchange term, in identical particle scattering, 209
Existence of scattering operator, 231-233
Excited states, 377
Exit channel, *see* Channel
Exponential operator, 212

Final state Hamiltonian, *see* Hamiltonian
Final state interaction, 337, 340
Final state form of state vector, *see* State vector
Final state potential, *see* Potential
Flux conservation, *see* Current, conservation of
Form factor, 347, 357, 369
Fourier expansion, 138, 140
Fourier transform, 169, 342, 344, 347
Fraunhofer scattering, 135
Fredholm determinant, 150, 155
Fredholm method, 149-157
  expression for phase shift, 154-156
  three-dimensional, 156-157
Free particle, 14
  comparison function, 73, 78
Free particle Hamiltonian, 188
Fresnel scattering, 135

Gallilean invariance, 255-261
Gallilean transformation operator, 256
Gauge invariance, 339
Generator of transformation, *see* Transformation
Green's function, 73 ff., 107-121
  asymptotic behavior, 118-121
  eigenfunction expansion, 125
  free particle, 121
  general form of, 112-114
  initial state, 366
  integral equation for, 134
  as inverse operator, 116
  many-particle, 121-126
    asymptotic behavior of, 123
    for motion in potential, 115-117, 124-126
  for motion in non-Hermitian potential, 120-121
  operator form of, 174-177
  outgoing-wave, 110-112, 114, 128, 191
  for particle with spin, 289-290
  radial, *see* Radial Green's function
  relation to radial Green's function, 114-115
  singularity of, 108
  symmetry of, 116
  symmetry property, 110
  time-dependent, 136-141, 212
  for unbounded domain, 110
Green's theorem, 139

Hamiltonian, 1
  final state, 333, 334, 336, 339
  initial state, 328, 352, 366
  hermiticity of, *see* Hermiticity
Hankel function, 123
Hartree potential, *see* Potential
Heisenberg equation of motion, 215
Heisenberg picture, 215 ff.
Heitler damping equation, *see* Damping equation
Hermitian matrix, 171
Hermiticity, 93, 173, 201, 238, 294, 326
  of K Matrix, *see* K Matrix
Hilbert space, *see* Vector space
Hypergeometric function, 64

Identical particles, 205-210
Impact parameter, 59
Impulse approximation, 341-350, 353, 356
  validity of, 348-350
Incident beam, 190
Index of refraction, *see* Refractive index
Inelastic scattering, 187 ff., 194, 331-332, 341, 347, 358, 364, 378
Ingoing-wave, 119
Ingoing-wave function, 119
Initial state Green's function, *see* Green's function
Initial state Hamiltonian, *see* Hamiltonian
Initial state potential, *see* Potential

SUBJECT INDEX

Integral, principal value of, see Principal value
Integral equation(s)
  with distorted wave, 130-131
  for scattering amplitude, 157-162
  of scattering theory, 127 ff.
  solution of, 127 ff.
  stationary state, 141
  time-dependent derivation of, 136-141
  for time-dependent wave function, 138-141
  for wave function, 127-136
Interaction picture, 212 ff., 214
Interference between Coulomb and nuclear scattering, see Coulomb interference
Invariance principles, 249 ff.
Invariance properties, 313
Inverse operator, 165

JWKB Approximation, 55-63
  classical limit, 59-63
  connection formulas, 57
  for Coulomb potential, 63

K Matrix, 34, 76, 227-248
  off energy shell, 239
  hermiticity of, 238, 240
KET, 165
Kinetic energy operator, 173, 332, 334, 352
Kronecker delta, 166

Laplacian operator in spherical coordinates, 26
Laue conditions, 358
Legendre expansion, 114, 265, see also Partial wave expansion
Logarithmic derivative, 35, 90, 100

Magnetic quantum number, 273
Many-body problem, 351
Many-body system, scattering by, see Scattering
Many-particle Green's function, see Green's function
Matrix, 170
Matrix element, 170, 182

Matrix methods in scattering theory, 307-316
Mean free path, 373
Mittag-Leffler theorem, 87-90
Mixed representation, 365
Momentum
  conservation of, 249-255
  representation, 168
  total, 251, 254, 259
Multiple scattering, 351-356

Newton's second law, 63
Normalization condition, 166
Normalization constant, 31-32
Normalization of vectors in vector space, 164 ff.

Operator, 165
  antilinear, see Antilinear
  antiunitary, see Antiunitary
  representation, of, see Representation
Operator equation for transition operator, 183
Operator formalism, 163 ff.
Operator form of scattering equations, 171-183
Optical potential, 363-379
  approximate form, 366-369
  approximations for, 377-379
  exact form of, 373-377
  local approximation to, 368
  multiple scattering expansion for, 377-378
Optical theorem, 9, 10, 22, 25, 33, 146, 183-186, 226, 235-237, 242, 293-295
Orbital angular momentum, see Angular momentum
Ordinary scattering, 187 ff., 329-332, 384
  S matrix for, see S Matrix
Orthogonal set, 85, 166
Orthogonality, 231, see also Scattering state vectors, orthogonality of
Orthogonality relation, 166
Orthonormality, 92
Orthonormality relation, 108
Outgoing wave boundary condition, 19, 124, 128, 189

Outgoing wave Green's function, *see* Green's function

Parabolic coordinates, 66
Parity
　conservation of, 185, 263-266
　intrinsic, 263
Partial wave expansion, 281-282, 292, *see also* Legendre expansion
Partial width, 383, 388
Pauli principle, *see* Identical particles
Pauli spin matrix, 270, 280
Periodic boundary conditions, *see* Boundary conditions
Permutation operator, 206
Perturbation theory, time dependent, 216
Phase amplitude method, 52-63
Phase shift, 29, 32, 383
　Born approximation, to, *see* Born approximation
　calculation of, 25-44
　effective range expansion of, 45-52
　　for charged particles, 51
　　validity of, 49-52
　Fredholm expression for, *see* Fredholm method
　high energy behavior, 40, 52-63
　integral expression for, 36-44
　low energy behavior, 40-52
　reality of, 33, 242, 294
　for square well, 35-36, 58
Pickup reaction, 187 ff., 334-336
Plane wave, 30
Plane wave expansion, 30, 241
Polarization, 280, 296, 299-307
　asymmetry relation, 313-316
　of incident beam, 299 ff.
　of scattered particles, 305-307, 313-315
Polarization tensor, 303
Polarization vector, 301-303
Positive energy bound states, 385
Post-prior paradox, 203
Potential
　complex, 326, 370
　distorting, 321 ff., 330
　final state, 333
　Hartree, 331

initial state, 333
local, 172
momentum dependent, 371-372
non-Hermitian, 120
nonlocal, 141, 173, 371
optical, *see* Optical potential
separable, 141-143, 153
spherically symmetric, 26
Potential matrix, *see* V Matrix
Potential operator, 172
Principal value of integral, 98, 105
Principal-value integral, 107
Probability amplitude, 167
Probability conservation, 185, 226, 233, *see also* Current conservation
Probability current, *see* Current
Probability current density, *see* Current density
Product space, 365
Projection, 167
Projection operator, 316-320, 378, 386
　for angular momentum, 318 ff.
　for spin, 317 ff.
Propagation, *see* Green's function
Pseudo potential, 183, 260

Radial Green's function
　analiticity property, 102
　boundary conditions, 82-83
　　general, 90-93
　closed form of, 79
　connection between eigenfunction expansion and closed form, 87-90
　definition, 79
　describing point to point propagation, 80
　discontinuity of derivative, 82
　eigenfunction expansions of, 84-107
　general form of, 93
　ingoing-wave, 106
　inhomogeneous part of, 98
　integral equation in terms of, 80
　Mittag-Leffler expansion of, 87-90
　outgoing-wave, 96, 103
　　in bounded domain, 93-95
　　for unbounded domain, 100-107
　singularity of, 90, 97, 105, 107

standing wave, 86, 95, 150
symmetry of, 83
unbounded domain, 95 ff.
Radial integral equation, 289-292, 298
Radial wave function, *see* Wave function
Radiative capture, 340
Raising and lowering operators, 262
Reaction matrix, *see* K Matrix
Rearrangement collision, 187 ff., 197-205, 226, 324-327, 332
Rearrangement collisions, 388
  S matrix for, *see* S Matrix
Rearrangement cross section, *see* Cross section
Reciprocal relation, 326, *see also* Scattering amplitude
Reciprocity, 84, 117, 135-136
Reduced mass, 2
Reflection invariance, 185, 263-266
Refractive index, 372
Regularity, *see* Boundary condition
Relative coordinate, 250, 328
Representation of operators, 169-171
Representation of vectors, 167 ff.
Representative, 167
Residual interaction, 328, 386
Residual potential, 78
Resonance energy, 388
Resonance parameters, relation to interactions, 383-389
Resonances, 379-389
Resolvent kernel, 153
Retarded Green's function, *see* Green's function, time-dependent
Reversal of motion, 267
Rotation invariance, 261-263
Rotation transformation operator, 262
Rutherford formula, 72

S Matrix, 34, 185, 192, 224, 227-248
  behavior under time reversal, 275
  diagonalization of, 244-248
  eigenstates of scattering operator, 245
  existence of, *see* Existence of scattering operator
  independent of total momentum, 259
  for ordinary processess, 230
  paramerization of, 277, 294
  for rearrangement collisions, 230-231
  spatial representation of, 251
  symmetric form of, 228-231
  symmetry of, 277
  unitarity of, 233-234, 294
Scalar product, 165
Scattered wave, 190
Scattering
  by collection of scatterers, 10, 361
  by Coulomb potential, 63-72
  cross section, *see* Cross section
  by hard sphere, 36
    low energy limit, 36
    relation to scattering length, 44
  by many-body system, 350-363
  by radial delta function potential, 80
  by shielded Coulomb potential, 68-72
  of spin-$\frac{1}{2}$ particle, *see* Spin, one-half
  by square well potential, 34-36
  by sum of two potentials, *see* Two-potential formula
Scattering amplitude, 7, 21, 32, 292-299
  coulomb, 67, 69
  forward, 370, *see also* Optical theorem
  integral equation for, *see* Integral equation
  integral expression for, 43, 129
  nonresonant, elastic, 382
  partial wave expansion, 131
  reciprocal relation for, 133
  reciprocity condition for, *see* Reciprocity
  spin-dependent, 284, 295-296, 298
Scattering experiment, elements of, 3-6
Scattering length, 44, 362, *see also* Effective range expansion
Scattering matrix, *see* S Matrix
Scattering operator, *see* S Matrix
Scattering state vector, *see* State vector
Schrödinger equation, 1, 172, 328
  integral form of, 73-79
  justification for use of time-independent, 21
  operator form of, 173
  radial, 27, 35
  regular solution, 35

solution of radial equation by means of Green's function, 81
with spin-orbit force, *see* Spin-orbit force
time-dependent, 17, 22, 190
time-independent, 21, 26
Schrödinger picture, 212 ff.
Semi classical approximation, 61
Separable potential, *see* Potential, separable
Shadow scattering, 22
Shape dependent parameter, 47, 49, *see also* Effective range expansion
Shape elastic scattering, 384
Shielded Coulomb potential, *see* Scattering by shielded Coulomb potential
Single scattering approximation, 356-359
Source at finite distance, 133-135
Source of incident particles, 3
Space reflection invariance, *see* Reflection invariance
Space translations, invariance under, *see* Translational invariance
Spatial representation, *see* Coordinate representation
Spherical Bessel functions, 28, 84
for large values of argument, 29
for small values of argument, 28
Spherical coordinates in 3n dimensions, 122
Spherical harmonics, 27
Spin, 278-320
one half, 279 ff., 292, 314, 319
Spin-dependent scattering amplitude, *see* Scattering amplitude
Spin eigenfunctions, 280, 286
orthogonality of, 282
Spin matrix, *see* Pauli spin matrix
Spin operators, 309-313
Spin-orbit force, 272, 281
eigenfunctions of, 283 ff.
Spin projection operator, *see* Projection operators
Spin space, 308
Spin variable, 286, 298
Spin-vector notation, 307 ff.
Square well, *see* Phase shift

State vector(s), 177-180
antisymmetrization of, *see* Antisymmetrization
explicit form, 178
final state form of, 198-202
ingoing-wave, 228
in interaction picture, 215
integral equation for, 177-180
orthogonality of, 178-180
Step function, 138
Stripping reaction, 334
Symmetry between initial and final states, *see* T Matrix, symmetry of
T Matrix, 34, 75, 158, 182, 193, 205, 247, 292-299, *see also* Transition amplitude
off energy shell, 159-162, 381
on energy shell, 161
in impulse approximation, *see* Impulse approximation
multiple scattering, expansion for, 355 ff., 378
phase relations implied by time reversal invariance, 275-276
reciprocal form of, 326
symmetry of, 202-205, 236
unitarity condition, 238, 240
Target, 4
with spin, 297-299
Three-dimensional Green's function, *see* Green's function
Time-dependent Green's function, *see* Green's function
Time-dependent Schrödinger equation, *see* Schrödinger equation
Time-dependent treatment of scattering, 211-226, *see also* Wave packet treatment of scattering
relation to time independent theory, 222-225
Time-independent, Schrödinger equation, *see* Schrödinger equation
Time-reversal invariance, 185, 236, 266-275, 334
Time-reversal transformation operator, 268-272

Time-reversed state, 268
  phase of, 272-274
Time translations, invariance under, 255
Time translation operator, 212, 216
  infinite limit of, 217-222
Total cross section, see Cross section
Total momentum, see Momentum
Transformation, generator of, 253
Transformation operators, 165 ff., 252-253, 256, 262, 268
Transition amplitude, 158, 182, see also T Matrix
Transition matrix, see T Matrix
Transition rate, 225
Translational invariance, 249-255, 332
Translational transformation operator, 253
Turning point, 52, 54
Two-body forces, 334
Two-potential formula, 40-44, 77, 80, 130-131, 148, 321-327, 373
  final state form, 326
  initial state form, 326
  partial wave form of, 79
  for rearrangement collisions, 325

Unit operator, 168
Unitarity, see S Matrix
Unitary operator, 171
Units, 196-197
Unpolarized beam, 303

V Matrix, 158
Vector addition coefficients, see Clebsch-Gordan coefficients
Vector space, 163 ff., 214, 365

Wave function, 172, see also State vector
  asymptotic form, 18
  boundary condition for scattering, 6
  for Coulomb potential, see Coulomb function
  distorted, 322 ff., 328
  expressed as scalar product, 168
  integral equation for, see Integral equation
  for motion of center of mass, 5
  partial wave expansion, 26
  radial, 30
    alternative forms for, 75-77
    normalization of, 37
  radial function, 27
    asymptotic behavior of, 28
  solution of integral equation for, 132-133
Wave matrix, 180 ff., 223, 350, 354, 374
  operator equation for, 180
  unitarity of, 181-182, 223
Wave number, 124
Wave operator, see Wave matrix
Wave packet, 13-25, 189-193, 219 ff., 227
  conditions imposed on, 17
  free particle, 14
  Gaussian, 16
Wave packet treatment of scattering, 13-25
WKB Approximation, see JWKB Approximation
Wronskian, 53-54, 74, 83

Zero range approximation, 362